Low Environmental Impact Polymers

Editors: Nick Tucker and Mark Johnson

rapra
TECHNOLOGY

Rapra Technology Limited

Shawbury, Shrewsbury, Shropshire, SY4 4NR, United Kingdom
Telephone: +44 (0)1939 250383 Fax: +44 (0)1939 251118
http://www.rapra.net

First Published in 2004 by

Rapra Technology Limited

Shawbury, Shrewsbury, Shropshire, SY4 4NR, UK

©2004, Rapra Technology Limited

A catalogue record for this book is available from the British Library.

Every effort has been made to contact copyright holders of any material reproduced within the text and the authors and publishers apologise if any have been overlooked.

ISBN: 1-85957-384-3

Typeset, printed and bound by Rapra Technology Limited
Cover printed by The Printing House, Crewe, Cheshire

Contents

Contributors

Martin P Ansell
Department of Engineering and Applied Science, University of Bath, Bath,
BA2 7AY, UK

Maurizio Avella
Istituto di Chimica e Tecnologia dei Polimeri - CNR, Olivetti Area - Building 70,
via Campi Flegrei, 34, 80078 – Pozzuoli (NA), Italy

James Barber
Metabolix, Inc., 21 Erie Street, Cambridge, MA 02139, USA

Tracy Bhamra
Department of Design and Technology, Loughborough University, Loughborough,
Leicestershire, LE11 3TU, UK

Roger van Erven
BiPP Biopolymer Products, PO Box 6241, 5600 HE Eindhoven, The Netherlands

Jean-Charles Gayet
BioMatera Inc., 3760 Panet, Jonquière (Québec), G7X 0E5, Canada

Mark Hughes
The BioComposites Centre, University of Wales, Bangor, UK

Mark Johnson
Warwick Manufacturing Group, The University of Warwick, Coventry, CV4 7AL, UK

Vicky Lofthouse
Department of Design and Technology, Loughborough University, Loughborough,
Leicestershire, LE11 3TU, UK

Mario Malinconico
Istituto di Chimica e Tecnologia dei Polimeri - CNR, Olivetti Area - Building 70,
via Campi Flegrei, 34, 80078 – Pozzuoli (NA), Italy

Laurent Masaro
6100 Deacon # 5N, Montreal (Québec), H3S 2V6, Canada

Marcia Miller
Metabolix, Inc., 21 Erie Street, Cambridge, MA 02139, USA

Leonard Y Mwaikambo
Department of Engineering Materials, University of Dar Es Salaam, PO Box 35131,
Dar Es Salaam, Tanzania

Mike O'Brien
Cargill Dow LLC, PO Box 5830, MS130, 12700 Whitewater Drive, Minnetonka, MN
55343, USA

Pierangelo Orlando
Istituto di Biochimica delle Proteine - CNR, via Pietro Castellino, 111, 80131 –
Naples, Italy

Dennis Price
25 Pantheon Road, Chandlers Ford, Eastleigh, Hants, SO53 2PD, UK

Anand R Sanadi
USDA Forest Service, Forest Products Laboratory, One Gifford Pinchot Drive,
Madison, WI 53726-2398, USA

Nick Tucker
Warwick Manufacturing Group, The University of Warwick, Coventry, CV4 7AL, UK

Robert Whitehouse
Metabolix, Inc., 21 Erie Street, Cambridge, MA 02139, USA

Bacterial Nomenclature

The following bacterial names have changed: the new names will be used throughout this book:

Comamonas acidovorans is now called *Delftia acidovoranas*

Alcaligenes eutrophus is now called *Wautersia eutropha*

Aureobacterium saperdae is now called *Microbacterium saperdae*

Low Environmental Impact Polymers

Preface

There is some irony in the presentation of limited life polymers. The previous history of polymer science has to a large extent been the struggle of the polymer scientist to produce materials with good environmental stability, resistant to ultra-violet light, extremes of pH and microbial attack. The result of success in these endeavours has been an anguished wail asking: 'how shall we be rid of these imperishable materials?' The answer is to tailor the properties of polymers so that they can enter the existing natural cycle of carbon circulation. If we are to truly enter into the spirit of this activity, such polymers should ideally, also be of recent biological origin. The widespread application and use of such materials is likely to have a striking and beneficial effect on the manufacturing economy.

The UK has seen a 75% reduction in manufacturing capacity over the last 25 years. The survivors are lean, agile, innovative and well aware of the forces of global competition. There is very little technological, competitive edge left in volume manufacturing. Factories in low wage areas no longer lag in technology. However, we are set for a new revolution in manufacturing, where the driving forces will not be those of simple financial bottom line measurements. We will have to incorporate social and environmental performance into the measures of the success of manufacturing industry – the triple bottom line. Current notions of recycling are just a stop-gap - we must develop profitable manufacturing methods that do not require expensive and inefficient processes. This will require a redirection of our economies to enable everyone to meet their basic needs and improve their quality of life while ensuring that the natural resources on which they depend are maintained and enhanced. New collaborations stretching along the whole length of the supply chain from the farmers through to manufacturers and end-users must be brokered. So said Professor Lord Kumar Bhattacharya at a recent European road-mapping event run at the Warwick Manufacturing Group to set the agenda for research development and industrial application of sustainable raw materials.

Warwick Manufacturing Group has a history of working with industry at the forefront of technical development, and part of the way this is done is to provide information for our industrial partners in an accessible form. It is with this end in mind that the editors conceived this book. We believe that low environmental impact polymers will find increasing application because of a combination of rising prices of fossil origin raw materials, and a collective desire to reduce the detrimental effects on the environment of manufacturing activity. It is therefore an appropriate time to present information that should be a starting point for the experienced polymer practitioner wanting to take up new materials, and an introduction to the possibilities of the application of such materials to designers, specifiers, end users, and waste managers.

Nick Tucker
Mark Johnson
June 2004

Acknowledgements

The editors gratefully acknowledge the contributions of our chapter authors, who have enabled us to span such a diverse topic. Thanks are also due to our editor Frances Powers and her team, particularly Claire Griffiths (editorial assistance) and Sandra Hall (typesetting and cover design).

Dr Gordon Smith of Warwick Manufacturing Group provides an environment that encourages us to thrive.

1 Synthesis of Polymers from Sustainable Resource Origin Raw Materials

Leonard Y. Mwaikambo

1.1 Introduction

In recent years the use of renewable resources as chemical feedstocks for the synthesis of polymeric materials has attracted considerable attention. The reason for such activity is due to the finite nature of traditional petrochemical derived compounds in addition to economic and environmental considerations. The use of fossil-based raw materials for the synthesis of industrial and domestic chemicals is limited due to the high-energy processing routes required. Fossil-based by-products are also responsible for the emission of carbon dioxide (CO_2), while biological break down of plastics releases carbon dioxide and methane, heat-trapping green house gases, which lead to a rise in global temperature. Similarly petrochemical based materials are often non-biodegradable, thus difficult to dispose at the end of their life cycle thus making them environmentally undesirable. While estimates of the stocks of traditional fossil-based resources often vary, it has been suggested recently that mineral oil stocks will be severely depleted in approximately 80 years, natural gas in 70 years and coal in 700 years, but the economic impact could hit much sooner [1]. Thus a key goal of the coming years will be the development of sustainable raw materials for the chemical industry that will replace current fossil-based feedstocks. However, both the emission of carbon dioxide as a by-product of the fossil-based processes and the biological breakdown of plastics releases CO_2 and methane. The challenge for researchers is to develop natural and man-made synthesis that would reduce the emission of gases such as CO_2. One way of overcoming the release of such toxic gases is to search for carbon dioxide neutral processes by utilising renewable materials as the resource for the raw materials for polymer synthesis. There have been a number of recent review articles, which address the economic as well as chemical feasibility of achieving a plant-based renewable chemical and materials industry.

Commercial polymer products derived from renewable resources are mainly carbohydrates, namely cellulose, which include virgin and regenerated celluloses, starch including the directly starch-based products where starch is used more or less unmodified and the more recently developed fermented starch polymers, where the starch first has been fermented and then polymerised and sugar-based materials. There are also a range of biodegradable polymers derived from fossil resources, such as polycaprolactone (PCL) and certain co-polyesters. There are also commercially biodegradable polymers, and these are often blends of fossil derived and renewable resources. Further more, there are the still very small groups of protein and fatty acid based polymers. This last group of polymers can be obtained by chemical or biochemical manipulation from a renewable compound, including processed natural oils such as castor oil, soy oil, rapeseed oil and euphorbia. The list in this last class of polymer feedstocks depends on the type of oil bearing plant seeds. However, whichever approach is being investigated any chemical or biochemical manipulation of an existing renewable resource should be accomplished in an environmentally friendly as well as cost effective manner. **Table 1.1** shows market price for biodegradable and non-biodegradable materials.

In practice, this means that where a renewable feedstock is being substituted for an existing petrochemical one, the overall process should compete in price as well as generate less harmful waste than the existing technology it is replacing. It is evident from **Table 1.1** that biopolymers are more expensive than fossil-based polymers except high amylase starch. It is common practice to blend high and low cost materials to reduce the production cost, as well as to obtain properties not attainable by one component.

Table 1.1 Market prices for biodegradable and non-biodegradable materials		
Material	**Description**	**Price, £/kg**
Nature works (Cargill Dow)	Polylactic acid	2.30-4.50 (38 five years ago)
Novon (Novon International)	43% starch, 50% synthetic polymer, 7% others	2.40-2.60
Mater-Bi (Novamont)	Starch-PCL/PVA blends	3.40-4.40
Biopol (Monsanto until 1999)*	PHB/PHBV	6.0-9.60
High amylose starch	Produced via selective breeding	0.90-1.00
Cellulose acetate	Chemically modified	2.40-3.20
Low/high density polyethylene	Derived from petrochemicals	0.50-0.60
Polystyrene	Derived from petrochemicals	0.60
* The Intellectual Property Rights are currently owned by Metabolix [2] PVA stands for polyvinyl acetate, PHB stands for polyhdroxybutyrate and PHBV stands for polyhydroxybutyrate-co-3-hydroxyvalerate		

It is worth noting that biopolymers were originally developed for their biodegradability, with other characteristics often being of secondary importance, however, focus has, over the preceding years, shifted to more refined and value-added products. All of these polymers will be discussed in this article.

Carbohydrate materials are already used in many different polymer applications and may be used directly, or modified into secondary products. Cellophane, which is considered to be biodegradable, was the first cellulose-based polymer to be made in 1908 by Jacques E Brandenberger from the plant-derived structural polysaccharide, cellulose. However, cellophane's inherent biodegradability proved not suitable for certain applications and its processing was difficulty because of inter- and intra-molecular hydrogen bonding. Cellulose melts could only be formed by derivatising the hydrogen groups on each cellulose monomer to prevent hydrogen bonding. However, biodegradability decreases as the number of these derivatised hydroxyl group increases because degradation by microbial enzymes relies heavily on steric induction from hydroxyl group. In fact, polymer materials derived from cellulose is a fairly old and established market segment, which can be further subdivided into viscose and cellulose acetate polymers. Research then shifted away from cellulose to a more processable biopolymer, starch.

The other class of renewable materials are those that can be used as a monomer directly in polymer production, namely cashew nut shell liquid (CNSL) and tannins. However, these are discussed in another chapter of this book. This chapter will discuss naturally occurring plant-based chemicals, and the chemical methods applicable for manipulating existing plant feedstocks to monomers suitable for the preparation of polymeric materials. Biodegradable synthetic-plant monomer blends will also be discussed.

1.2 Carbohydrates as Renewable Resources

The generic term 'carbohydrate' includes monosaccharides, oligosaccharides and polysaccharides as well as substances derived from monosaccharides by reduction of the carbonyl group, by oxidation of one or more terminal groups to carboxylic acids, or by replacement of one or more hydroxyl group(s) by a hydrogen atom, an amino group, a thiol group or similar heteroatomic groups. It also includes derivatives of the compounds mentioned previously. Polysaccharide (glycan) is the name given to a macromolecule consisting of a large number of monosaccharide (glycose) residues joined to each other by glycosidic linkages. They form a heterogeneous group of polymers of different length and composition. Both α- and β-glycosidic linkages exist and these linkages may be located between the C_1 or C_2 of one sugar residue and the C_2, C_3, C_4, C_5 or C_6 of the second residue (**Figure 1.1**). A branched sugar results if more than two types of linkage are present in a single molecule. A polysaccharide may consist of one (homopolymer) or several types of monomers (heteropolymer) allowing production of an indefinite number of different types of polysaccharides namely plant and algae, animal and microbial based that are biomaterials. These biomaterials can either be used as they are (cellulose), depolymerised (regenerated cellulose) or polymerised such as hemicellulose polymers, polylactic acid and starch-based polymers. Other cellulose-based biopolymers are a modification of depolymerised cellulose namely diacetate and triacetate. A brief description of the synthesis of polysaccharide-based biopolymers (natural and man made) is given next.

1.2.1 Cellulose

Cellulose is a skeletal polysaccharide ubiquitous in the plant kingdom and one of the commonest naturally occurring crystalline polymers and comprises about 40% of all organic matter [3]. It is usually found in fibrous forms, and functions as the reinforcement for the amorphous lignin and hemicelluloses (discussed later) resulting in the composite structure found in all plants. The primary structure of cellulose is essentially a regular unbranched linear sequences of 1→4 linked β-D-glucose [4] (**Figure 1.1**). Neighbouring chains may form hydrogen bonds leading to the formation of microfibrils.

Examples of the naturally synthesised polysaccharide biopolymers obtained in the form of fibres are cotton, kapok and akund [5, 6]. Other polysaccharide biopolymers that are in fibrous form are the bast (hemp, flax, jute) and leaf fibres (sisal and henequen) [6]. Wood is another type of polysaccharide fibre.

Algae produces alginate, which is a polyuronide copolymer containing two building units namely β-D-mannuronic acid and α-L-guluronic acid, both of which are glycosidically linked 1→4 and occur in domains of blocks. The animal polysaccharides may be conveniently divided into two groups namely, the one encompassing the connecting tissue hyaluronate, the chondroitin sulfates, dermatan sulfate and keratan sulfate. These are all known to be linear polysaccharides with alternating 1→3 diequatorial and 1→4

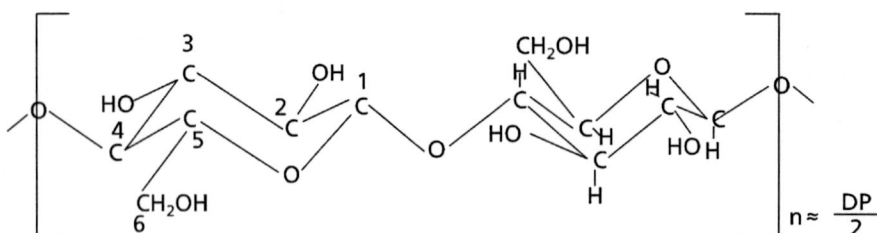

Figure 1.1 Schematic diagram of cellulose (DP is the degree of polymerisation)

diequatorial glycosidic linkages. A wide range of microorganisms produce the microbial polysaccharides.

However, polysaccharide cellulose is sometimes degraded (depolymerised) to produce regenerated cellulose. The process entails soaking cellulose in a base which is mainly sodium hydroxide followed by a reaction with carbon disulfide to give sodium xanthate as shown in **Figure 1.2.**

Sodium cellulose xanthate is then passed into an aqueous coagulating bath containing sulphuric acid and salts, usually sodium and zinc sulfates [5], although other salts such as bicarbonates are also being used. Carbon disulfide is lost producing regenerated cellulose with a degree of polymerisation of around 600. Products obtained from this process are in the form of fibre (viscose rayon) or sheets of cellophane with essentially the original cellulose chemical structure shown in **Figure 1.2.**

The development of thermoplastics based on cellulose and its derivatives dates back to the 19th century when work by Braconnot and Schönbein in the preparation of cellulose nitrate was followed by the introduction of camphor as a plasticiser in obtaining thermoplastic cellulose nitrate [6]. This early cellulose-based polymer, which could only be produced by moulding into film, was later followed by production of injection moulded cellulose acetate [7].

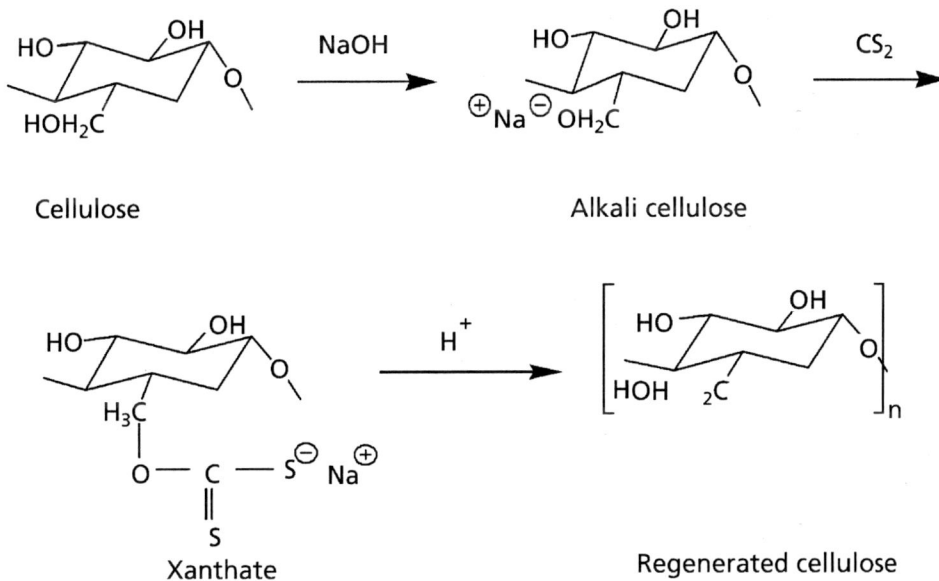

Figure 1.2 Conversion of cellulose to regenerated cellulose (viscose) using caustic soda

5

The process to produce cellulose acetate involves the de-polymerisation of the cellulose molecule and is called acetylation. Acetylation was first attempted by Schutzenberger in 1865 [8]. However, the reaction was found to be unstable and latter attempts used acetic acid and acetic anhydride producing triacetate commonly used for cigarette filters. In the early 1900s diacetate polymer was produced in which two hydroxyl groups are acetylated per ring leaving one hydroxyl group for hydrogen bonding. By controlling the replacement of hydroxyl groups by acetyl groups, at least 74% and not more than 92% of the hydroxyl groups are acetylated [9]. The availability of one hydroxyl group might be of great benefit for the development of composite materials using polar charged resin-matrices such as epoxies and phenolics. The diacetate is made in a two step process in which acetylation with acetic anhydride containing sulfuric acid as a catalyst gives the triacetate, and then controlled partial hydrolysis with dilute aqueous sulfuric acid gives the diacetate (**Figure 1.3**).

The development for cellulose diacetate (**Figure 1.3**) was exploited for the manufacture of 'canvas' wings aeroplanes used during the first World War. This is because cellulose diacetate dissolves is soluble and this characteristic leads to its use to strengthen and windproof canvas, ethyl ethanoate solvated cellulose diacetate is painted onto the canvas panels and by allowing the solvent to evaporate a strong composite structure is formed. Cellulose diacetate became the major thermoplastic moulding material in the early part of the twentieth century and is still used for plastic combs and toothbrushes. It is also used for apparel and photocopying transparent papers where they exhibit high thermal resistance. The benefit of cellulose is that it is inherently biodegradable and environmentally friendly and continuously renewable.

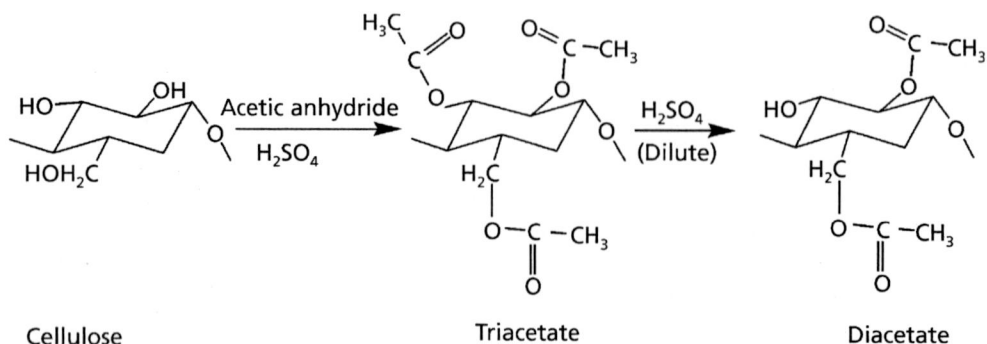

Figure 1.3 Conversion of cellulose to regenerated cellulose (triacetate and diacetate) using acetic anhydride in sulfuric acid medium

1.2.2 Starch

Starch is a particular form of cellulose and is a biopolymer of anhydroglucose units linked by α→4 linkages. It is an abundant (318 million metric tonnes) naturally occurring biodegradable polymer made up mainly of two polysaccharides namely amylose (molecular weight of up to 2,000,000) and amylopectin (100-400,000,000). Amylose is a linear 1→4 linked α-D-glucan (**Figure 1.4**) that occurs naturally in the crystalline form in starch granules [4] and amylopectin is a α-1,6- branched α-1,4 glucan polymer (**Figure 1.5**).

The amylose (**Figure 1.4**) and amylopectin (**Figure 1.5**) molecules are in an ordered arrangement within the starch granule and this gives crystallinity to the granule. Starch granules exhibit hydrophilic properties and strong inter-molecular association via hydrogen bonding due to the hydroxyl groups on the granule surface. This strong hydrogen

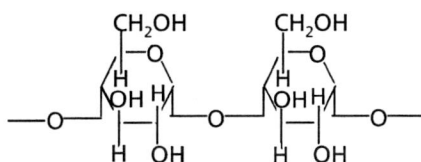

Figure 1.4 Linear α-1,4-glucan–amylose (200 to 2000 anhydroglucose units)

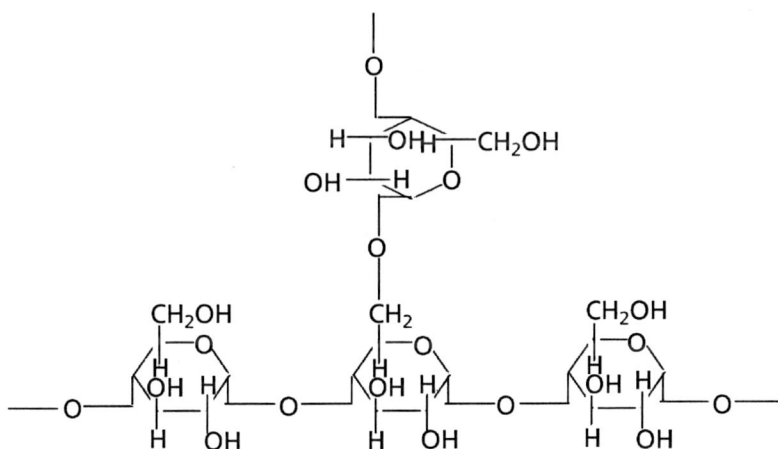

Figure 1.5 Branched polymer – amylopectin (α-1,4-glucan with 1,6-glucosidic linked branches containing 20-30 anhydroglucose units)

bonding association and crystallisation leads to poor thermal processing since the melting temperature is higher than the thermal decomposition temperature and degradation sets in before thermal melting. The hydrophilicity and thermal sensitivity renders the starch molecule unsuitable for thermoplastic applications. Starch can be used alone or compounded with synthetic polymers in amounts exceeding 50%.

However, it is believed that starch is not a true thermoplastic polymer but in the presence of plasticisers such as water or propan-1,2,3-triol (glycerol) and at higher temperatures (90-180 °C) and under shear, it readily melts and flows. 'Thermoplastic' starch has two main disadvantages when compared to most plastics currently in use. It is water-soluble and therefore exhibits poor environmental stability, poor mechanical properties and processability [10]. To improve some of these problems various physical and chemical modifications of starch molecules have been considered including blending, chemical derivatisation and graft copolymerisation [11, 12]. Its water resistance can be improved by mixing it with certain synthetic polymers and adding crosslinking agents such as calcium and zirconium salts and lignin [13].

Some of the early work in this field involved graft copolymerisation synthesis such as starch-*g*-polymethylacrylate and starch-*g*-polystyrene. This process entails generating free radicals on starch and reacting these free radicals with their respective vinyl monomers. The starch-based copolymers obtained can be injection moulded or extruded into films with properties similar to low density polyethylene [14]. However, these copolymers with vinyl polymer branches also have limited biodegradability because of the presence of non-degradable branch units although the properties are acceptable for end use applications. Carvalho and co-workers [15] report that the first starch plastics in the market were starch blended polyethylene. They were only bio-disintegrable and not completely biodegradable within a limited time frame. Data show that only the surface starch biodegraded leaving behind a recalcitrant polyethylene material. In the granular state it was used as filler for polyolefins and as a component in synthetic polymer blends. Most of the starch-based copolymers were made with the view that they could be composted. Companies such as Novamont, Novon International, National Starch and Chemical market a range of commercial bioplastics based upon de-structured thermoplastic starch [15, 16]. This starch can be regarded as a replacement for polystyrene pellets in packaging materials.

The modification of starch hydroxyl groups by esterification to form starch esters of an appropriate degree of substitution (ranging between 1.5-3) imparts thermoplasticity and water resistance [17]. Unmodified starch shows no thermal transition except the onset of thermal degradation at around 260 °C. Starch acetate with degree of substitution of 1.5 shows a sharp glass transition temperature (T_g) at 155 °C and starch propionate with the same degree of substitution would exhibit a T_g of 128 °C [16]. Plasticisers such as glycerol triacetate and diethyl succinate are completely miscible with starch esters (**Figure 1.6**) and can be used to improve processing. The water resistance of starch esters is greatly improved compared to the unmodified starch. The starch ester resin reinforced with plant fibres has been found to possess mechanical properties comparable to general-purpose polystyrene [18].

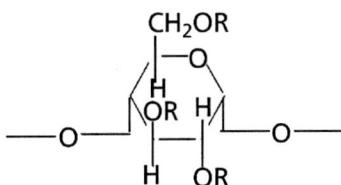

Figure 1.6 Starch triester

Where R (**Figure 1.6**) represents $-COOCH_3$ and $-CO(CH_2)_nCH_3$, n = 2-18. A random copolymer of mono, di and tri-substituted starch ester is shown in **Figure 1.7**.

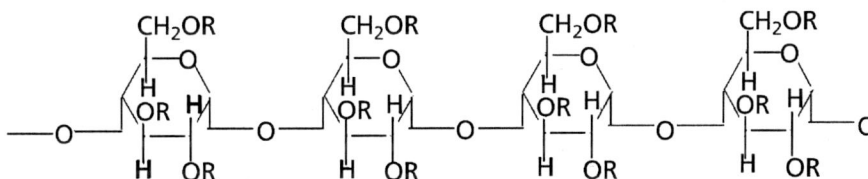

Figure 1.7 Random copolymer of mono, di and tri-substituted starch ester

Commercialised biodegradable plastics based on starch esters (**Figure 1.7**) and blends of starch with aliphatic polyesters are being marketed [18]. Appropriately formulated starch esters with plasticisers and other additives provide resin compositions, which can be used to make injection moulded products and for direct lamination onto Kraft paper. Starch acetates with a degree of substitution of up to around 2.5 undergo complete and rapid biodegradation. For example, about 70% of the carbon of the starch triacetates can be converted to carbon dioxide at 58 °C in 45 days [18].

The problems posed by the inclusion of the vinyl groups on the starch molecule, such as low level of biodegradation, resulted in a search for other chemical units that would impart complete degradation of the starch-based polymers. Research into obtaining chemical derivates of starch brought about the synthesis of a completely biodegradable polymer based on starch. A copolymer of starch and ε-caprolactone monomer has been synthesised in the presence of water (as a swelling agent) to gelatinise the starch granule to obtain starch-polycaprolactone (**Figure 1.8**) [10].

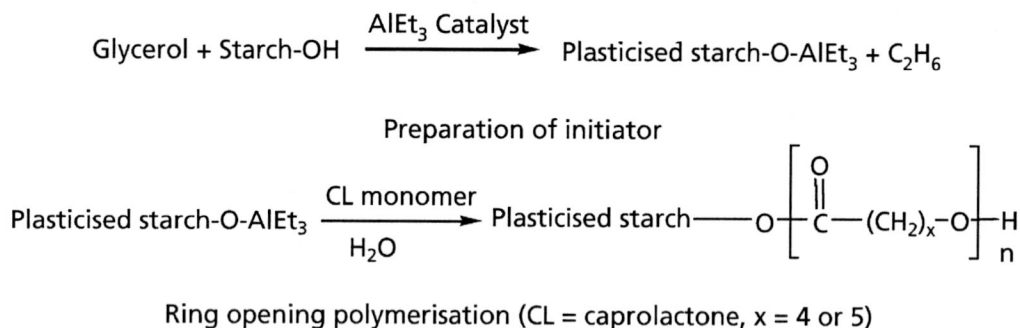

Glycerol + Starch-OH $\xrightarrow{\text{AlEt}_3 \text{ Catalyst}}$ Plasticised starch-O-AlEt$_3$ + C$_2$H$_6$

Preparation of initiator

Plasticised starch-O-AlEt$_3$ $\xrightarrow[\text{H}_2\text{O}]{\text{CL monomer}}$ Plasticised starch $-$O$-\left[\overset{\overset{\displaystyle O}{\|}}{C}-(CH_2)_x-O\right]_n$H

Ring opening polymerisation (CL = caprolactone, x = 4 or 5)

Figure 1.8 Grafting polycaprolactone onto starch

The starch-polycaprolactone copolymer (**Figure 1.8**) is completely biodegradable and has been used for lawn and leaf collection compost bags and agricultural mulch film [18]. In addition to the biopolymers discussed there are biomolecules that can be polymerised to produce biodegradable polymers namely lactic acid.

1.2.3 Hemicelluloses

Hemicellulose is the second most abundant polysaccharide after cellulose, comprising one-fourth to one-third of most plant materials [19], and in the past twenty years they have been used as feedstock for the production of sugars. The term 'hemicellulose' was coined in the 19th century [20]. Hemicelluloses are mostly heteropolysaccharides classified according to the sugar residues present namely arabinans, galactans, mannans and xylans, and they are either linear or branched polymers. The β-1→4-D-xylan is the most abundant hemicellulose (**Figure 1.9**) built from β-1→4-linked D-xylopyranosyl residue, which forms the linear backbone of the polymer. It is the second biggest constituent after cellulose in plant fibres and constitutes between 30-40% of the dry weight of plants.

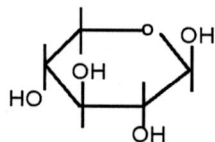

Figure 1.9 β-D-Xylopyranose (xylose)

The hemicelluloses (**Figure 1.9**) are not present in the cell wall as distinct bundles as is the case for α-cellulose, but instead appear as an individual molecule. They are more closely associated with lignin than cellulose, and exist in an amorphous state. The amorphous state of the hemicelluloses is due to the presence of many side groups, which prevent the close association between molecules required for the formation of crystalline regions.

Hydrophobic polymers have been synthesised using hemicelluloses through etherification or esterification of hydroxyl groups [19, 21, 22]. Hemicelluloses were extracted from maize bran via synthesised ring opening followed by laurylamine linkages with cyanoborohydride, both steps were carried out in water to produce a polymer. The polymer was extruded into a film that showed hydrophobic characteristics.

1.2.4 Polylactic acid

Lactic acid is produced through the fermentation of sugar feedstocks obtained from sugar beet or sugar cane, or from the conversion of starch from corn, potato peel, or another starch source. It can be polymerised to produce polylactic acid (PLA), which is already being used in commercial applications in drug encapsulation and in biodegradable medical devices. In 1932 Carothers produced a low molecular weight product by heating lactic acid under vacuum. An example is given of the hydrolysis of cornstarch, which yields simple sugars that can be readily fermented into lactic acid. The L-lactic acid is produced by bacterial fermentation of corn sugar (D-glucose). The process involves step polymerisation of lactic acid (α-hydroxyl acid) to produce poly (L-lactide) (PLLA). Direct condensation routes have also been reported by Mitsui Chemicals (Japan), which results in a high molecular weight PLLA polymer. The two polymerisation approaches are illustrated in **Figure 1.10**.

The direct condensation involves the removal of water of condensation by the use of solvents under high vacuum and temperatures. The resulting polymer can be used as is, or coupled using isocyanates, epoxides or peroxide to produce a range of molecular weights. Another route is to remove water under milder conditions, without solvent, to produce a cyclic intermediate dimmer, referred to as lactide (**Figure 1.10**).

This monomer is readily purified under vacuum distillation. Ring opening polymerisation of the dimmer is accomplished under heat, again without the need for solvent. By controlling the purity of the dimer it is possible to produce a wide range of molecular weights. Copolymerisation of poly lactic acid (PLA) (**Figure 1.10**) produces polymers with a wide range of properties from glassy and tough, to flexible materials. However, PLA is still more expensive than conventional plastics and exhibits low T_g (about 60 °C).

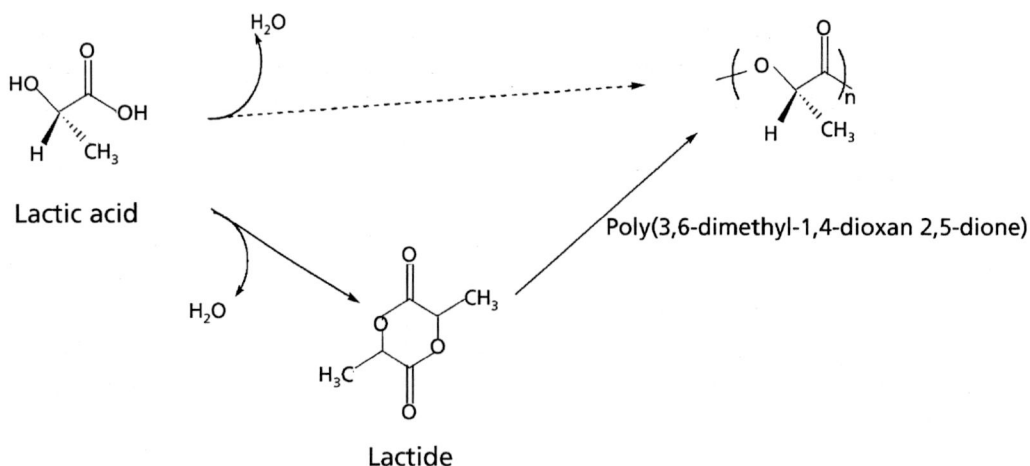

Figure 1.10 Polymerisation of polylactic acid

Copolymerisation with diglycidyl ether of bisphenol-A and other stiffer polymers improves the T_g hence the rigidity of PLA. Nevertheless the versatile characteristics of PLA have made it readily convertible into a variety of products such as fibre, using conventional melt-spinning processes, and mono component and bicomponent, continuous and staple fibres of various types are easily produced. Other end uses for PLA-based products are medical sutures and food packaging.

1.2.5 Polyhydroxyalkanoates (PHA)

To date, two types of polyhydroxyalkanoates (PHA) polymers have been developed, namely polyhydroxybutyrate (PHB) and polyhydroxyvalerates (PHV). These are based on fermented sugars (sucrose, gluctose and lactose) with different starch crops as starting materials. PHB was discovered as a constituent of bacterial cells in 1926 by Lemoigne [23] and it has properties of biodegradable thermoplastics and elastomers [24, 25]. Hocking and co-workers [11], and Jesudason and co-workers [26], prepared PHB polymers by polymerisation of racemic β-butyrolactone in the presence of AlMe₃/water catalyst and separated them into isotactic, atactic and syndiotactic fractions and the three fractions showed varying degradation characteristics [27, 28]. The development of PHB was meant to be as a substitute for polyvinyl chloride (PVC), polyethylene (PE) and polypropylene (PP) polymers. Polyhydroxyalkanoates are polyoxo-esters produced intracellularly by a wide variety of bacteria for the purpose of carbon and energy storage when they are placed in an environment of nutrient limitation and are therefore bio-renewable [29, 30]. Copolymers derived from hydroxybutyrate (HB) and hydroxyvalerates (HV) have been prepared that exhibit thermoplastic characteristics suitable for clinical applications and are biodegradable [28, 31, 32]. PHA sourced PHB is reported to exhibit properties similar to those of PP but its properties can be modified by blending or inclusion of other monomers [31, 33]. Other studies have indicated the possibility of producing PHA from palm oil.

1.3 Oils and Fats as Chemical Feedstocks

Oils and fats are the most important chemical feedstocks, using mainly the carboxy functionality of fatty acids, although recently research has been concentrating extensively on the fatty compounds for the selective functionalisation of the alkyl chain.

Oils and fats of plant and animal origin (not discussed in this chapter) constitute about 90% of the current renewable raw materials for their use based on the fatty acid carboxy group, while less than (10%) involve transformations of the alkyl chain. The chemical components present in natural oils are triglycerides and glyceroltriesters of saturated and unsaturated fatty acids. These oils and fats also possess two reactive sites, the double bonds of the unsaturated fatty acids and the carboxyl group, linking the acid to the glycerol. This implies that the mode of reaction involving the conversion of vegetable oil triglycerides to reactive monomers would vary depending on the degree of unsaturation and the functional groups present such as alkenes and epoxy groups. **Table 1.2** shows some of the vegetable oils and their components.

Table 1.2 Fatty acid composition in vegetable oils [34, 35], wt%					
Oil source	Oleic	Linoleic	Linolenic	Ricinoleic	Erucic
Flaxseed (linseed)	12-34	14-26	35-65	-	0
Rapeseed	55-65	18-25	8-11.5	-	0
Castor	3.0	4.2	-	85-90	41-52
Soybean	20-25	50-56	7-10	-	-
Grapeseed	-	66-72	-	-	-
Hazelnut	77-81	-	-	-	-
Sesame	33-47	33-48	-	-	-
Walnut		67-72	16-25	-	-

Natural oils vary in their physical properties even though they are made up of the same or similar fatty acids. This is because individual oils vary over a relatively large range in the proportions of the component fatty acids (**Table 1.2**) and there is some variation in the structure of the individual component triglycerides.

Table 1.2 shows varying quantities of fatty acids within and amongst different type sources. Oleic, linoleic and linolenic fatty acid are the most common and appear in large quantities with linoleic acid appearing the most followed by oleic and linolenic acid. Eruric acid is rare and is found in rapeseed oil only. **Figure 1.11** and **Figure 1.12** show alkenes only and alkene epoxy groups found in natural oils.

The functionalisation of triglycerides to reactive monomers is usually through the epoxidation route by the *in situ* performic acid procedure. The epoxidation process is carried out using appropriate concentrations of hydrogen peroxide, formic acid and tungsten (as catalyst medium). Temperature and time are crucial in these processes in order to obtain the desired composition. **Figure 1.13** shows a typical example of epoxidised double bonds of triglycerides that follows this route.

Chemo-enzymic epoxidation is applied because it suppresses, *in situ*, undesirable ring opening of the epoxide. The chemo-enzymic process initially involves the conversion of the unsaturated fatty acid or ester into unsaturated percarboxylic acid by a lipase-catalysed reaction with hydrogen peroxide, which is then self-epoxidised in an essentially intermolecular reaction [36-38]. The formation of mono and diglycerides is suppressed

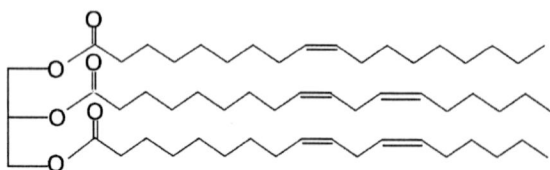

Figure 1.11 Typical fatty acids of plant oil

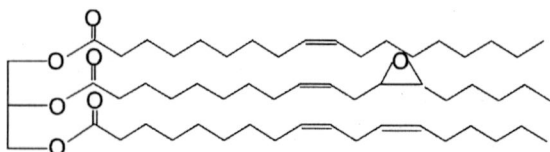

Figure 1.12 The general structure of epoxidised triglycerides

by adding about 5 mol% free fatty acid. Soybean and other vegetable oils have been oxidised by this method with conversion of over 90% monomer. **Figure 1.14** shows the conversion of vegetable oils into epoxidised triglyceride using the chemo-enzymic process.

Even with the highly unsaturated linseed oil the conversion is maintained [40]. The biocatalyst involved in this process lipase from *Candida antarctica* can be used several times.

When vegetable oils are subjected to perhydrolysis they are likewise epoxidised by the peroxy fatty acid formed in **Figure 1.14**. However, several other methods have been used to produce epoxides from unsaturated fatty acids using dioxiranes, hydrogen peroxide/methyl

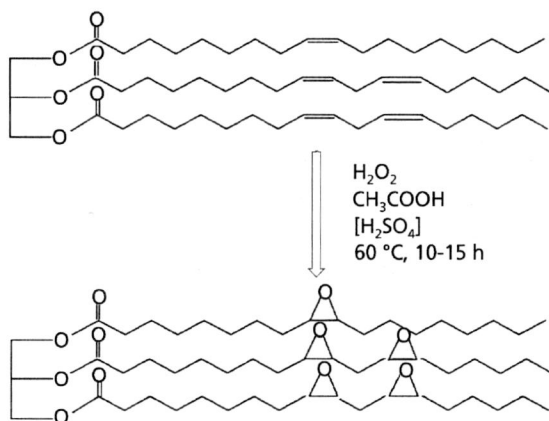

Figure 1.13 Epoxidised plant oil by peracetic acid *in situ*

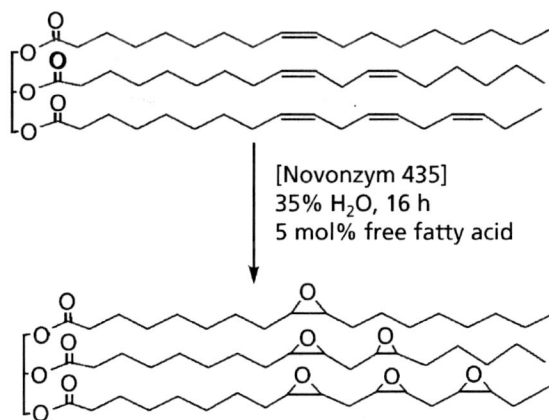

Figure 1.14 Chemo-enzymic epoxidation of vegetable oils [39]

trioxorhenium and hydrogen peroxide-peroxyphosphotungstenates [40]. These processes are performed in an organic solvent, which would need to be disposed of or recycled with a corresponding increase in the cost of the process.

However, the *in situ* epoxidisation involving this natural catalyst does occur without use of the enzyme following the rules of the Prileshajev epoxidation [40]. Epoxidised vegetable oils are currently used mainly as PVC stabilisers, while new applications have been discovered with photochemically initiated cationic curing [41].

The synthesis process to obtain monomeric materials from fatty acids entails the epoxidation of the alkenes and, depending on the monomer type required, this is followed by hydroxylation. Most natural oils can be used for polyol production but the type of polyol and the suitability of the polyol for its intended application are determined by the fatty acid composition. Work is in progress at the University of Warwick to epoxidise triglycerides of rapeseed and euphorbia oil and by using traditional methods to produce epoxies it is expected that vegetable oil epoxy polymers will be produced. These will further be used as resin-matrices for the manufacture of plant fibre composites.

1.3.1 Hydroxylation (Ring Opening) of Vegetable Oil

Epoxidised triglycerides can be used as crosslinking agents. However, most novel processes will require a second step converting the epoxidised triglycerides to low or highly hydroxylated polyol. For instance Sherringham and co-workers [42] have shown that *Euphorbia lagascae* can undergo two separate hydroxylation processes, one resulting in low hydroxylation and the other in high hydroxylation as shown in **Figure 1.15**. The level of hydroxylation is dependent on the catalyst and reaction temperature.

The synthesis processes shown in **Figure 1.15** are carried out *in situ* and the level of functionalisation will depend on whether the triglycerides contain either the alkenes only or the alkenes and epoxy groups. It will also depend on the amount of unsaturation and the number of epoxy groups present in the particular fatty acid. Sherringham and co-workers [42] have shown that the two processes can be reduced to only one process to cut down the cost by carefully controlling the processing temperature and catalyst. **Figure 1.15** also shows low hydroxylation through route (a) to (b) and high hydroxylation is obtained by carrying out *in situ* step (a) to (c) then (d).

Vegetable oil alkenes are therefore, useful sites that can undergo chemical modifications and under appropriate and tailored experimental conditions are converted to highly reactive groups such as hydroxyl groups. These are then used to produce resins for industrial applications.

Figure 1.16 shows a typical example of a scheme used to modify vegetable oil such as rapeseed and euphorbia oil. However, the same scheme can be applied to all types of vegetable oil such as those in **Table 1.2**.

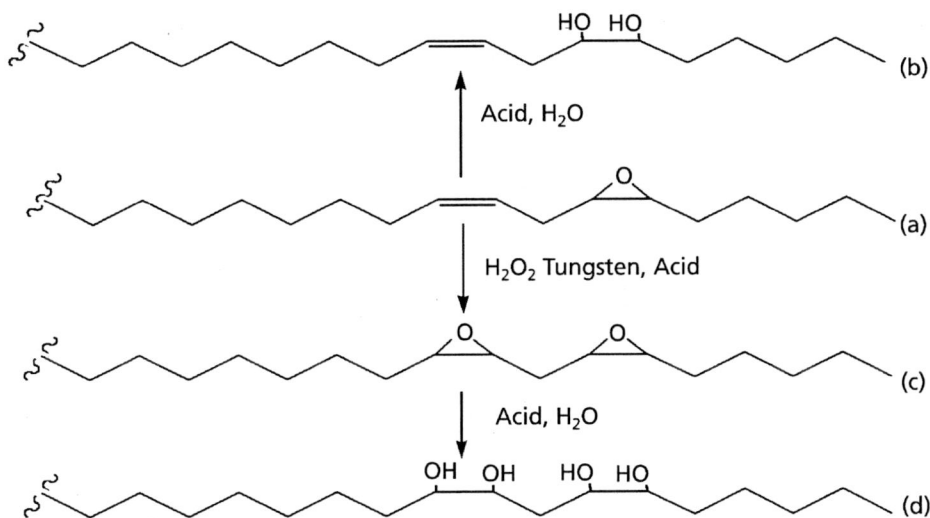

Figure 1.15 Epoxidation and hydroxylation of vernolic acid in *Euphorbia lagascae*

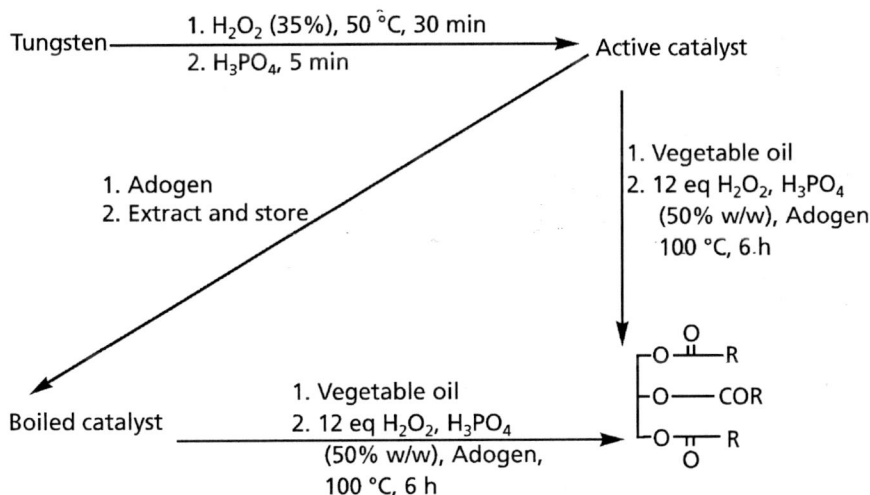

Figure 1.16 General scheme for chemical modification of vegetable oils to obtain reactive monomers

One of the methods applied to convert rapeseed oil into epoxide uses the Venturello-Ishii catalytic method (described by Noyori and co-workers [43]) in conjunction with hydrogen peroxide to effect conversion of unsaturated oil to the corresponding epoxide, which is then converted *in situ* to polyols. The reaction can be accomplished without the use of a solvent. This is because the active form of the catalyst is able to dissolve both in the aqueous phase and in the fatty acid phase of the reaction mixture, as long as the reaction mixture is properly emulsified. This catalyst can be prepared from tungsten powder, hydrogen peroxide and phosphoric acid, followed by addition of a phase transfer catalyst such as Adogen 464 [methyltrialkyl (C_8-C_{10}) ammonium chloride]. The catalyst can be extracted into an organic solvent such as dichloromethane and stored until needed, but typically it is added directly to the reaction. Reaction times vary from around an hour at 100 °C, to a few days at room temperature, with reasonably clean reactions. The catalyst can be recovered at the end of the reaction and recycled, thus avoiding further contaminating waste. A second stage is required to produce the polyols from the polyepoxides. The reaction is acid catalysed ring opening of the epoxide and uses about 50% phosphoric acid (by weight, with respect to the natural oil) at a reaction temperature of 100 °C and with vigorous stirring. This transformation can be accomplished relatively easily in organic solvents mixed with water, with a much lower reaction temperature and acid concentration. The overall process is shown in **Figure 1.16**.

There are problems in recovering the desired polyols from the acidic waste, however in contrast the aqueous phosphoric acid-catalysed process allows the product to be easily separated from the acid phase, which is re-usable as is the Venturello-Ishii catalyst. If the product is to be fully hydroxylated, then the two processes outlined above can be combined in a one-pot procedure, thus reducing the need for an additional work-up between the two steps.

1.3.2 Vegetable Oils as Feedstocks for Polyurethane Polymers

Figure 1.17 (a and b) shows the conversional technique used to polymerise polyols with di-isocyanate (4, 4′ methylene bis(diphenyl di-isocyanate) (MDI) with or without catalysts to produce a polyurethane polymer. However, different catalysts, accelerators and initiators have been explored to polymerise the monomers for the production of polyurethane. Toluene 2, 4 di-isocyanate (TDI) is also used to make polyurethanes.

Accelerators such as cobalt naphthenate, aromatic tertiary amines and free radical initiators such as methyl ethyl ketone peroxide, benzoyl peroxide, cumyl hydroperoxide are some of the common additives in the polyol-isocyanate polymerisation process. The choice of initiators and accelerators depends on the reactivity of the polymer and the temperature and the time desired for the cure reaction. The polymerisation of polyol

Isocyanate group

$$O=C=N-\bigcirc-CH_2-\bigcirc-N=C=O$$

(a) A di-isocyanate

$$H-O-CH_2-CH_2-O-H$$

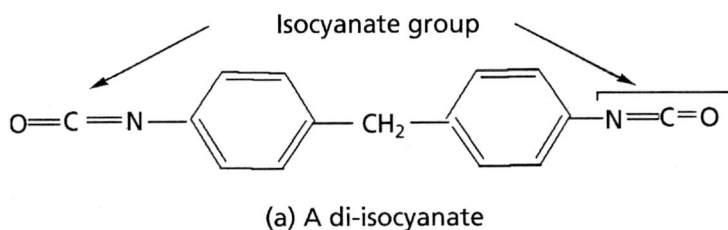

(b) A diol/polyol group

Figure 1.17 (a) Isocyanate and (b) diol or polyol group

(**Figure 1.17b**) into polyurethane can lead to either thermoplastic or thermoset resin-matrices.

Low crosslinking results in thermoplastics, for example types of elastomeric polyurethanes have been produced from rapeseed oil with low hydroxyl groups. High crosslinking results in thermoset materials implying that the starting monomer had a high content of hydroxyl groups. The structural versatility of the natural oil triglycerides allows the extent of chemical modification to be altered in a way that leads to a product with a wide range of properties. For example, resin viscosity can be altered by reducing the level of hydroxylation for the intended level of crosslinking which will hinder improvement in the thermal and mechanical properties of the cured resin, while at the same time it will alter the hydrophilicity and biodegradation of the polyurethane. This is an important property for the vegetable oil-based polyurethane as biodegradation but not hydrophilicity is the most sought after characteristic.

Current research work is directed at utilising vegetable oil triglycerides for the preparation of polymeric materials including composite manufacture. Triglycerides are therefore chemically modified as discussed previously into polyols, which are then used as polymer precursors. The degree of hydroxylation of particular triglycerids depends on the conversion of the double bonds to exorane groups. Optimised epoxidation provides room for optimal conversion to hydroxyl groups (polyols), which then become the reactive sites for the formation of polyurethanes. The reaction of diols with diisocyanate to give polyurethanes is often carried out in the presence of a catalyst. If the hydroxyl monomer has more than two hydroxyl groups then crosslinking of the polymer can occur. The most commonly used diisocyanates are TDI and MDI. For a given polyol-diisocynate system, the higher the hydroxyl content of the polyol, the greater the degree of crosslinking and hence the more rigid the polyurethane formed. Conversely, the longer the diol component, the more flexible the polyurethane formed. By altering the hydroxyl content of the feedstocks it is possible to prepare both very flexible elastomers and very rigid polyurethanes. It is also possible to use the highly hydroxylated materials in the preparation of rigid lightweight foams.

Polyurethanes have been used for a large variety of applications from flexible comfort foams for use as seat cushions and bedding and the like, to the matrices of high performance composites. Other applications include coatings, adhesives and sealants and use in rapid prototyping of pre-production/low production runs of commercial products. The production of polyurethane from renewable source will also result in increased agricultural and industrial economies. **Figure 1.18** shows the most general scheme applied to functionalise vegetable oils.

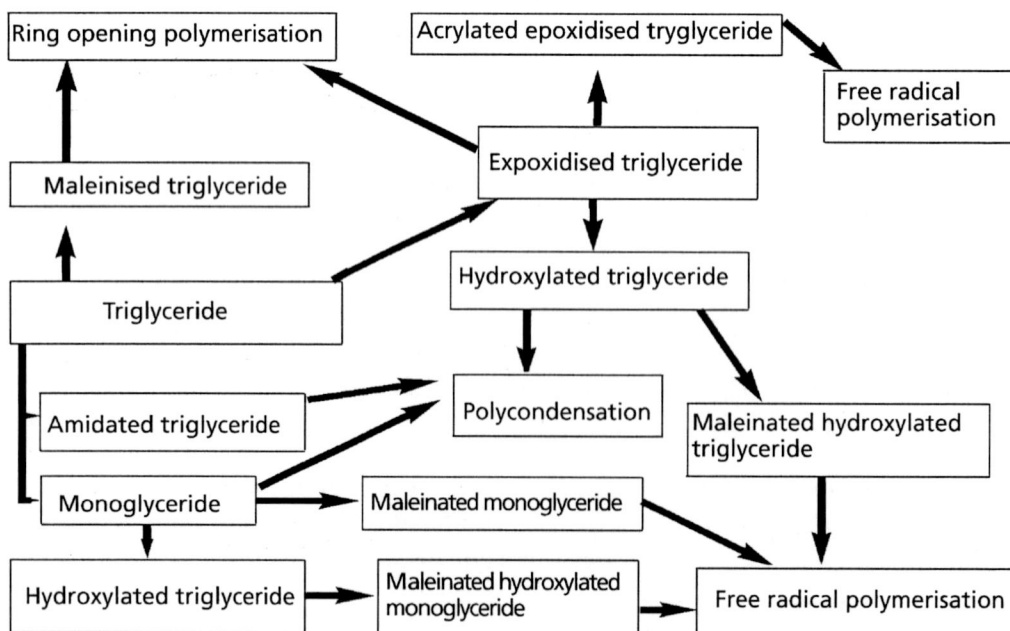

Figure 1.18 *In situ* functionalisation and polymerisation of fatty acid of plant origin

Figure 1.18 indicates the diversity of the products that can be obtained by functionalising soybean oil triglycerides to reactive monomers, which is used in resin manufacture. However, it worth noting that the reaction pathways shown in **Figure 1.18** would also apply to the synthesis of any other naturally occurring triglyceride.

1.4 Conclusions

This chapter has shown that naturally occurring polymeric polysaccharides can be used in their virgin form to produce materials for industrial applications. It has also been demonstrated that these same polysaccharide materials can undergo physical and chemical modifications to produce useful polymers such as regenerated and acetylated celluloses with the potential for industrial end uses.

In addition to polysaccharides, vegetable oils are currently offering another renewable source of industrial raw materials that can be used to produce items ranging from plasticisers (epoxidised triglycerides) to resinous materials such as polyurethane.

However, whereas, the market cost of some of the polysaccharides are low enough to compete with fossil-based products such as cellulosic fibres and amylase starch (**Table 1.1**) the market cost of chemically modified natural resources such as oil/fat-based products is not worth the investment. There is, therefore, need for viable technologies that would reduce the production costs of the plant oils and fats-based naturally occurring materials. The production of polymers from non-fossil renewable and biodegradable natural resources will have economic and environmental advantages.

References

1.   T.U. Gerngross and S.C. Slater, *Scientific American*, 2000, No.8, 36.

2.   M. Johnson, *Personal Communication*, 2002.

3.   Y. Cao and H. Tan, *Carbohydrate Research*, 2002, **337**, 14, 1291.

4.   E. Atkins, *Applied Fibre Science, Volume 3*, Ed., F. Happey, Academic Press, London, UK, 1979.

5.   D.R. Perry, *Identification of Textile Materials*, The Textile Institute, Manara Printing Services, London, UK, 1975.

6.   R.H. Kirby, *Vegetable Fibres*, Interscience Publishers, London, UK, 1963.

7.   H. Warth, R. Mülhaupt and J. Schätzle, *Journal of Applied Polymer Science*, 1997, **64**, 2, 231.

8.   D. Walton and P. Lorimer, *Polymers*, Oxford Chemistry Primers No.85, Oxford Science Publications, Oxford, UK, 2000.

9.   C.A. Farnfield and P.J. Alvey, *Textile Terms and Definitions*, 7th Edition, The Textile Institute, Manchester, UK, 1975.

10. E-J. Choi, C-H. Kim and J-K. Park, *Macromolecules*, 1999, **32**, 22, 7402.

11. P.J. Hocking, R.H. Marchessault, M.R. Timmins, T.M. Scherer, R.W. Lenz and R.C. Fuller, *Macromolekulare Chemie - Rapid Communications*, 1994, **15**, 6, 447.

12. G.F. Fanta and E.B. Bargley in *Encyclopedia of Polymer Science Technology*, *Supplement 2*, Eds., J.S. Trefil, P. Ceruzzi and H.J. Morowitz, Wiley Interscience, New York, NY, USA, 1977, p.665.

13. R.L. Shogren, J.W. Lawton, K.F. Tiefenbacher and L. Chen, *Journal of Applied Polymer Science*, 1998, **68**, 13, 2129

14. J.L. Willet, *Journal of Applied Polymer Science*, 1994, **54**, 1685.

15. A.J.F. Carvalho, A.A.S. Curvelo and J.A.M. Agnelli, Thermoplastic starch composites, in *Proceedings from the Third International Symposium on Natural Polymers and Composites*, ISNaPol/2000, Eds., L.H.C. Mattoso, A. Leao and E. Frollini, Sao Pedro, Brazil, 2000, 212-217.

16. R. Narayan, M. Kotnis, H. Tanaka and N. Miyachi, inventors; Evercom, Inc., assignee; US Patent 5728824, 1998.

17. R. Narayan, Microfibre reinforced biodegradable starch ester composites with enhanced shock absorbance and processability, U.S. Patent allowed, 1997.

18. S. Bloembergen and R. Narayan, inventors; Evercom, Inc., assignee; US Patent 5,462,983, 1995.

19. E. Fredon, R. Granet, R. Zerrouki, P. Krausz, L. Saulnier, J.F. Thibault, J. Rosier and C. Petit, *Carbohydrate Polymers*, 2002, **49**, 1, 1.

20. J. Puls and K. Poutanen in *Enzyme Systems for Lignocellulose Degradation*, Ed., M.P. Coughlan, Elsevier Applied Science, New York, NY, USA. 1989.

21. M. Vincendon, *Journal of Applied Polymer Science*, 1998, **67**, 1, 455.

22. R. Sun, J.M. Fang, L. Mott and J. Bolton, *Holzforschung*, 1999, 53, 3, 253

23. *Biomaterials: Novel Materials from Biological Sources*, Ed., D. Byrom, Stockton Press, New York, NY, USA, 1991.

24. P.J. Barham, A. Keller, E.L. Otun and P.A. Holmes, *Journal of Materials Science*, 1984, **19**, 9, 2781.

25. T.L. Bluhm, G.K. Hamer, R.H. Marchessault, C.A. Fyfe and R.P. Veregin, *Macromolecules*, 1986, **19**, 11, 2871.

26. J.J. Jesudason, R.H. Marchessault and T. Saito, *Journal of Environmental Polymer Degradation*, 1993, **1**, 2, 89.

27. C. Jaimes, R. Dobreva-Schué, O. Giani-Beaune, F. Schué, W. Amass and A. Amass, *Polymer International*, 1999, **48**, 1, 23.

28. Y. Jia, T.J. Kappock, T. Frick, A.J. Sinskey and J. Stubbe, *Biochemistry*, 2000, **39**, 14, 3927.

29. U. Müh, A.J. Sinskey, D.P. Kirby, W.S. Lane and J. Stubbe, *Biochemistry*, 1999, **38**, 826 826.

30. A. Steinbüchel in *Biomaterials, Novel Materials from Biological Sources*, Ed., D Byrom, Macmillan Publishers Ltd., Basingstoke, UK, 1991.

31. L.G. Griffith, *Acta Materialia*, 2000, **48**, 1, 263.

32. S. Slater, T.A. Mitsky, K.L. Houmiel, M. Hao, S.E. Reiser, N.B. Taylor, M. Tran, H.E. Valentin, D.J. Rodroguez, D.A. Stone, S.R. Padgette, G. Kishore and K.J. Gruys, *Nature Biotechnology*, 1999, **17**, 10, 1011.

33. M. Yasin, S.J. Holland and B.J. Tighe, *Biomaterials*, 1990, **11**, 7, 451.

34. B.R.T. Keene in *Proceedings of The Development of Euphorbia lagascae as a New Oil Crop within the European Community (PL98/4460), Concerted Action: Workshop 1: Summary of Presentations*, Winchester, UK, 1999, p68.

35. C. Bastioli in *Starch Polymer Composites in Degradable Polymers*, Eds., G. Scott and D. Gilead, Chapman and Hall, Cambridge, UK, 1995.

36. F. Björkling, H. Frykman, S.E. Godtfredsen and O. Kirk, *Tetrahedron*, 1992, **48**, 22, 4587.

37. M.R.G. Klass and S. Warwel, *Journal of Molecular Catalysis A*, 1997, **117**, 1-3, 311.

38. S. Warwel and M.R.G. Klass, *Journal of Molecular Catalysis B*, 1995, **1**, 1 29.

39. M.R.G. Klass and S Warwel, *Industrial Crops and Products*, 1999, **9**, 2, 125.

40. J.O. Metzger, U. Biermann, W. Friedt, S. Lang, W. Lühs, G. Machmüller, R. Klass, H.J. Schäfer and M.P. Schneider, *Angewandt Chemie International*, 2000, **39**, 2206.

41. J.V. Crivello and R. Narayan, *Chemistry of Materials*, 1992, **4**, 3, 692.

42. J.A. Sherringham, A.J. Clark and B.R.T. Keene, *Lipid Technology*, 2000, **12**, 129.

43. R. Noyori, M. Aoki and K. Sato, *Chemical Communications*, 2003, 1997.

2 Chemistry and Biology of Polymer Degradation

Maurizio Avella, Mario Malinconico and Pierangelo Orlando

2.1 Introduction

Many synthetic polymers are produced and utilised because they are resistant to chemical and physical degradation. Synthetic polymers were developed for durability and resistance to all forms of degradation. Polymers are widely accepted for these properties and they are relatively inexpensive to make compared to the other products that polymers replace such as glass, paper and cardboard. These conveniences enhance the quality and comfort of life in modern society.

One major drawback to using polymers is the problem with disposal. Since they are somewhat resistant to degradation, polymers tend to accumulate in what is today's most popular disposal system, the landfill. This leads to questions about what effects polymers have on the environment, whether they biodegrade at all, and if they do, what effect the products of biodegradation have on the environment.

There are several different types of degradation that can occur in the environment. These include biodegradation, photodegradation, oxidation, and hydrolysis [1]. Often, laypeople will lump all of these processes together and call them biodegradation. However, most chemists and biologists will agree that biodegradation involves enzymically promoted breakdown caused by living organisms, usually micro-organisms [2].

The challenge posed to chemists and biochemists is to determine to what extent polymers biodegrade. Several chemists, biochemists, polymer manufacturers and independent testing groups have taken up this challenge. However, there is some disagreement as to what constitutes a biodegradable polymer. Polymer manufacturers are willing to accept breakdown of a polymer structure into smaller fragments that still contain the polymer. Conversely, environmentalists will not stop short of complete degradation into the monomer units or mineralisation into products usable by organisms.

This conflict has created a need for a standard system for determining the degree to which a polymer can biodegrade. One proposed system includes the following: measuring physico-mechanical changes (changes in morphology and physical properties), chemical changes and products formed, and weight loss. When comparing the degree to which different polymers biodegrade, several factors must be taken into consideration. The first of these factors is the environment. Polymers may be tested in a natural or simulated environment. Simulated testing environments can be normal or accelerated. These are utilised to determine whether the polymer begins to degrade following disposal or while it is still in its intended use [3].

The next aspect of biodegradable polymers, to consider, is polymer concentration. The polymer may have a high or low mass weight compared to its volume. In addition, the form of the polymer may range from a thin film to a thick building structure. Finally, the environmental effects of the polymer must be considered. A polymer that biodegrades is of little value if the products that form are found to contaminate water supplies or be toxic to living organisms in the environment. One group that utilises this system of comparison is Krupp and Jewell at Cornell University in Ithaca, New York, USA. Their study was conducted on thin films of various polymers that contained varying amounts of starch. In 1989, these polymers were claimed to be biodegradable by their respective manufacturers [4]. The reason starch is added to the polymers is that some micro-organisms utilise starch, a polymer of glucose, as a nutrient source and secrete enzymes to break it down for consumption. These enzymes act on the polymer, provided the polymer could be broken down.

Starch is added to polymers in two basic ways. The first method of starch addition is the attachment of acrylic acid segments at various locations on the polymer chain. The acid segments hydrogen bond with the starch, allowing the mixture of components. The second way starch is added is to graft a polyethylene (PE) molecule to a starch molecule. This grafted molecule readily reacts with pure PE or starch to produce a copolymer [5]. PE films that contained starch show no evidence of anything other than biodegradation of the starch component. In addition, based upon mass loss, most of the materials tested were affected identically by the aerobic and anaerobic bacteria. The polyhydroxybutyrate and polyhydroxyvalerate (PHB/PHV) film readily biodegraded. After four weeks of exposure in the bioreactors, the film had a mass loss of 90-100%. At the same time, little mass was lost after being exposed to the hot water bath. In addition, the film exerted a biological oxygen demand (BOD) that was 61% of the theoretical BOD (compared to around 70% for a cellulose sorghum mixture) [6].

This study, along with others, points to some conclusions about the biodegradability of polymers. The first conclusion is that naturally occurring polymers biodegrade and chemically modified natural polymers may biodegrade but this depends on the extent of modification. Next, synthetic addition polymers, like those described in the beginning, with carbon as the only atom in the backbone do not biodegrade at molecular weights

(M_w) above 500. If an addition polymer contains atoms other than carbon in the backbone, it may biodegrade depending on any attached functional groups. Synthetic condensation polymers are generally biodegradable to different extents depending on chain coupling (ester > ether> amide > urethane), morphology (amorphous > crystalline), molecular weight (lower > higher), and hydrophilic polymers degrade faster than hydrophobic. However, if a polymer is water soluble, that does not necessarily mean that it is biodegradable [7].

The biodegradation of these polymers proceeds by hydrolysis and oxidation. The presence of hydrolysable and/or oxidisable linkages in the polymer main chain, the presence of suitable substituents, correct stereoconfiguration, balance of hydrophobicity and hydrophilicity, and conformational flexibility contribute to the biodegradability of the polymers [3]. Biodegradable polymers may be divided into three classes:

(a) natural polymers originating from plant or animal resources, (e.g., cellulose, starch, protein, collagen, etc.),

(b) biosynthetic polymers produced by fermentation processes by microorganisms, (e.g., poly-hydroxy alkanoates),

(c) certain synthetic polymers possessing the biodegradable properties explained earlier, (e.g., polycaprolactone (PCL) and polylactic acid (PLA)).

Natural polymers like cellulose and starch are not thermally processable unless modified. Cellulose has a degradation temperature below its melting temperature and hence cannot be processed in the melt. Moreover, because of its complex morphology of crystalline regions and hydrogen bonding, cellulose is difficult to dissolve in common solvents. Processing with the help of a suitable solvent has been possible with *N*-methylmorpholine *N*-oxide (N-MMNO) and water [8] or dimethylacetamide (DMAc)/lithium chloride (LiCl) [9, 10]. Hence cellulose can be spun into lyotropic liquid crystalline fibres, but the process is complicated and dangerous due to rigorous solvent handling protocols [11]. However, modified cellulose, such as cellulose esters, can be melt processed as a thermoplastic polymer, since their melting temperature can be significantly reduced below their degradation point [12]. Similarly, modified starch has been incorporated into polymer blend systems to provide biodegradability [13]. On the other side, PHA are high molecular weight polymers produced by bacteria or other microorganisms.

A variety of PHA can be synthesised by bacteria depending on the species of bacterium and the compounds supplied as carbon sources for growth. PHA are biodegradable and biocompatible thermoplastics which can be melted and moulded [14]. These polymers have attracted attention because of their potential use in both medical and industrial applications owing to their properties of biocompatibility and biodegradability. The other group of biodegradable polymers that has been of interest includes the synthetic polymers like PLA and PCL. However, in the case of these polymers, morphology affects the degradation process [15]. Amorphous regions are preferentially degraded first due to their greater surface compared to the spherulites. A similar effect of morphology on degradation has been observed in the case of more crystalline poly (alpha-hydroxy acids). However, varying the stereoisomer ratio of D- or L-lactate units, PLA can be made semicrystalline or fully amorphous [16].

The variability of properties and performances of the polymers previously mentioned has prompted research groups to investigate the possibility of studying their behaviour in blends and composites. One of the aims of this present chapter is to report on the state-of-the-art of such research.

The blending approach is a fundamental tool in many areas of macromolecular material in view of their technological applications. In principle, blending can be considered as a form of compounding and this may allow many converters to make their own trials, i.e., try to modify some deficiencies of a polymer by blending it in an extruder with a second polymer. Blending is more economic than copolymerisation, which is an alternative method to modify the performance of a given polymer [17]. The only serious limitation for blending is the mutual repulsion of different polymers [18]. In fact, unlike low molecular weight compounds, different polymers have a high tendency to demix, due to repulsive entropic forces. Only the creation of specific interactions, like polar interactions or the addition of polymeric emulsifiers, may increase the enthalpic contribution to mix, and render many polymer blends stable, compatible and technologically attractive [18]. The same considerations hold for composites, where the filler and the matrix must have a stable interface to allow a good dispersion and, hence, a strong mechanical response. The above requirements must, of course, be fulfilled when the blend components belong to the family of biodegradable polymers or when the filler is a natural fibre. This challenge has been faced by polymer chemistry and will be the subject of the following sections.

2.2 Microbial Degradation of Natural and Synthetic Polyesters

One-fifth by volume of solid urban waste, in the modern economy, is disposable materials manufactured from synthetic polymers that are barely degradable, such as PE, polypropylene (PP) or polystyrene. Often, these materials are collected in landfill sites or subjected to incineration in municipal plants, with increasing costs and biological hazards. Only a few thermoplastic materials are subject to differential collection and industrial recycling. Furthermore, some of these materials come directly into the environment. Consequently, pollution by plastic materials has become a serious problem and has stimulated, during the last decade, the increased interest in environmentally biodegradable polymers [19].

The biodegradation of these materials is a process that involves chemo-organotrophic prokaryotic (bacteria) and eukaryotic (fungi and protozoa) microorganisms, capable of excreting enzymes (depolymerases) that degrade the polymeric matrix and/or utilise water-soluble oligomers released as nutrients for the production of energy, biosynthesis precursors and energy-storing materials. Conceptually, the biodegradation of plastic materials must be distinguished from the ageing process that normally takes place in the environment by photochemical degradation and non-specific biological attacks, in that this kind of degradation can be accompanied, or not, by a complete bio-assimilation of the compounds released. Furthermore, this type of degradation could give rise to a 'stealth' form of pollution by releasing into the environment cytotoxic or phytotoxic substances, as demonstrated for some kind of polyesters [20]. On the contrary, in the case of biodegradable polymers these mechanisms contribute in a synergistic mode to the degradation cycle by making the inner part of the material more accessible to the water and consequently, to the enzymes and colonising microorganisms.

To promise that the biodegradability of plastic materials is a necessary, but not sufficient condition, for the solution of the pollution problem and spontaneous environmental biodegradation should, potentially, be limited to the material that 'inevitably' comes into the environment. The biodegradation process would preferentially take place in industries specialising in biotechnological waste treatment in plants for microbial biomass and biopolymer production or for anaerobic biogas formation. In fact, to achieve the greatest efficiency in the biodegradation process, it is necessary to utilise fermentative plans, selected or genetically modified microorganisms and, at the same time, to warrant the biocontainment. Aerobic biodegradation is a fast, robust process that yields production of biomass and of commercially interesting products, but in liquid media it incurs high costs related to culture aeration, agitation and cooling. Anaerobic fermentation is a slow process, with requirements for the maximum conversion efficiency the selection of the appropriate microbial strains, the stabilisation of the culture after inoculum and the incurring of costs related to the extraction of products. Furthermore it yields products of low value, such as fatty acids or alcohol, or commercially interesting, but of great volume, such as fossil fuels, consequently, it is economically viable if fermentable wastes are utilised [21-23].

Water is the natural vehicle for the extracellular enzymes that hydrolyse at the interfacial surface of water-lipid hydrophobic polymers such as polyesters. The characteristics that influence the biodegradation cycle greatly are the hydrophilic grade of the polymeric matrix, the presence in the principal, or lateral chains, of chemical bonds that involve oxygen or nitrogen atoms, the flexibility of the chain and a low level of crystallinity. Thus, these typical features of a biodegradable polymer strongly limit their industrial applications. The research effort in the last decade has been (and it will be more so in the future) to consolidate both these demands. Consequently, many biodegradable and thermoplastic polyesters and *co*-polyesters of natural origin, such as the microbial PHB and PHA [6, 23], or of synthetic origin such as PLA, poly(L-lactide) (PLLA) and PCL, have been studied and utilised for industrial or bio-medical applications. Their blends, that provide better material properties, are under study (see next chapter).

Microorganisms are able to utilise, for their metabolic purposes, all natural polymers; 'if *Nature* makes it (or recognises it), *Nature* will eat it' and aliphatic polyesters and *co*-polyesters, recognised as common microbial energy storing materials, are easily biodegraded. Aromatic polyesters such as polyethylene terephthalate (PET) and polyhydroxybutrate (PBT) are quite insensitive to biodegradation, confirming that extracellular depolymerases don't hydrolyse the aromatic ester bond. On the contrary, random *co*-polyesters containing aliphatic and aromatic monomers like *co*-polyester of terephthalic and adipic acids with 1,4 butandiol (BTA) are biodegradable in relation to the length of the aromatic sequence, in that aromatic monomers and short aromatic chains can be internalised in the microbial cell and further metabolised [24].

Extracellular depolymerases can be distinguished in endo or exo forms if the hydrolysis involves the internal bond or proceeds from the ends of polymer chain, respectively (see **Figure 2.1**). Polyester depolymerases have been isolated and characterised from the *in vitro* culture of pre-selected microorganisms, at optimum pH and salt conditions and in the presence of polyester the as sole energy and carbon source. In nature, aerobic and anaerobic microorganisms can utilise the water-soluble oligomers released in a relation of commensalisms with the degrading microorganisms [25]. A schematic representation of the metabolic destiny of aliphatic polyesters and of the establishing microbial consortia (in relation to the presence of fast-degraded polysaccharides such as starch and cellulose) is depicted in **Figure 2.2**. The hydroxyl-acids released from PHA and PCL come into the general catabolic pathway of the 'β-oxidation of fatty acids' in which the carboxylic chain is progressively shortened by two carbon units to forms acetyl-coenzyme A (acetyl-SCoA). Acetyl-SCoA is a branch point of metabolism in that:

i) in the aerobic metabolism it enters into the Krebs cycle where it is mineralised into CO_2 and H_2O for energy production and formation of precursors for biosynthetic reactions;

ii) in anaerobic acidic fermentation it can be converted to acetate and butyrate;

iii) in anaerobic methanogen fermentation it is utilised for auxotrophic assimilation of CO_2 by reductive carboxylation to pyruvate;

iv) it is utilised as a precursor for fatty acid synthesis and consequently of aliphatic polyesters and *co*-polyesters in PHA synthesising bacteria.

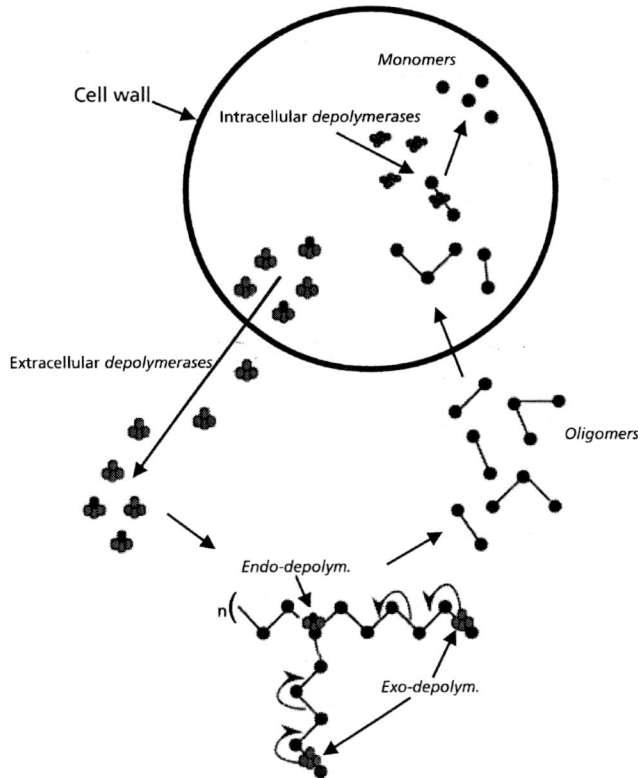

Figure 2.1 Action mechanism of extracellular depolymerisations

L-lactate released from PLLA can be converted to pyruvate, another branch point of metabolism, or to propionate *via* acrylil-SCoA formation. The presence of fast-degraded polysaccharides, *via* pyruvate formation, slow down polyester degradation, in observance of the general law that regulates biodegradation: microbial opportunism. **Figure 2.3** illustrates a schematic diagram of the most common tests of biodegradability, these will be analysed in detail for various kinds of polyesters.

2.2.1 Polyhydroxyalkanoates

Literature focused on the physiological and engineering aspects of microbial polyester (in particular PHB) and *co*-polyester production, including complete description of PHA synthesis and degradation pathways for relevant micro-organisms, detection, analysis

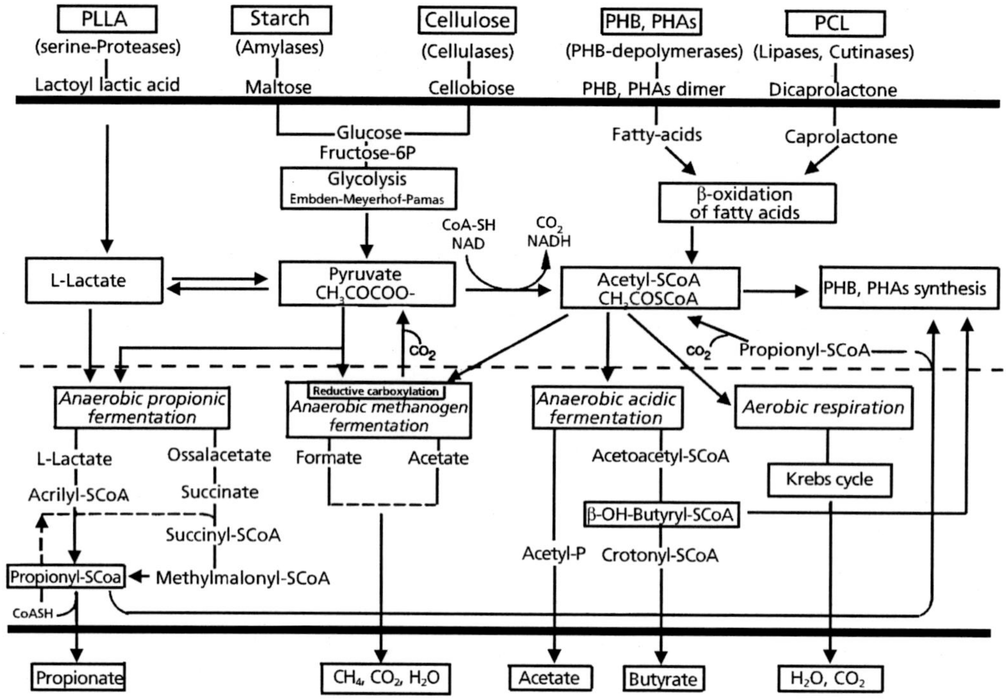

Figure 2.2 Metabolic destiny of aliphatic polyesters and establishing microbial consortia

and extraction of PHA, strategies and plans for fermentative production have been published [6, 23]. Furthermore, almost all the enzymes involved in PHA synthesis and degradation have been characterised and the related genes cloned [6, 23, 26]. It is interesting to note that the biosynthesis and composition of *co*-polyesters such as PHB/PHV and poly(3-hydroxybutyrate-*co*-4-hydroxybutyrate) is influenced, and can be directed, by the choice of the substrates utilised as a source of carbon in the culture medium, but only to a relative extent [6]. In *Pseudomonas oleovorans*, for example [6], by varying the size of fatty acids or of the precursor alkanes in the culture medium, random *co*-polyesters with monomer units ranging to C_6-C_{12} are obtained but, only for the fatty acids C_6-C_9 is a direct correlation in monomer composition (ranging from 100% to 65%) achieved. On the contrary, the amount of C_{10} monomer is the highest when 3-hydroxybutyrate (57%), butyrate (33%), dodecanoate (32%), hexadecanoate (30%) are used instead of decanoate (17%), thus indicating that fatty acid size influences, to a limited extent, the PHA synthetic pathways and the monomer composition of PHA.

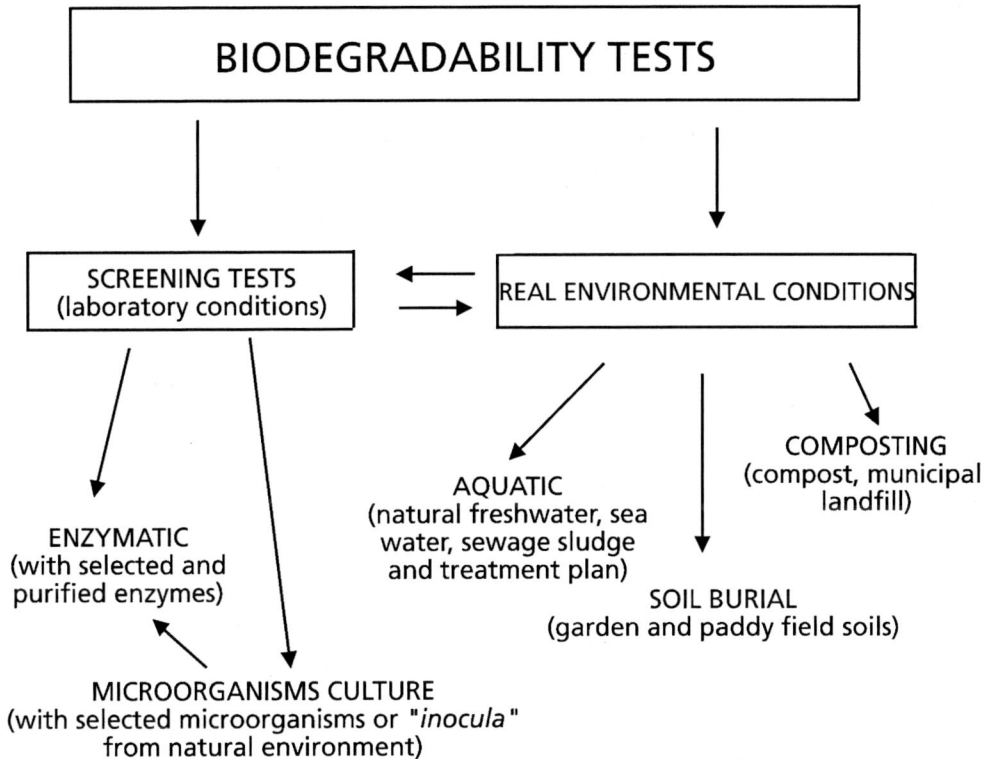

Figure 2.3 Common biodegradability tests

PHA films and *co*-polyester blends are swiftly degraded in both aerobic and anaerobic laboratory conditions [6, 22, 23]. Mergaert and co-workers, have investigated PHA biodegradation in natural environments [27] and in soil [28] and have evaluated the influence of abiotic and biotic factors in the degradation. They observed in tensile test pieces made of PHB and PHB/PHV buried in soil, a surface erosion that was enhanced by increasing incubation temperatures [28]. Sample molecular weights decreased to the same extent in abiotic and biotic conditions and were unaffected at lower temperatures. This suggests that the most relevant degradation was due to abiotic hydrolysis. PHA film aerobic degraders, and were isolated and quantified in garden and paddy field soils [29]. Recently microbial PHB degraders were isolated from soil and a PHB/co-PHA-depolymerase was characterised from an *Arthrobacter sp.* strain by Asano and co-workers [30]. Furthermore, fungal contribution to *in situ* biodegradation of PHB/PHV films in soil has been demonstrated [31]: fungi grew very rapidly along the film's surface and expanded their hyphae in a three-dimensional manner. Anaerobic degradation in municipal landfill conditions are reputed to be unfavourable [32] but anaerobic methanogen fermentation of PHB and PHB/PHV films has been demonstrated [22]. Probably, the biodegradation rate in landfills depends strongly on the microbial population present and upon temperature [33].

Aerobic and anaerobic degradation of PHA films by sewage sludge, *in situ* or in laboratory culture conditions, has been demonstrated and microorganisms isolated and characterised [34-37]. Recently, Khan and co-workers [37] have isolated from a sewage treatment plant a new chemoorganotrophic denitrifying bacterium from the *Comamonadaceae* family, capable of aerobic and anaerobic degradation of PHB and PHB/PHV. An high rate of NO_3^- reduction to N_2, without appreciable accumulation of nitrite and nitrous oxide, was observed in the oxidation of PHB/PHV.

Biodegradation *in situ* of PHB and co-PHA samples has been studied in natural waters under real-life conditions. These studies revealed the biodiversity of the microflora involved in biodegradation [38]. Imam and co-workers [39], have demonstrated biodegradation of plastic blends made from corn-starch and PHB/PHV, in tropical coastal waters. The rate of degradation correlated with the amount of starch present, the degradation of starch being two orders of magnitude higher than that of PHB/PHV degradation. Incorporation of polyethylene oxide (PEO) in the blends slightly retarded the degradation rate. Mabrouk and co-workers [40] have isolated a marine *Streptomyces sp*, able to grow in a simple mineral liquid medium with PHB and co-PHA as the sole source of carbon. Excreted PHB-depolymerase expression was repressed by the presence of simple carbon sources. Kasuya and co-workers [41] have purified from a *Marinobacter sp*. bacterium, isolated from marine benthos, drawn at 1165 m in depth, a PHB-depolymerase capable of hydrolysing poly(3-hydroxypropionate) and poly(4-hydroxybutyrate).

Recently a rapid and sensitive screening, based on DNA amplification by DNA-polymerase chain reaction (PCR) has been established to evaluate potential PHB-degrading bacteria [42]. The PCR primers have been designed on the basis of DNA sequence homology for known PHB-depolymerase fibronectine type III linker domains. Furthermore a sensitive and quantitative evaluation of depolymerase-directed PHB degradation and of enzyme-substrate binding kinetics has been performed by a quartz crystal microbalance technique [43].

PHB-depolymerases present a bifunctional organisation composed of a hydrophobic, carboxyl-terminal solid-surface binding domain and a catalytic domain (containing a penta-peptide lipase like box), separated by a linker region [44-47]. The structure of this model can explain the decrease in degrading activity observed in the presence of an excess of enzyme, and the failure in hydrolysing some co-PHA. In fact, to exert catalytic activity the hydrolytic domain has to fold down and gain surface access. Consequently, if the surface is saturated by enzyme molecules, or the polymeric structure is too rigid to allow this mechanism, the hydrolysis of the ester bond can't take place. Structural effects of *in vitro* PHB and *co*-PHA crystal degradation by *Alcaligenes faecalis* PHB-depolymerase, has been studied by Iwata and co-workers [48] and Abe and co-workers [49]. The enzyme preferentially degrades the amorphous phase of polymer chains, and the degradation of the crystalline phase progresses from the edges of crystalline lamellar stacks to yield narrow cracks along long axes and crystal fragments. Recently, it has been demonstrated for two deletion mutants of *Pseudomonas stutzeri* PHB-depolymerase, lacking the substrate binding domain and the linker region, that hydrolytic activity is present only for water-soluble oligomers, thus indicating that catalytic activity is independent from the substrate-binding domain and the linker region and, conversely, that these regions are fundamental for the hydrolysis of solid samples at the lipid-water interface [50]. Furthermore, analysis of water-soluble oligomers released during hydrolysis indicates that PHB-depolymerase recognises, as a substrate, at least two monomer units, but trimer and tetramer derivatives are two orders of magnitude higher in the initial phase of hydrolysis. Kasuya and co-workers [51] have investigated the substrate specificity of three PHB-depolymerases purified from *A. faecalis*, *P. stutzeri*, and *Delftis acidovorans*, classified A or B, with respect to the different localisation of the lipase-box in the catalytic domain. The enzymes showed a relatively narrow substrate specificity for polyester films, in that hydrolysed, but at a different extent, the same kind of polyesters, [i.e., poly(3-hydroxybutyrate), poly(3-hydroxypropionate), poly(4-hydroxybutyrate), poly(ethylene-succinate) and poly(ethylene-adipate)]. The enzymes were able to bind, but not to hydrolyse, structurally correlated polyesters that present:

i)   a side chain greater than methyl group, [i.e., poly(3-hydroxyoctanoate) and poly(3-hydroxypentanoate)];

ii) a number of carbon and oxygen atoms into a backbone between two carbonyl groups greater than 4, [i.e., poly(butylene succinate), poly(butylene adipate), poly(-6-hydroxyhexanoate), poly(-5- hydroxypentanoate] or less than 3 atoms, [i.e., poly(-2-hydroxypropionate]. These findings suggest a similar conformational organisation of the active site in the catalytic-domain of PHB-depolymerases.

2.2.2 Synthetic Polyesters

PCL *in vitro* is hydrolysed by various lipases [52, 53] and is degraded by bacteria and fungi widely distributed in nature [54-57]. An highly efficient PCL-depolymerase has been purified and characterised from a mutant of the bacterium *Alcaligenes faecalis*, constitutive for the enzyme expression [58]. Also this PCL-depolymerase resembles a kind of lipase, in that the enzyme is able to hydrolyse triglycerides, but not PHB. Liu and co-workers [59] have reported that *Pseudomonas* lipase can hydrolyse both amorphous and crystalline PCL domains in films of PCL/PLLA blends, but cannot degrade PLLA. Similar results have been reported by Tsuji and co-workers [60] for the hydrolysis of PCL/PLLA blends by *Rhizophus arrhizus* lipase. Murphy and co-workers [61] have demonstrated that a PCL-depolymerase from the phytopathogen *Fusarium moniliforme* is a cutinase (EC 3.1.1.74), the enzyme deputed to hydrolyse the cutin, the polyester component of the plant cuticle. In conclusion, enzymes from the two major classes of excreted esterases (EC 3.1), lipases and cutinases, with different efficiency, related to factors such as molecular weight, crystallinity and porosity, are able to degrade PCL, and its blends. The non-specificity of enzymic attack can explain the environmental biodegradability of PCL. Recently, aerobic degradation of PCL in a culture of active sludge suspension has been evaluated using ^{3}H radiolabelled-PCL [62]. The onset of biodegradation occurs after a few hours, earlier than previously reported by respirometry evaluation, and it is completed after roughly 72 days, producing tritiated water (80-90%) and biomass. It is also interesting to note that *Candida antarctica* lipase, also in an immobilised form, is able, in water-containing toluene, to catalyse both the degradation of PCL into cyclic dicaprolactone and the inverse polymerisation reaction [63, 64]. These results are relevant in order to establish a potential biotechnological recycling of PCL-based plastics.

PLA and PLLA are biocompatible and bioabsorbable polymers developed for medical applications such as sutures, plates and screws used in internal fixation of fractures, or matrices for drug delivery. More recently, these materials have become potentially utilisable as biodegradable plastics engineered from renewable raw materials (L-lactic acid can be efficiently produced by waste-polysaccharide fermentation), in applications such as packaging and agricultural mulching films. This perspective has stimulated the study of the environmental impact and the biodegradation of PLLA and its blends. It is known that PLA and PLLA in natural environments are subjected to an abiotic autocatalytic hydrolysis, that increases with the increasing crystallinity of the polymeric matrix, proceeds preferentially from the core of the material, is enhanced by water, temperature and alkaline pH, and induces the hydrolytic removal of the chains in the amorphous regions, thus increasing crystallinity in the residual polymeric matrix [65-67]. Lactic acid and water-soluble oligomers, released from abiotic degradation, are quickly bio-assimilated from many microorganisms. Due to autocatalytic hydrolysis, the biotic contribution to degradation has been difficult to evaluate, i.e., whether the biotic process is a true biodegradation or a bio-assimilation of soluble products. Torres and co-workers [68] have screened a collection of filamentous fungi for their ability to utilise D,L-lactic acid and racemic oligomers as sole carbon and energy sources. Furthermore, they isolated five strains of wild-type filamentous fungi when plates of racemic PLA were buried in the soil. Biodegradation of racemic PLA by *Pseudomonas putida* [69] and *Amycolatopsis sp.* [70] has been reported. More recently, biotic and abiotic degradation of PLLA, in aquatic aerobic headspace [71] and in compost [72], in both aerobic and anaerobic thermophilic conditions, has been studied. Rapid (5 weeks in biotic environment) biodegradation of PLLA films to fine powder, by mixed culture of compost microorganisms, has been reported by Hakkarainen and co-workers [73]: the films in the corresponding abiotic medium still looked intact. Gallet and co-workers [74] reported that PLLA films, buried in an outdoor-soil environment in south Finland, were subjected during the first year to abiotic auto-hydrolysis; during the second year microorganisms assimilated the small products of degradation. In fact, the lactide content in PLLA samples did not change during the two years of observation; after 20 months lactic acid and lactoyl lactic acid appeared as result of auto-hydrolysis and after 24 months the amount of lactoyl lactic acid decreased, as result of bio-assimilation.

In vitro studies have been performed to evaluate potential attacks by excreted enzymes and to determine the correlation of hydrolysis to the structural features [75-80]. It has been reported that proteinase-K from *Tritirachium album* [77-80], the lipase from *Rhizophus delemer* [80] and polyester/polyurethane-depolymerase [81] are able to hydrolyse PLLA releasing L-lactic acid. The microbial and enzymic attacks seems preferentially restricted to the amorphous region at the free and tie chains of the polymeric matrix, the crystalline domain and the amorphous structure present in the folding surface being barely degradable. More recently [82], a PLLA-depolymerase from *Amycolatopsis sp* strain K 104-1 that degrades to lactic acid high molecular weight crystalline PLLA, both in emulsions and solid films, has been purified and characterised. For the catalytic properties *Amycolaptosis sp*-depolymerase seems to be, like proteinase K, a serine protease. In fact, the enzyme also degrades casein and fibrin but do not hydrolyse collagen, tributyrin and PCL (substrates of lipases) or PHB.

In conclusion, it is interesting to mention that the biodegradability and composting of polyesters and co-polyesters are very interesting from an engineering point of view, such as aliphatic polyesters containing aromatic constituents [22] and polyester/polyurethanes [81] which have recently been reviewed.

2.3 Biodegradable Blends and Composites: Preparation, Characterisation and Properties

In the following section some case studies on the design criteria for the preparation of biodegradable blends and composites are reported, together with their characterisation and analysis of the most relevant properties.

2.3.1 Microbial Polyesters

The first microbial polyester that has attracted interest from a technological point of view was PHB and its copolymers with valerate (PHBV). PHB was introduced in the early 1980s by ICI as an innovative thermoplastic material obtained through a biotechnological approach [83, 84]. It is produced by the fermentation of sugars by bacteria such as *Wautersia eutropha* [85]. High yields and high purity are obtained.

It has been found that PHB is thermoplastic with a high degree of crystallinity and a well-defined melting point (T_m) (around 180 °C). At relatively low undercooling (temperature between melting and crystallisation), PHB crystallises from the melt, giving rise to the formation of large spherulites (see **Figure 2.4**). The nucleation behaviour, the crystallisation and the morphology of PHB have been studied by Barham and co-workers [86]. Unlike other thermoplastic polymers, such as PP and PE, PHB is truly biodegradable and highly biocompatible.

Films of PHB show gas barrier properties comparable to polyvinyl chloride and PET. The combination of all these properties indicates that PHB may compete with commodity polymers in the packaging industry, especially in areas where non-biodegradable plastic items, due to environmental pollution, are not allowed. PHB can be injection moulded and extruded providing care is taken to minimise melt temperatures and residence time. It is melt unstable and degrades to crotonic acid if kept for a relatively long time at temperatures of only a few degrees above its melting point. Injection-moulded PHB bars show a high crystallinity with a brittle behaviour, especially at temperatures below the glass transition temperature (T_g). It can be concluded that PHB suffers two limitations in its use: a very narrow processing window, and a relatively low impact resistance.

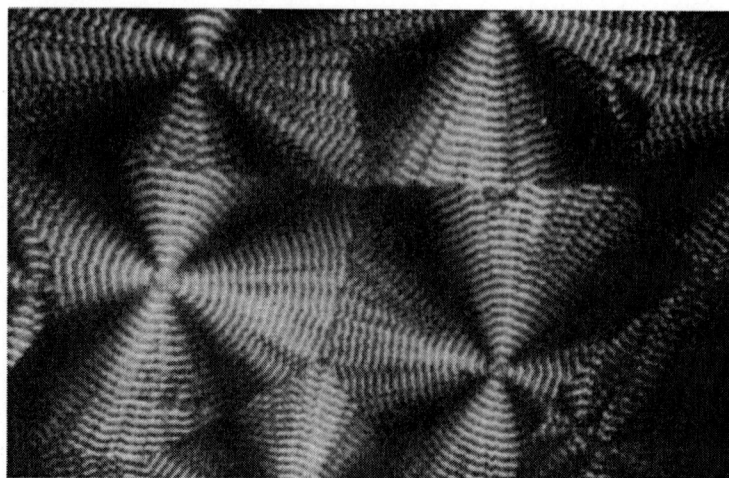

Figure 2.4 Optical micrographs of PHB spherulites

By the proper choice of culture media, it is possible to induce the biosynthesis of hydroxybutyrate-hydroxyvalerate (PHBV) [87, 88]. The presence of valerate reduces the melting temperature of the polymer, thus improving the processing window of such polyesters, which are otherwise limited by their thermal degradation, which occurs above 200 °C. Moreover, valerate induces ductility in the highly rigid, crystalline PHB.

Microbial polyesters are rapidly degraded in this environment, the rate decreasing with increasing crystallinity. At constant overall crystallinity, the rate of degradation also depends on the average size of the spherulites.

2.3.2 PHB and PHBV Blend with other Polymers Blends

PHB and its copolymers have been mixed with a variety of polymers, having very different characteristics; biodegradable and non-biodegradable, amorphous or crystalline with both different melting points and T_g in order to improve their processability and low impact resistance. Miscible blends have been formed with PEO [89-91], poly(epichlorohydrin) (PECH) [92] and poly(vinylacetate) [93]. Partly miscible blends are formed with poly(methylmethacrylate) [94, 95], while compatible blends have been obtained with PCL [96], ethylene-propylene rubber [93], polybutylacrylate (PBA) [97, 98], and polysaccharides [99].

In Section 3.3.2.1, the results relative to some microbial polyester-based systems are discussed, with particular emphasis placed on the influence of the enzymic degradation upon the final properties of the prepared materials.

2.3.2.1 Poly(hydroxy butyrate)/Poly(epichlorohydrin) (PHB/PECH) Blends

Atactic PECH (aPECH), i.e., poly(oxy-2-chloromethyl-ethylene), is an uncrystallisable polymer that has been blended with PHB by solution casting from dichloromethane, in a wide range of compositions [92]. The thermal and microscopic analysis of the blends has shown a single T_g. Moreover, the melting point of the PHB decreases with blending and the interaction parameter results in a negative value. These findings suggest that PHB and PECH form a miscible blend in the amorphous phase.

Microscopy has not shown phase separation even in the solid state. As a matter of fact, after the crystallisation of the PHB there has been no segregation of the PECH component in interspherulitic contact zones, and separate intraspherulitic regions of PECH have not been revealed, suggesting that the uncrystallised component is incorporated in the interlamellar or interfibrillar regions of PHB spherulites. Moreover, the growth rate of spherulites, at constant crystallisation temperature, decrease with the increasing of PECH percentage. The addition of the PECH to PHB also causes a reduction of the overall crystallisation rate calculated by applying the Avrami equation [100]:

$$X_t = 1 \exp(-kt^n)$$

Where X_t is the crystallinity developed at the time t, k is the kinetic constant of the growth and n is a parameter depending on the geometry of the growing crystals and on the nucleation process.

Small angle X-ray scattering studies on PHB/PECH blends [33] provided information about the localisation of the amorphous component in the spherulitic structure of the crystalline polymer. The scattering observed for the blends resulted from the superposition of the scattering due to the crystalline regions (made up by alternate stacking of lamellae and thin amorphous layers) and from the amorphous PECH inhomogeneity placed outside them. The result showed that the PECH molecules are dispersed at the molecular level in interfibrillar zones, where they can assume a random-coil conformation. Moreover, the blends were annealed, (i.e., kept for long time at a temperature between the crystallisation and the melting temperatures), in order to investigate the influence of this thermal treatment on the crystalline structure. It has been assessed that the annealing treatment promotes a general perfecting and rearrangement of the sample morphology, enhancing the crystallinity and the crystal dimensions of PHB in the pure state and in the blends, likely favouring the trend of the PECH molecules to assume a globular conformation.

Enzymtic and bacterial degradation of the blends was also investigated utilising the *Microbacterium saperdae* to degrade PHB/PECH blends of different formulations [101]. The culture growth was followed by spectrophotometric measurements of the optical density at 540 nm. The polymer degradation was determined by measuring the weight loss of the films after bacterial growth. The experimental procedure required periodic removal of samples, washing with distilled water several times and drying to reach a constant weight. The growth rate of *Aureobacterium saperdae* was found to decrease by increasing the PECH content in the blend, and to be zero when the percentage of PECH is 60 wt.% (see **Figure 2.5**). This compromised degradation of PHB by using *Aureobacterium saperdae* in the blends has been attributed to the reduced accessibility of the PHB to the bacteria. Finally during the bacterial growth only PHB was metabolised, whereas neither degradation or abiotic release of aPECH was detected for blend films.

2.3.2.2 Poly(hydroxy butyrate)/Poly-ε-caprolactone (PHB/PCL) Blends

Blends of poly-ε-caprolactone and PHB or PHBV are very interesting due to their technical properties and their inherent biocompatibility and biodegradability. Kumagai and Doi

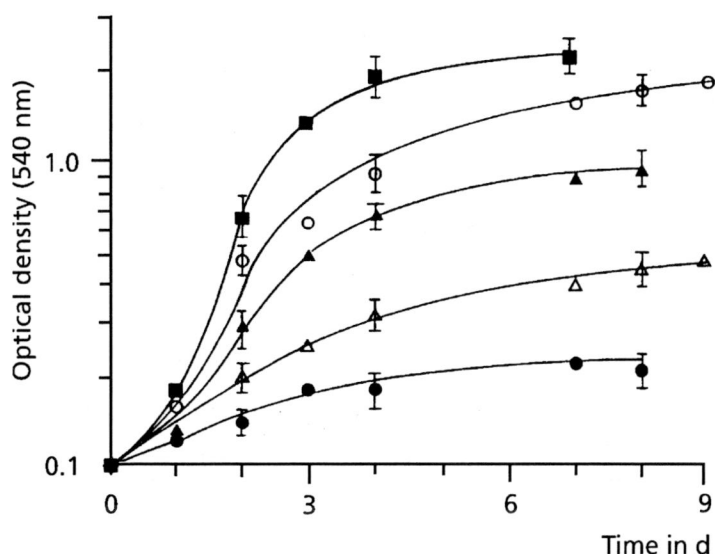

Figure 2.5 Growth curves of *A. saperdae* cultures of PHB/aPECH blends

[102] have investigated the miscibility, morphology and biodegradability of such blends. These authors, in agreement with subsequent studies performed by Gassner and Owen [103], found that PHB and PCL are immiscible in the amorphous state. On the contrary, miscibility has been reported by McCarthy and co-workers [96], due to the very low molecular weight of the utilised PCL.

Although PHB and PCL are immiscible on the molecular scale, a small amount of solubilisation of one component into the other has been suggested, which explains the reduction of the melting point in the blends [96].

In general, PCL acts as a polymeric plasticiser, i.e., it lowers the T_g and hence the elastic modulus, making a material that is more flexible. From dynamic-mechanical measurements, it has been inferred that for a PCL content of 60% and above, the PCL phase forms a continuous matrix, with PHB spherulites embedded in it. The mechanical behaviour of the blends is then dominated by the ductile PCL matrix. Instead, for compositions with less than 60% of PCL, the PHB phase becomes continuous. However the inclusion of soft PCL does not catastrophically lower the rigidity of the sample.

In general, in mechanical blends of PCL and polyhydroxyalkanoates no synergistic effects have been found that could be exploited to obtain a material with better performance than those of the individual components. To induce compatibilisation the approach of reactive blending of PCL and PHBV has been undertaken [104]. The two polymers have been melt mixed by adding peroxide, (i.e., dibenzoylperoxide and dicumylperoxide). According to the type of peroxide, two different temperatures were used in the blend preparation. The same blend compositions have been prepared in absence of peroxide for comparative purposes. Change of the thermal properties of the PHBV (in particular the apparent T_m) suggested that DCPO induces some structural change in PHBV. As a matter of fact, PHBV becomes more soluble in chlorinated solvents after DCPO treatment, indicating the formation of crosslinks by radical reactions. The same phenomenon is present in PHBV matrix blends, whilst it is absent in PCL-matrix blends, as the peroxide appears to have no influence on the PCL melting point in addition to its solubility characteristics.

Beyond calorimetric analysis, spectroscopic and morphological investigations of the reactive PHBV/PCL blends have shown the existence of a graft copolymeric phase in the interfacial regions between the polymers. Furthermore, the decrease of the elastic modulus of the blends obtained with DCPO has been correlated to the larger plastic deformation of PCL particles in such blends.

Biodegradation studies on injection moulded PHBV/PCL samples were performed under municipal solid waste composting conditions according to the Laboratory Scale Composting Test Method [105]. All samples had completely degraded after composting for 21 days. Moreover, it has been observed that the biodegradation increases rapidly with the increasing PCL content of the blends. This acceleration may be caused by the decrease of crystallinity in the whole samples and/or by the modification of the surface.

2.3.2.3 PHBV/Poly(butyl acrylate) Blends

An attractive route to the impact modification of PHBV by a reactive blending method involves the use of acrylate rubbers, such as PBA. A method in which the powders of PHB (or PHBV), as obtained from the bacterial polymerisation and subsequent purification, are thoroughly mixed with proper amounts of acrylate monomers and free-radical initiator has been developed by Martuscelli [97], resulting in blends which a minor elastomeric phase is intimately dispersed in the polyester matrix. Blends were prepared by mixing 70 g of polyester powder with 30 g of acrylic monomer, into which 60 mg of benzoyl peroxide (0.2 wt.% of acrylic monomer) are dissolved. The mixture is gently stirred for 24 hours at room temperature. Subsequently, this homogeneous mixture is warmed to 90-100 °C and allowed to remain at this temperature under stirring, for more than 24 hours, to permit acrylate polymerisation.

The impact properties obtained by Charpy impact tests on sharply notched samples of PHB and PHBV containing 4% of valerate (PHBV4) and on their blends with PBA, demonstrated that a positive influence is exerted by the rubber on the fracture toughness of microbial polyesters. The effect is particularly marked at temperatures close to, and above, room temperature. In fact, while PHB and PHBV4 are brittle at room temperature, their blends are much more ductile. The enhancement of properties is less pronounced for PHBV4, because the addition of valerate comonomer is already effective in the induction of ductility [98].

Swelling experiments have been carried out on several blends to investigate the possible formation of crosslinks between PHBV and PBA. It is conceivable that PHBV can undergo similar reactions leading to graft copolymers in the presence of butyl acrylate monomers and/or growing PBA macroradicals. The microbial degradation of plain PHBV and PHBV/PBA blends has been also investigated [98].

Aureobacterium saperdae cultures, where the only carbon source was the polymeric sample, were used to degrade pure PHBV4 and PHBV4/PBA blends (80/20 and 70/30 weight ratios). The microorganism was precultured overnight on 0.1% Lennox LB broth, about 3.5 cm^3 samples of this culture were used to inoculate 500 cm^3 flasks containing 100 cm^3 of mineral medium (mineral medium composition: 1 mg/cm^3 NH$_4$Cl, 0.5 mg/cm^3 MgSO$_4$.7H$_2$O and 0.005 mg/cm^3 CaCl$_2$ 2H$_2$O in 66 nM KH$_2$PO$_4$ (pH = 6.8). The polymeric samples were added to the flasks and incubated at 30 °C, with shaking, for 15 days. In addition, control experiments were run to verify chemical hydrolyses of polymeric samples immersed in mineral medium and no weight loss was found after 15 days. Samples at different stages of polymer degradation were used to perform various morphological analysis. The percentage of polymer degradation was determined by measuring the weight loss of the sample during bacterial attack. Having preliminary checked the non-biodegradability of the PBA phase, the weight loss was normalised on the PHBV content, thus obtaining the percent of degradation in blends. Polymer samples were removed from the culture medium at different stages, washed several times with distilled water and dried under vacuum to a constant weight. The thickness of the polymeric samples was measured before and after the bacterial attack. Since decrease of the thickness during the biodegradation corresponds to the percentage of weight loss, the polymer erosion must take place via surface dissolution. Scanning electron microscopy analysis of the surface of pure PHBV4 after bacterial attack showed homogeneous superficial erosion caused by the degradative enzymes, while no change took place inside the sample. During bacterial degradation of the PHBV4/PBA blend (80/20), pieces of the PBA component released in the culture were macroscopically visible. As a consequence of the bacterial attack, the PHBV4 present on the surface was eroded and pieces of the dispersed PBA component were released, allowing new PHBV4 zones to be accessible to the degradative enzymes (see **Figure 2.6**).

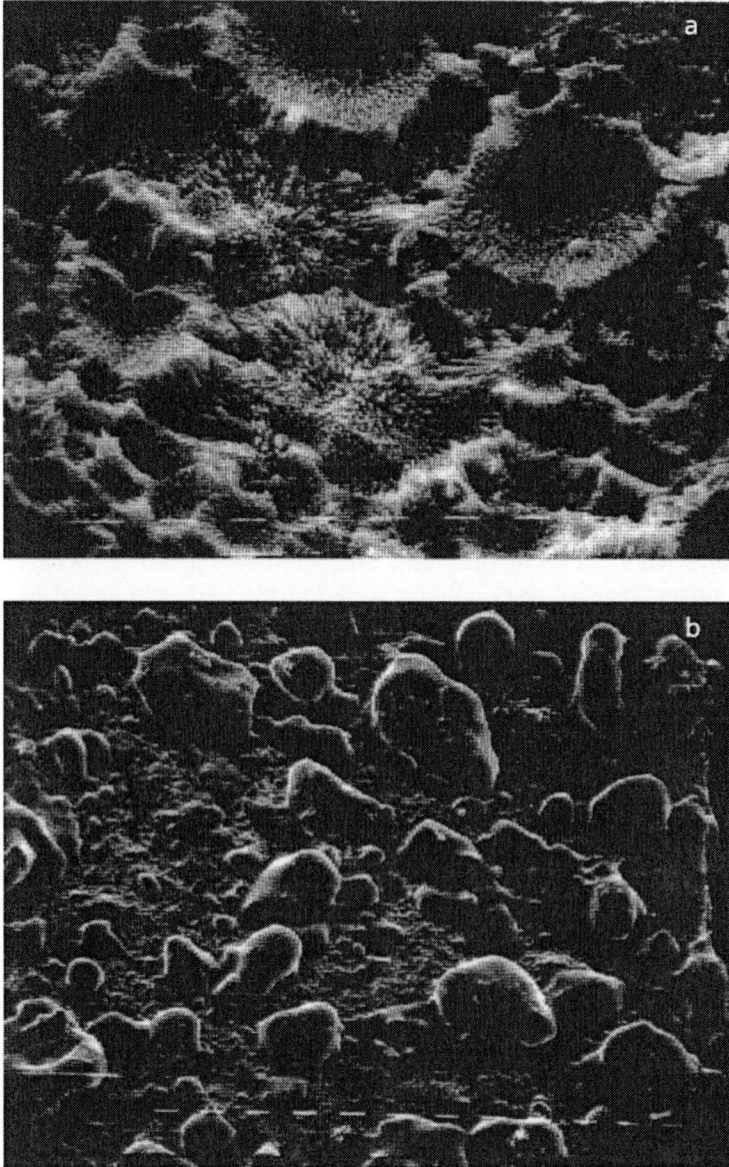

Figure 2.6 SEM micrographs of PHBV/PBA samples after microbial attack

2.3.2.4 PHBV/Polysaccharide Blends

Poly(hydroxybutyrate-*co*-valerate) has been blended with cellulose and starch derivatives [99, 106]. Thermoplastic cellulose esters such as cellulose acetate butyrate (CAB) and cellulose acetate propionate have been found to form miscible blends with PHBV in the amorphous phase. These blends show that to offer miscibility as the dominant criteria a single T_g should be applied. As a matter of fact, PHBV/CAB blends with up to 50% of CAB exhibit only one low T_g both in dynamic scanning calorimetry measurements and dynamic mechanical analysis (DMA). At intermediate compositions, ranging between 50% and 70% of CAB, a broad T_g ideally separable in the two next steps by a flexural point, has been found. Finally, for blend compositions with a CAB percentage larger than 70%, a single T_g at higher temperature is detectable. The higher T_g values follow the Fox equation in the range 0%-50% of PHBV This situation fits the case of the partial miscibility of PHB and CAB, with a separation of pure PHBV at CAB contents less than 30%. However, a more careful analysis leads to a different explanation. Indeed, the value (ΔC_p, being C_p the heat capacity) of the T_g permits the calculation of the composition of the unknown multiphase by means of the relationships:

$$\Delta C_1 = x^1_A \, \Delta C_A + (1 - x^1_A) \, \Delta C_B$$

$$\Delta C_2 = x^2_A \, \Delta C_A + (1 - x^2_A) \, \Delta C_B$$

Where ΔC_1 and ΔC_2 are the values of the T_g steps related to phase 1 and 2, AC_A and ΔC_B are the specific heat capacity changes corresponding to the T_g of the pure components A and B, and x^1_A and x^2_A are the molar fraction of the component A in the phases 1 and 2, respectively.

The DC_p value is associated with the low temperature T_g much more than that expected on the basis of the PHBV content and it transpires that both components should contribute to the low temperature T_g of the blends. Thus the hypothesis of partial miscibility of PHBV and CAB with the separation of a pure phase of PHBV must be excluded.

The behaviour of PHBV/CAB blends has been elucidated by carbon ^{13}C-nuclear magnetic resonance studies [99]. These studies demonstrated that PHBV and CAB are miscible in the amorphous phase and two T_g are revealed owing to the existence of dynamic heterogeneity. This heterogeneity reflects different molecular mobilities of the blend components, even if the latter experience equivalent or average free volumes. However, the equivalence of free volumes does not necessarily imply a single glass T_g, as demonstrated by Miller and co-workers [107].

When CAB is the major component, the PHBV crystallisation is completely inhibited and no trace of crystallinity is revealed even after months of blend storage at room temperature. Blends that are richer in PHBV crystallise with storage at ambient temperature, becoming opaque. The mechanical properties of blends containing 20%-50% of PHBV reflect, of course, the amorphous character of such blends. Indeed, in the range 20%-50% of PHBV, the elastic modulus and the tensile strength decrease with increasing PHBV percentage. The tear strength is almost constant, while the elongation at break increases remarkably. Above 50% of PHBV the elastic modulus, the tensile strength and the tear strength increase while the elongation at break drops significantly. In particular, a synergistic effect is found for PHBV/CAB compositions above 70% of PHBV.

PHB and PHBV copolymer has also been combined with starch [108, 109], which is a inexpensive biodegradable filler produced in surplus for food needs. Owen and Koller [108] have investigated the structure and the mechanical properties of melt-pressed sheets of PHB and PHBV filled with various amounts of maize starch granules. No further component such as a compatibilisers were used. The addition of the starch to the PHB and PHBV causes a decrease in breaking, strain and stress and an increase in the elastic modulus. Thus the PHB becomes even more brittle by the addition of starch. This increases the crystallinity content of the matrices without changing the nucleation density of the spherulites. Shogren [109] has shown that PEO-coated granular starch causes a large improvement in the tensile properties of PHBV/starch composites over uncoated starch. The introduction of a PEO interfacial layer seems to enhance the degree of adhesion between the starch and the PHBV and/or increase the toughness and resistance to crack propagation around starch granules.

Biodegradation studies on PHBV blended with native cornstarch, and with cornstarch pre-coated with PEO have been carried out by Imam and co-workers [110]. The weight loss of the samples was measured and the deterioration in tensile strength tested. The extent and rate of weight loss were similar in both pure PHBV and in PHBV containing starch, whilst the weight loss was lowest in PHBV blends prepared with PEO-coated starch. The rate of deterioration of the mechanical properties was highest for pure PHBV and less for PHBV/PEO-coated starch. By means of Fourier transform infrared spectroscopy, it has been found that more extensive starch degradation occurs as the starch content increases, while the PHBV in the blends becomes less susceptible to the enzymic attack.

2.3.3 Polycaprolactone (PCL)

The increasing usage of polymers in disposable items as packaging creates obvious problems of garbage disposal and demand for their degradation. Aliphatic polyesters currently constitute the most attractive class of artificial polymers, which can be degraded by contact with living organisms. PCL is an important member of the aliphatic polyester family, known to be susceptible to biological degradation. It is investigated mainly in the context of its application for use in drug delivery systems and packaging materials. There are literature reports on the degradation of aliphatic polyesters in a living environment [111]. This degradation can result from enzymic attack or from simple hydrolysis of ester bonds or both. The degradation rate depends on moisture level, nutrient supply, temperature and pH.

2.3.4 Starch/Polymer Blends

To increase biodegradability whilst simultaneously lowering the cost and preserving resources, it is possible to blend polymeric materials with natural products. Among these, the starches; being one of the biggest (in the EU, 5 million tons/year from corn) easily recovered and chemically-biologically modifiable, are playing an increasing role in the development of new environmentally sound polymeric materials.

Starch is a semicrystalline polymer stored in granules as a reserve in most plants (see **Figure 2.7**). It is composed of repeating 1,4-α-D glucopyranosyl units: amylose and amylopectin. The amylose is almost linear, in which the repeating units are linked by α (1-4) linkages; the amylopectin has an α (1-4) linked backbone and approximately 5% α (1-6) linked branches. The relative amounts of amylose and amylopectin depend upon

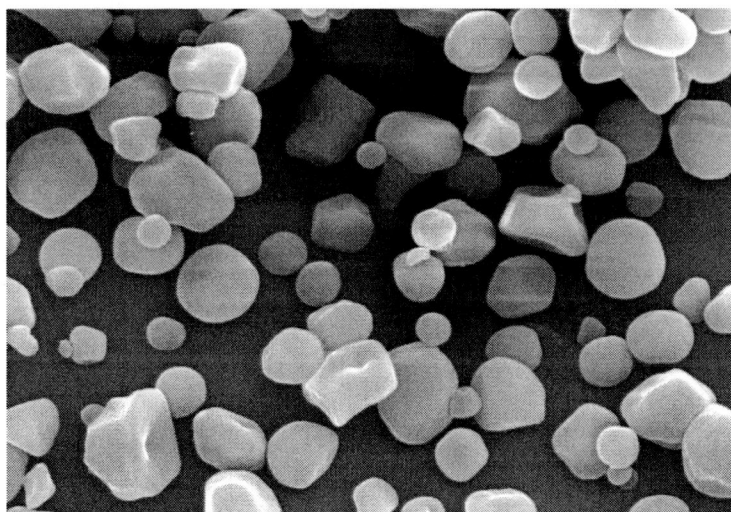

Figure 2.7 SEM micrographs of starch amylomaize powder

the plant source. Corn starch granules typically contain approximately 70% amylopectin and 30% amylose. Several studies focus upon the development of starch-based materials, for the reasons mentioned above. However the hydrophilicity of starch is responsible for an incompatibility with most hydrophobic polymers. Therefore, the improvement of the reactive interface between polymers and starch can play a critical role in ensuring that the properties of each component contribute to the bulk properties and in obtaining composite materials with good physical, mechanical and biological properties. The compatibilisation between polymers can occur by blending a preformed compatibilising agent or inducing, in appropriate conditions, reactions between the functional groups of the two incompatible polymers in order to obtain the *in situ* formation of an interfacial agent.

2.3.5 Polyesters/High Amylose Starch Composites by Reactive Blending

Two biodegradable thermoplastic polyesters PCL and poly (3-hydroxybutyrate-*co*-valerate containing 5% by mole of hydroxyvalerate (HV)] were blended with a maize starch with a high content of amylose (70%) using a reactive blending approach to promote interactions between the two phases.

For PCL/starch composites, the compatibilisation was induced by adding an anhydride functionalised PCL phase to the two components. The influence of this phase and its effect on the morphology and properties of the composites was examined. As far as PHBV/starch composites, the improvement of the interfacial adhesion was promoted by means of reactively blending the two components, and adding a small amount of an organic peroxide.

For both systems, mechanical composites were also prepared to compare their properties to those of the compatibilised materials. Thermal, morphological and mechanical tests were also carried out. Finally, biodegradation analyses were performed on the systems to evaluate the influence of the starch on the rate of the polyester's biodegradation.

2.3.5.1 Preparation of PHBV/Starch and PCL/Starch Composites

The well known hydrophilicity of starch renders it incompatible to a wide range of polyesters (which are hydrophobic in nature), such as PHB and PCL. The nature of the starch, its unmelted components, and its insolubility in the polyester matrices permits us to classify our systems as composites and not blends. Thus, the compatibilisation between starch and polyester could be evaluated only by recovering starch granules from the polymeric matrices. Then, in order to promote compatibilisation between the polyester (PHB or PCL) and the starch, two different reactive preparation methodologies were used. In the case of PCL-based composites it was possible to prepare a compatibilising agent by using a PCL with a lower molecular weight.

Being as PHBV has particular properties, such as high molecular weight, it was not possible to utilise the same preparation methodology. In this case the reactive blending methodology was preferred. PHBV was mixed in the melt with high-amylose starch using a Brabender-like apparatus operating at 190 °C for 15 minutes and at 32 rpm. For reactive blending, 2% of the bis (*tert*-butylperoxyisopropyl)benzene was also added in the same operation. Composites containing 80% and 70% by weight of PHBV were prepared with and without the addition of peroxide. The samples and codes are listed in **Table 2.1**.

PCL was mixed in the melt with high-amylose starch using a Brabender-like apparatus, but operating at 80 °C for 15 minutes and at 32 rpm. For the compatibilised composites, a pre-compatibilising agent was added simultaneously to the PCL and starch phases in the melting mixer at the same operating conditions.

This pre-compatibilisation agent was obtained using the following experimental routine: in a round bottom flask equipped with refrigeration and nitrogen inlets, PCL with low molecular weight (20000 Da) and pyromellitic anhydride (APM) were added. The APM was added in an excess molar ratio (20:1) to the PCL. The APM was dissolved in a small amount of tetrahydrofuran and the solution was added to the melted PCL. The reaction was carried out at 110 °C for 24 hours under nitrogen flux. The final product was dissolved in acetone and precipitated by water. In this way it removed the APM because of its solubility in the water/acetone mixture. Probably, the presence of water is responsible for the hydrolysis of the non-reacted anhydride group. Different amounts of pre-compatibiliser agent were used; the compositions and codes of the prepared composites are detailed in **Table 2.1**.

2.3.5.2 PHBV/Starch Composites

The compatibilisation between PHBV and starch is improved by using a particular type of blending called 'reactive blending'. This methodology consists of conferring chemical reactivity upon the polymers that are to be blended, thus promoting their compatibilisation. The major qualification in the use of the reactive blending methodology is the presence of reactive groups on the backbone chains of both polymers in order to create strong interactions between the two incompatible polymers that are responsible for a lower interfacial energy and a more stable morphology.

PHBV is characterised by the presence of carbonyl groups in the backbone chains, these constitute the sites of potential physical interactions with starch hydroxyl groups. The

Table 2.1 Description of composition of starch-based composites and relative codes		
Composite composition (w/w%)	Weight percentage of the pre-compatibiliser (%)	CODE
PHBV/Starch 100/0	0	PHBV 100
PHBV/ Starch 80/20	0	PHBV 80
PHBV/Starch 70/30	0	PHBV 70
PHBV/Starch 100/0	2	PHBV 100 R
PHBV/Starch 80/20	2	PHBV 80 R
PHBV/Starch 70/30	2	PHBV 70 R
PCL/Starch 100/0	0	PCL 100
PCL/Starch 90/10	0	PCL 90
PCL/Starch 70/30	0	PCL 70
PCL/Starch 50/50	0	PCL 50
PCL/Starch 100/0	2.5	PCL 100 L
PCL/Starch 90/10	2.5	PCL 90 L
PCL/Starch 70/30	2.5	PCL 70 L
PCL/Starch 50/50	2.5	PCL 50 L
PCL/Starch 100/0	5	PCL 100 M
PCL/Starch 90/10	5	PCL 90 M
PCL/Starch 70/30	5	PCL 70 M
PCL/Starch 50/50	5	PCL 50 M
PCL/Starch 100/0	10	PCL 100 H
PCL/Starch 90/10	10	PCL 90 H
PCL/Starch 70/30	10	PCL 70 H
PCL/Starch 50/50	10	PCL 50 H

chemical interaction between the two components was induced by adding 2% by weight of peroxide during the blending. This organic peroxide has a half-life of roughly 30 minutes at a temperature of 190 °C. The preparation temperature of the composites was about 190 °C. In this way it is possible to prevent a reaction that is too quick, keep the blending homogenous and avoiding the complete crosslinking of the PHBV. The starch used had a high amylose content. This choice allows a greater number of hydroxyl groups being present on the surface that is responsible for a stronger interaction with the carboxylic groups of the PHBV. Moreover, high amylose starch is characterised by a lower granule size that allows good dispersion of the starch into PHBV.

The composites so obtained are characterised by thermal properties similar to those of PHBV. This result confirms that the chemical interactions between PHBV and starch are limited to a few polymer chains sufficient to promote compatibilisation between the two components.

2.3.5.3 PCL/Starch Composites

The compatibilisation between starch and PCL was promoted by adding a compatibilising agent during the blending of the two components. This agent was prepared by the chemical modification of the end-groups of a low molecular weight PCL (M_w = 20000 Da) with reactive groups that are able to react with the -OH groups of the starch. It is preferential to have a PCL with a low molecular weight which has a higher concentration of reactive hydroxyl end groups. This chemical modification was performed as described in **Scheme 2.1**, which is the reaction of PCL with an excess of a dianhydride, the APM. The excess of APM permits the attainment of high conversions and the avoidance of undesirable chain-extension reactions between modified and unmodified PCL. The reaction was performed in bulk, using pyridine as a catalyst at a temperature of 110 °C, for 24 hours. These process parameters were selected on the basis of reaction monitoring performed by the use of infrared (IR) spectroscopy.

In these conditions the IR analysis has shown the decrease of the bandwidth relative to the PCL hydroxyl end groups stretching ($\sim$3630 cm^{-1}) and the appearance of a band at 3200 cm^{-1} due to pH stretching of carboxylic groups. The intensity of this latter band increased after the washing of the product because of a partial hydrolysis of the anhydride groups present on the PCL backbone.

2.3.5.4 Biodegradation Analysis

The biodegradability of the neat thermoplastic polyesters and starch-based composites was investigated by a compost simulation test. This test was a modification of the ASTM D5338

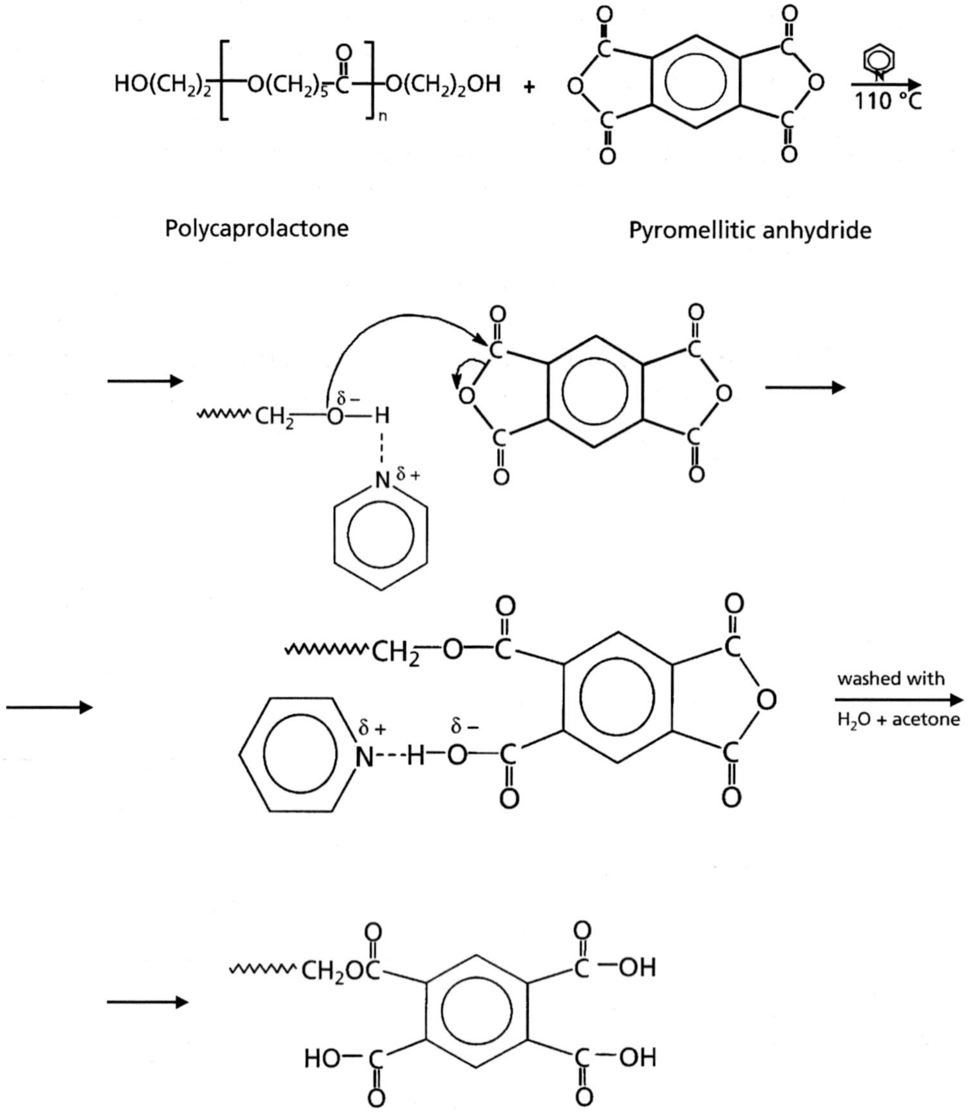

Polycaprolactone   Pyromellitic anhydride

washed with

H_2O + acetone

Scheme 2.1 Reaction of PCL with an excess of APM

procedure of controlled composting [112]. A 3 litre reactor was filled with 2 kg of mature compost and maintained at a temperature of 55 °C ± 2 °C. The system was continuously aerated with previously water-saturated and heated, pressurised air. Test specimens of the same initial shape, (i.e., the same exposed surface to biodegradation), were buried inside the reactors. The samples were withdrawn from the compost reactors at different times, washed with distilled water and dried at 60 °C to constant weight and the disintegration of the materials was evaluated (referred as % of weight loss). The compost used in this test was kindly supplied by Centro Ricerche Produzioni Animali (CRPA) of Reggio Emilia, Italy. It was produced at the Composting Plant of Limidi of Soliera, Modena, Italy, from a mixture of residual sludges (municipal sewage), grass and wood chips. The compost was stored indoors for several months. Prior to use, the compost humidity was adjusted to 60%.

PHBV, PCL, PHBV/Starch 70/30 (compatibilised and non-compatible), PCL/Starch 70/30 (compatibilised and non-compatible) samples were submitted to the compost test and the percentage of weight loss was evaluated at different times during the biodegradation process. In **Table 2.2** the results are summarised. From these findings it is possible to observe the following:

1. PCL polymers present, at the conditions used, a faster biodegradation rate than the PHBV.

2. The presence of a starch phase is responsible for a faster biodegradation for both the systems for neat polymers.

3. After 20 days of incubation both PCL/Starch and PHBV/Starch composites are almost completely biodegraded.

4. The compatibilisation procedures do not affect the biodegradability properties: as a matter of fact PHBV/Starch and PCL/Starch compatibilised and non-compatibilised composites show the same percentage weight loss.

Table 2.2 Percentage of weight loss of neat PCL, neat PHBV, PCL/Starch and PHBV/Starch mixtures in a simulated compost test		
Samples	Time of incubation Ten days	Time of incubation Twenty days
Neat PCL	35%	60%
PCL/Starch 70/30 NR	>75%	>90%
PCL /Starch 70/30 R	>75%	>90%
Neat PHBV	17 %	41%
PHBV/Starch 70/30 NR	*	100%
PHBV/Starch 70/30 R	*	100%
* Samples completely deteriorated NR: not reacted R: reacted		

2.3.6 PCL/PVOH

An investigation of PCL, poly(vinylalcohol) (PVOH) and their blends have been carried out by De Kesel and co-workers [111]. The materials were incubated in the presence of a pure strain of microorganism isolated from an industrial composting facility. In these conditions pure PCL films are completely assimilated over periods of 600-800 hours. Pure PVOH is not degraded, even over much longer exposure times. Unexpectedly, the blends, even PCL rich, are not altered (neither weight loss, oxygen uptake nor microorganism growth) in the presence of these microorganisms. It has been shown that inactivation of the strain by PVOH does not occur and is not responsible for this phenomenon. PVOH, even when present in small amounts in the incubation medium, adsorbs on the surface of PCL or blended films and so the PCL phase is then inaccessible to micro-organisms.

2.3.7 Polylactide (PLA)

Polylactide and its copolymers have long been studied and used for various medical applications, from sutures, bone fracture nails and supports for controlled drug delivery. Only recently have there been serious attempts at bringing the polymers to large-scale production for a variety of markets. Process development is still needed to bring the costs to competitive levels, and much effort has to be put into tailoring the polymer properties. They should suit existing processing equipment and conditions and impart desired properties to the final products, from touch to mechanical strength. In this Chapter it is shown that poly(L-lactide) can be produced, stabilised and processed into block films, thin fibres and non-woven fabrics by common methods, that the products possess attractive properties, and that these properties can be tailored by modification of the resin with plasticisers. Polylactides are completely biodegradable in natural systems when water and microorganisms are present. When modifying the properties by additives, care has to be taken that all additives are at least as rapidly degradable as the polymer matrix, and that they are totally inert and harmless.

2.3.8 PLA/Bionolle

In order to improve the mechanical behaviour of PLA a series of blends consisting of polylactic acid (PLA) and aliphatic succinate polyester (Bionolle) have been prepared and characterised [113]. The results of the mechanical tests have shown that using more than 20 wt% Bionolle can significantly increase the toughness of the PLA, increase the elongation at break (more than 200%) and increase the impact strength (more than 70 J/m). These properties were not significantly affected by the ageing behaviour of PLA for more than two months. DMA results show that Bionolle reduces the elastic modulus of the blends between −15 °C and 60 °C.

Soil degradation rates of the PLA/Bionolle blends also increase with increasing Bionolle content. However, for blends with less than 30 wt% of Bionolle, the degradation rates do not significantly increase. Enzymic degradation rates of the blends are higher than for those of the two polymers, and these rates increase with increasing PLA content. Composting biodegradation rates increase with increasing Bionolle content. The biodegradation tests, in soil, showed that the biodegradation rate of the polyester was extremely fast, while the rate for PLA was relatively slow. After degrading for 45 days, Bionolle degraded by almost 100%, while PLA only degraded by about 14%. For the blends with 70 and 50 wt% Bionolle A30/B70 and A50/B50, respectively, the degradation rate is relatively fast. After 45 days, the A30/B70 blend degraded by about 77%; the A50/B50 blend degraded about 65%. These values are equal to those expected on the basis of additivity rule. However, for blends with less than 30 wt% Bionolle, the degradation percentage values are less than those expected by the rule of mixtures.

The degradation testing results at 55 °C show that Bionolle had a very high weight loss rate, while PLA showed a slight weight loss after being tested for 20 days. For PLA/Bionolle blends, the weight loss rates are between the rates of the two parent polymers and rises with increasing Bionolle content. Even in poisonous composting conditions (without bacteria), Bionolle and PLA/Bionolle 50/50 blend still have relatively high degradation rates.

2.4 Conclusions

This Chapter shows that chemistry and biology are both involved in the research effort to identify innovative materials for many applications in which a sustainable management of resources requires a diversification of technological options. Still many questions remain open to discussion, related to testing methods, performances and economic sustainability.

It is a fact that, in the last few decades, an enormous amount of industrial attempts to produce and commercialise biodegradable polymers, blends and composites has been attempted. Most of them are listed in **Table 3.3**. Some of the products shown are still alive, some survive, some are already dead. However, their existence testifies that diversification is necessary to create a possible and better future for mankind, and research efforts are still necessary to rationalise the different options.

Acknowledgements

The authors wish to thank Dr. L. Calandrelli for technical assistance.

Table 2.3 Inventory of biodegradable materials		
Commercial Name	**Composition**	**Producer**
MATER-BI ZI01U	Starch + PCL	Novamont
MATER-BI YI01U	Starch + acetate derivatives	Novamont
MATER-BI ZF03U/A	Starch + PCL	Novamont
BIOPOL	PHBV	Monsanto Europe SA
EASTAR BIO 14766	Aliphatic/aromatic copolyester	Eastman Chemical Europe
ECOFLEX	Aliph/aromatic Polyester	BASF
BAK 2195	Poly-ester-amide	Bayer
	Adipic acid bisphenol A	Dow
BIONOLLE	Poly(butylene adipate-*co*-butylene succinate-*co*-ethylene adipate-*co*-ethylene succinate)	Showa Denko Europe
	PCL + Estane	Petroplast
ACEPLAST RT	Cellulose acetate DS = 2.25	Acetati
ACEPLAST LS	Cellulose acetate DS = 2.4	Acetati
ECOPLA	PLA	Cargill
TONE P787	PCL	Union Carbide
FASAL	Sawdust	Ifa
ELVANOL 71-30	PVA insoluble high DH	DuPont
GOHSENOL GH 23	PVA soluble med.ium DH	Nippon Ghosei
BIOSOLO	PE	Indaco
NATURGRADE+	PE	Ecostar
EPI CP530	PE	Epi Environmental Products Inc.
EPI CP560	PE	Epi Environmental Products Inc.
EPI CP 590	PE	Epi Environmental Products Inc.

Table 2.3 Continued		
Commercial Name	**Composition**	**Producer**
SKY GREEN	Poly(butylene adipate-*co*-butylene succinate-*co*-ethylene adipate-*co*-ethylene succinate)	Sunkyong Industries
PHOENIX PL. NV	PE	Phoenix Plastics
BIOSAN	Cellophane	
PARAGON	Starch	Avebe
BIOTEC	Starch + PCL	Biotec
FLUNTERAPLAST	Starch (?)	
ECOSTAR (NOVON)	Starch + (?)	Novon International (USA)
BIOCETA	Cellulose acetate DS = 2.0	Mazucchelli 1849 Spa
CAPRA 650	PCL	Solvay Interox (UK)
BIOMER	PHA	Biomer
	PLA	Neste Oy
	Poly(butylene terephthalate) derivative	General Electric Plastics
KYTEX	Chitine/chitosan	Marine Commodities
POLYVIOL	Poly(vinyl alcohol)	Wacker
MOWIOL	Poly(vinyl alcohol)	Hoechst
VINNAPAS	Poly(vinyl acetate)	Wacker
SOARNOL	Poly(ethylene vinyl alcohol)	Elf Atochem
EVATANE	Poly(ethylene vinyl acetate)	Elf Atochem
ESCORENE	Poly(ethylene vinyl acetate)	Exxon
POLYOX	Poly(ethylene oxide)	Union Carbide
	Poly(vinyl alcohol)	Idroplast
		Primalco
BIOFLEX	PCL + starch	Biotec GmbH & CoKG

Table 2.3 Continued		
Commercial Name	**Composition**	**Producer**
BIOPAC	Starch	Biopac
GREENPAC ECOFOAM	Starch	National Starch & Chemical Company
PAPER FOAM	Paper	PSP Papierschaum Pries GmbH
		Chronopol
		Dupont
		National Starch
		Planet Polymer
		Japan Corn Starch
BIOPOL	PHB/HV 35 or 150 µm	Safta (Italy)
COMPOPHAN	35 or 115 µm	Van Leer Packaging (D)
CELLOPHANE	Cellulose pure, 25 or 42 µm	UCB Films
NOBELON	Polyester aliphatic	Société Nationale des Poudres et Explosifs
Mulch paper	Paper	Ken Bar Inc.
DEPOSA	PLA	Fiberweb
	Biopolymers + starch	Fardis Belgium
PLANTCOBIO	Natural fibres	Plantco France
TEXON	Cellulose fibres	Texon International
Technilin	Non woven flax fibres	Technilin
Polynat	Biopolymers + natural fibres	Polynat
Greenfix	Biopolymers + natural fibres	Greenfix France Industrie
Sequana	Mulch paper	Ahlstrom Paper Group
	Biopolymers + starch	SCA Vivadour
	Mulch paper	Arjo Wiggins, Centre de recherche d'Apprieu

Table 2.3 Continued		
Commercial Name	**Composition**	**Producer**
BIOPAPIER	Mulch paper, and non woven materials (natural fibres)	Azcos
Biofilm	Polyester Aliph/arom	Barbier
	Polyester + starch	Ulice
BIO Polyane	?	Sté Prosyn Polyane
DS: degree of substitution *(?): there are undisclosed components in the composition*		

References

1. G. Swift in *Agricultural and Synthetic Polymers: Biodegradability and Utilisation*, Eds., J.E. Glass and G. Swift, ACS Symposium Series No. 433, ACS, Washington, DC, USA, 1990, 2.

2. R. Narayan in *Degradable Materials: Perspectives, Issues and Opportunities*, Eds, S.A. Barenberg, J.L Brash. R. Narayan and A.E Redpath, CRC Press, Boca Raton, FL, USA, 1990, 1.

3. S.J. Huang and P.G. Edelman in *Degradable Polymers: Principles and Applications*, Eds., G. Scott and D. Gilead, Chapman & Hall, London, UK, 1995, 18.

4. L. Krupp and W. Jewell, *Environmental Scientific Technology*, 1992, **26**, 193.

5. S.M Goheen and R.P Wool, *Journal of Applied Polymer Science*, 1991, **42**, 2691.

6. Y. Doi, *Microbial Polyesters*, VCH Publishers, New York, NY, USA, 1990.

7. G.M. Chapman in *Polymers from Agricultural Coproducts*, Eds., M.L. Fishman, R.B. Friedman and S.J. Huang, ACS Symposium Series No. 575, ACS, Washington, DC, USA, 1994, 29.

8. H. Chanzy, A. Peguy, S. Chaunis and P. Monzie, *Journal of Polymer Science, Polymer Physics Edition*, 1980, **18**, 1137.

9. A.F. Turbak, A. El-Kafrawy, F.W. Snyder and A.B. Auerback, inventors; International Telephone and Telegraph Corporation, assignee; US 4,302,252, 1981.

10. A.F. Turbak, A. El-Kafrawy, F.W. Snyder and A.B. Auerback, inventors; International Telephone and Telegraph Corporation, assignee; US 4,352,770, 1982.

11. V. Dave and W.G. Glasser, *Polymer*, 1997, **38**, 9, 2121.

12. J.E. Sealey, G. Samaranayake, J.G. Todd and W.G. Glasser, *Journal of Polymer Science: Part B: Polymer Physics*, 1996, **34**, 1613.

13. W.J. Maddever and P.D. Campbell in *Degradable Materials: Perspectives, Issues and Opportunities*, Eds., S.A. Barengerg, J.L. Brash, R. Narayan and A.E. Redpath, CRC Press, Boca Raton, FL, USA, 1990, 237.

14. D.F. Gilmore, R.C. Fuller and R. Lenz in *Degradable Materials: Perspectives, Issues and Opportunities*, Eds., S.A. Barengerg, J.L. Brash, R. Narayan and A.E. Redpath, CRC Press, Boca Raton, FL, USA, 1990, 481.

15. P. Jarret, W.J. Cook and J.P. Bell, *Polymer Preprints*, 1981, **22**, 2, 351.

16. P. Christel, F. Chabot, J.L. Leray, C. Morin and M. Vert, *Proceedings of Biomaterials*, 1980, 271.

17. F. Kollinsky and G. Markert in *Proceedings of Multicomponent Polymer Systems*, Houston, TX, USA, 1971, 175.

18. D.R. Paul, *Polymer Blends, Volume 2*, Eds., D.R. Paul and S. Newman, Academic Press, New York, NY, USA, 1978, p.12-35.

19. A.C. Albertsson and S. Karlsson, *Macromolecular Symposia*, 1999, **144**, 1.

20. M.N. Kim, B.Y. Lee, I.M. Lee and J.S. Yoon, *Journal of Environmental Science and Health*, 2001, **36**, 4, 447.

21. C.F. Phelps and P.H. Clarke, *Biotechnology*, The Biochemical Society, London, UK, 1983.

22. D.M. Abou-Zeid, R-J. Müller and W-D. Deckwer, *Journal of Biotechnology*, 2001, **86**, 113.

23. G. Braunegg, G. Lefebvre and K.F. Genser, *Journal of Biotechnology*, 1998, **65**, 127.

24. R.J. Müller, I. Kleeberg and W-D. Deckwer, *Journal of Biotechnology*, 2001, **86**, 87.

25. G. Gottschalk, *Bacterial Metabolism*, Springer-Verlag, New York, NY, USA, 1986.

26. G.W. Huisman, E. Wonink, R. Meima, B. Kazemier, P. Terpstra and B. Witholt, *Journal of Biological Chemistry*, 1991, **266**, 4, 2191.

27. J. Mergaert, C. Anderson, A. Wouters, J. Swings and K. Kersters, *FEMS Microbiology Review*, 1992, **9**, 2-4, 317.

28. J. Mergaert, A. Webb, C. Anderson, A. Wouters and J. Swings, *Applied Environmental Microbiology*, 1993, **59**, 10, 3233.

29. C. Song, U. Uchida, S. Ono, C. Shimasaki and M. Inoue, *Bioscience, Biotechnology and Biochemistry*, 2001, **65**, 5, 1214.

30. Y. Asano and S. Watanabe, *Bioscience, Biotechnology and Biochemistry*, 2001, **65**, 5, 1191

31. B.I. Sang, K. Hori, Y. Tanji and H. Unno, *Applied Microbial Biotechnology*, 2002, **58**, 2, 241.

32. M. Day, K. Shaw and D. Cooney, *Journal of Environmental Polymer Degradation*, 1994, **2**, 121.

33. J. Mergaert, C. Anderson, A. Wouters and J. Swings, *Journal of Environmental Polymer Degradation*, 1994, **2**, 177.

34. D.F. Gilmore, S. Antoun, R.W. Lenz and R.C. Fuller, *Journal of Environmental Polymer Degradation*, 1993, **1**, 269.

35. B.H. Briese, D. Jendrossek and H.G. Schlegel, *FEMS Microbiology Letters*, 1994, **117**, 1, 107.

36. J. Biedermann, A.J. Owen, K.T. Schloe, F. Gassner and R. Sussmuth, *Canadian Journal of Microbiology*, 1997, **43**, 6, 561.

37. S.T. Khan and A. Hiraishi, *FEMS Microbiology Letters*, 2001, **205**, 253.

38. J. Mergaert, A. Wouters, C. Anderson and J. Swings, *Canadian Journal of Microbiology*, 1995, **41**, 1, 154.

39. S.H. Imam, S.H. Gordon, R.L. Shogren, T.R. Tosteson, N.S. Govind and R.V. Greene, *Applied Environmental Microbiology*, 1999, **65**, 2, 431.

40. M.M. Mabrouk and S.A. Sabry, *Microbiological Research*, 2001, **156**, 4, 323.

41. K. Kasuya, H. Mitomo, M. Nakahara, A. Akiba, T. Kudo and Y. Doi, *Biomacromolecules*, 2000, **1**, 2, 194.

42. K. Sei, M. Nakao, K. Mori, M. Ike, T. Kohno and M. Fujita, *Applied Microbial Biotechnology*, 2001, **55**, 6, 801.

43. K. Yamashita, Y. Aoyagi, H. Abe and Y. Doi, *Biomacromolecules*, 2001, **2**, 1, 25.

44. A. Behrends, B. Klingbeil and D. Jendrossek, *FEMS Microbiology Letters*, 1996, **143**, 2-3, 191.

45. C.K. Hansen, *FEBS Letters*, 1992, **305**, 2, 91.

46. D. Jendrossek, M. Backhaus and M. Andermann, *Canadian Journal of Microbiology*, 1995, **41**, 1, 160.

47. D. Jendrossek, A. Frisse, A. Behrends, M. Andermann, H.D. Kratzin, T. Stanislawski and H.G. Schlegel, *Journal of Bacteriology*, 1995, **177**, 3, 596.

48. T. Iwata, Y. Doi, S. Nakayama, H. Sasatsuki and S. Teramachi, *International Journal of Biological Macromolecules*, 1999, **25**, 1-3, 169.

49. H. Abe and Y. Doi, *International Journal of Biological Macromolecules*, 1999, **25**, 1-3, 185.

50. T. Hiraishi, T. Ohura, S. Ito, K. Kasuya and Y. Doi, *Biomacromolecules*, 2000, **1**, 3, 320.

51. K. Kasuya, T. Ohura, K. Masuda and Y. Doi, *International Journal of Biological Macromolecules*, 1999, **24**, 4, 329.

52. Y. Tokiwa and T. Suzuki, *Nature*, 1977, **270**, 76.

53. K-E. Jaeger, A. Steinbüchel and D. Jendrossek, *Applied Environmental Microbiology*, 1995, **61**, 8, 3113.

54. C.V. Benedict, J.A. Cameron and S.J. Huang, *Journal of Applied Polymer Science*, 1983, **28**, 335.

55. C.V. Benedict, W.J. Cook, P. Jarrett, J.A. Cameron, S.J. Huang and J.P. Bell, *Journal of Applied Polymer Science*, 1983, **28**, 327.

56. Y. Oda, H. Asari, T. Urakami and K. Tonomura, *Journal of Fermentation Bioengineering*, 1995, **80**, 265.

57. H. Nishida and Y. Tokiwa, *Journal of Environmental Polymer Degradation*, 1993, **1**, 227.

58. Y. Oda, N. Oida, T. Urakami and K. Tonomura, *FEMS Microbiology Letters*, 1997, **152**, 339.

59. L. Liu, S. Li, H. Garreau and M. Vert, *Biomacromolecules*, 2000, **1**, 3, 350.

60. H. Tsuji and T. Ischizaka, *International Journal of Biological Macromolecules*, 2001, **29**, 83.

61. C.A. Murphy, J.A. Cameron, S.J. Huang and R.T. Vinopal, *Applied Environmental Microbiology*, 1996, **62**, 2, 456.

62. S. Ponsart, J. Coudane, B. Saulnier, J.L. Morgat and M. Vert, *Biomacromolecules*, 2001, **2**, 2, 373.

63. H. Ebata, K. Toshima and S. Matsumura, *Biomacromolecules*, 2000, **1**, 4, 511.

64. R.A. Gross, B. Kalra and A. Kumar, *Applied Microbial Biotechnology*, 2001, **55**, 6, 655.

65. D. Cam, S.H. Hyon and Y. Ikada, *Biomaterials*, 1995, **16**, 11, 833.

66. S.J. de Jong, E.R. Arias, D.T.S. Rijkers, C.F. van Nostrum, J.J. Kettenes-van den Bosch and W.E. Hennink, *Polymer*, 2001, **42**, 2795.

67. H. Tsuji, *Polymer*, 2002, **43**, 1789.

68. A. Torres, S.M. Li, S. Roussos and M. Vert, *Applied Environmental Microbiology*, 1996, **62**, 7, 2393.

69. A. Torres, S.M. Li, S. Roussos and M. Vert, *Journal of Environmental Polymer Degradation*, 1996, **4**, 213.

70. H. Pranamuda, Y. Tokiwa and H. Tanaka, *Applied Environmental Microbiology*, 1997, **63**, 4 1637.

71. S. Karjomaa, T. Suortti, R. Lempiäinen, J-F. Selin and M. Itävaara, *Polymer Degradation and Stability*, 1998, **59**, 333.

72. M. Itävaara, S. Karjomaa and J-F. Selin, *Chemosphere*, 2002, **46**, 879.

73. M. Hakkarainen, S. Karlsson and A-C. Albertsson, *Polymer*, 2000, **41**, 2331.

74. G. Gallet, R. Lempiäinen and S. Karlsson, *Polymer Degradation and Stability*, 2001, **71**, 147.

75. R.T. Mc Donald, S.P. McCarthy and R.A. Gross, *Macromolecules*, 1996, **29**, 7356.

76. H. Tsuji and S. Miyauchi, *Polymer Degradation and Stability*, 2001, **71**, 415.

77. H. Tsuji and S. Miyauchi, *Biomacromolecules*, 2001, **2**, 2, 597.

78. Y. Kikkawa, H. Abe, T. Iwata, Y. Inoue and Y. Doi, *Biomacromolecules*, 2002, **3**, 2, 350.

79. S.M. Reeve, S.P. McCarthy, M.J. Downey and R.A. Gross, *Macromolecules*, 1994, **27**, 825

80. D.F. Williams, *Engineering in Medicine*, 1981, **10**, 5.

81. T. Nakajima-Kambe, Y. Shigeno-Akutsu, N. Nomura, F. Onuma and T. Nakahara, *Applied Microbial Biotechnology*, 1999, **51**, 2, 134.

82. K. Nakamura, T. Tomita, N. Abe and Y. Kamio, *Applied Environmental Microbiology*, 2001, **67**, 1, 345.

83. M. Leimogne, *Annales Institute Pasteur*, 1925, **39**, 144.

84. J. Merrick, *Photosynthetic Bacteria*, 1978, **199**, 219.

85. H.G. Shlegel, H. Gottschalk and R. Von Bartha, *Nature (London)*, 1961, **191**, 463.

86. P.J. Bahram, A. Keller, E.L. Otun and P.A. Holmes, *Journal of Materials Science*, 1984, **194**, 2781.

87. D. Byron, *Trends in Biotechnology*, 1978, **5**, 246.

88. P.A. Holmes, *Physics in Technology*, 1985, **16**, 32.

89. M. Avella and E. Martuscelli, *Polymer,* 1988, **29,**1731.

90. M. Avella, E. Martuscelli and P. Greco, *Polymer,* 1991, **32,** 9.

91. M. Avella, E. Martuscelli and M. Raimo, *Polymer,* 1993, **34,** 3234.

92. P. Sadocco, C. Bulli, G. Elegir, A. Seves and E. Martuscelli, *Die Makromolekulare Chemie,* 1993, **194,** 2675.

93. P. Greco and E. Martuscelli, *Polymer,* 1989, **30,** 1475.

94. Lotti, M. Pizzoli, G. Ceccorulli and M. Scandola, *Polymer,* 1993, **34,** 4935.

95. J.S. Yoon, C.S. Choi, S.J. Maing, H.J. Choi, H. Lee and S.J. Choi, *European Polymer Journal,* 1993, **29,** 1359.

96. M. Gada, R.A. Gross and S.P. McCarthy in *Biodegradable Plastics and Polymers,* Eds., Y. Doi and K. Fukuda, Elsevier Science, Amsterdam, NL, 1994, 479.

97. M. Avella, B. Immirzi, M. Malinconico, E. Martuscelli, G. Orsello, A .Pudia and G. Ragosta, *Die Angewandte Makromolekulare Chemie,* 1993, **205,** 151.

98. M. Avella, L. Calandrelli, B. Immirzi, M. Malinconico, E. Martuscelli, B. Pascucci and P. Sadocco, *Journal of Environmental Polymer Degradation,* 1995, **3,** 49.

99. C.M. Buchanan, S.C. Gedon, A.W. White and M.D. Wood, *Macromolecules,* 1992, **25,** 7373.

100. M. Avrami, *Journal of Chemical Physics,* 1939, **7,** 1103.

101. P. Sadocco, M. Canetti, A. Seves and E. Martuscelli, *Polymer,* 1993, **34,** 3368.

102. Y. Kumagai and Y. Doi, *Polymer Degradation and Stability,* 1992, **36,** 241.

103. F. Gassner and A.J. Owen, *Polymer,* 1994, **35,** 2233.

104. M. Avella, B. Immirzi, M. Malinconico, E. Martuscelli and M.G. Volpe, *Polymer International,* 1996, **39,** 191.

105. J. Gu, D.T. Eberiel, S.P. McCarthy and R. Gross, *Journal of Environmental Polymer Degradation,* 1993, **3,** 1, **143.**

106. N. Lotti and M. Scandola, *Polymer Bulletin,* 1992, **29,** 407.

107. J.B. Miller, K.J. McGrath, C.M. Roland, C.A. Trask and A.N. Garroway, *Macromolecules,* 1990, **23**, 4543.

108. I. Koller and J. Owen, *Polymer International,* 1996, **39**, 175.

109. R.L. Shogren, *Journal of Environmental Polymer Degradation,* 1993, **3**, 1, 75.

110. S.H. Imam, S.H. Gordon, R.L. Shogren and R.V. Greene, *Journal of Environmental Polymer Degradation,* 1993, **3**, 1, 205.

111. C. De Kesel, C. Vander Wauven and C. David, *Polymer Degradation and Stability,* 1997, **55**, 107.

112. ASTM D5338-98e1, *Standard Test Method for Determining Aerobic Biodegradation of Plastic Materials Under Controlled Composting Conditions,* 1998.

113. S.P. McCarthy, *Macromolecular Symposia,* 1999, **144**, 63.

3 Quantifying the Range of Properties in Natural Raw Material Origin Polymers and Fibres

Mark Hughes

3.1 Introduction

Sustainable development, defined by the Brundtland Report [1] as *development which meets the needs of the present without compromising the ability of future generations to meet their own needs*, is increasingly becoming a priority of governments and businesses. But how can we satisfy society's demand for materials (and energy) in a sustainable fashion as it continues to grow?

Many of our modern day materials are based upon essentially non-renewable resources and are, to all intents and purposes, unsustainable in the long-term. Perhaps the most prominent of these resources is oil and whilst a proportion is converted into functional materials such as plastics and resins, the greater part is burnt to produce energy.

Renewable and potentially sustainable alternatives to fossil reserves are industrial crops. Industrial crops can form the basis of a variety of functional materials. For instance, a variety of commonly available vegetable oils can be used as the starting point for resins; starches can form plastics and fibre crops like flax and hemp can be used in the fabrication of structural composite materials. The environmental benefits of these materials are well documented – renewability, biodegradability and low embodied energy, for example. With these potential benefits, why are these materials not in widespread use?

A recent study has highlighted that one of the principal stumbling blocks hindering the wider adoption of materials derived from natural, renewable resources is a lack of reliable property information [2] and goes onto suggest that more readily available information would promote usage. One of the principal reasons why this lack of information is so apparent is that there is often enormous variation in the properties of materials of natural origin. Quantifying the range of these properties is essential, if reliable technical information is to be made available to designers and specifiers wishing to use natural raw material origin polymers, fibres and composite materials based thereon.

This chapter sets out to explore some of the factors that give rise to this variability in polymeric materials of natural origin, how the some of these properties are measured and discusses the practical implications of this variability. It is important to understand how this variability arises in the first instance, as it gives a fuller appreciation and understanding of these materials and their behaviour, as well as their limitations. For this reason, a significant portion of what follows is a discussion of the factors that give rise to variability in natural fibres and raw material polymers.

The properties of natural polymer fibres (wood fibre and long vegetable fibres) as well as polymers synthesised from renewable raw materials will be considered. Whilst the properties of wood and wood fibres have been well researched and documented over the years, a class of plant fibres known as 'long vegetable fibres' are currently receiving a significant amount of attention in the research and industrial communities. This interest has arisen because the properties of these fibres make them attractive as potential alternatives to synthetic organic and inorganic (principally glass) fibres. Potential applications for these fibres might include non-woven structures for horticultural and soil stabilisation projects, or even in more technically demanding roles such as reinforcement in composite materials. A subset of long vegetable fibres known as 'bast' fibres is of particular importance because of their excellent mechanical properties. Bast fibres are currently enjoying considerable attention as alternative fibre crops in the UK and elsewhere. Whilst these fibres had been used for many centuries for apparel textiles (excellent quality linen was produced at the time of the Pharaohs) as well as for industrial applications such as ropes and cordage, their use had been in decline in recent decades due to the prevalence of synthetic fibres. Nevertheless, in recent years renewed interest has been shown, in view of their potential to help mitigate carbon dioxide emissions when used as alternatives to glass fibre. For similar environmental reasons, there has been a drive to produce a range of thermosetting and thermoplastic polymers from renewable biomass. This drive has resulted in the limited commercialisation of resins and plastics from renewable feedstock and research continues apace in this area. In marrying together both polymers derived from natural resources as well as natural fibres, useful composite materials may be formed. These may, in time, provide a sustainable alternative to many of the materials used in, for instance, transportation and construction.

3.2 Properties

In dealing with the properties of materials, it is useful to consider briefly what is actually meant by 'property' and which of these properties should be measured. In general, this chapter will concern itself with the better known physical and mechanical properties such as 'density' or 'stiffness'. Frequently, when dealing with natural materials, especially fibres, it is difficult to know whether it is a true material property, such as Young's modulus, that is being measured, or whether it is actually the property of a structure. As will be discussed in Section 4.8, this can lead to difficulties in the evaluation of the properties of natural plant fibres.

When considering the properties of natural raw material origin polymers, it is difficult to envisage a situation where the ultimate mechanical, physical or biological properties are not ultimately linked to the starting materials – vegetable oils or other plant extracts – all of which highly complex and variable in their own right.

When measuring the physical and mechanical properties of either natural raw material origin polymers or fibres for industrial applications, it is natural to consider standard tests, such as those developed and published by various standards organisations. These include; the International Organisation for Standardisation (ISO), European Committee for Standardisation (CEN), British Standards Institution (BSI), Deutsches Institut für Normung e.V. (DIN), or the American Society for Testing and Materials (ASTM). Standards will, therefore, have been developed for similar materials and with suitable adaptation (if necessary) can be used to measure the properties of natural raw material origin polymers and fibres. Some tests, devised for 'natural materials' such as textile fibres, will have been developed around non-industrial or non-technical applications, e.g., BS EN ISO 5079:1996 [3]. This often gives rise to different terminology, making it difficult for engineers or designers to utilise property data developed for alternative applications. Where standard test protocols do not exist, then it is desirable to look to the published literature for techniques that have been used successfully to measure material properties.

When we speak of 'low environmental impact polymers', the ability of the material to biodegrade at the end of its lifecycle is of significance. Few tests have been specifically developed to assess the biodegradability of natural raw material origin polymers and fibres. Again, standard tests have been developed for many similar materials and these may often be adapted to assess biological attack in biopolymers. As will be discussed in Section 3.9.3.3, recently, advances have been made in the development of norms for composting and biodegradation of biopolymers and these have now been accepted into the standards (e.g. ISO 14855 [4]; ASTM D6400-99e1 [5]).

3.3 Variability in Natural Origin Materials

In many ways, natural raw material origin polymers and fibres can be thought of as a 'half-way house" between wholly synthetic materials, in which the raw materials, manufacturing and properties are all tightly controlled, and naturally occurring materials which are manipulated by mankind to suit his own ends. Perhaps the best known and almost certainly the most widely used material in this latter category is wood. In the UK alone some 40 million cubic metres of wood product (Wood Raw Material Equivalent) are consumed annually [6]. In its most basic form sawn timber is highly variable when compared to materials such as steel. The properties of wood are affected by a number of factors including, for example, species, density, the occurrence of defects such as knots and its moisture content [7]. As wood is processed further, much of this 'natural' variability can removed as the defects and heterogeneity is 'engineered' out of the wood [8]. In the search for more and more reliable structural materials, wood is often 'reconstituted' into products such as 'glulam' (laminated laths of wood bonded with an adhesive) or laminated veneer lumber (LVL) which possesses greater homogeneity in properties than sawn timber. It is interesting to note that taking the process of 'engineering' still further leads to more homogeneous materials such as paper and fibreboard, which possess highly reproducible properties. It is worthy of note, that despite the natural variability of sawn wood products, designers are able to work with these materials because of the grading systems and standards that have been developed.

3.4   The Influence of the Chemistry and Structure of Natural Origin Fibres Upon Their Properties

There are many forms of natural fibre obtained from plant material. Wood is perhaps the best known and most widely used. There is, nevertheless, a large group of non-wood fibres which come under the general heading 'industrial crops' which are of particular interest at the present time because of their potential to replace synthetic fibre. Fibres, such as cotton, sisal, coir, flax and hemp are well known to mankind and have been used for millennia for textiles and cordage. These long vegetable fibres are obtained from various parts of the plant and can be classified accordingly. Cotton and kapok are, for example, seed hairs, whilst sisal is a leaf fibre. Flax, hemp and jute belong to the group of fibres known as bast fibres. Flax and hemp, in particular, are currently receiving a significant amount of attention in the UK and Europe in view of their potential as reinforcement in polymer matrix composites (PMC).

The basic chemical building blocks for all the so-called 'lignocellulosic' fibres are similar, whether we are referring to a refined wood fibre, or a flax fibre. In a similar fashion the organisation of the structural elements in the cell wall are similar. In the following sections much of the discussion will relate to wood fibre since a very large body of work has been accumulated in this area. Nevertheless, the same principals apply to all lignocellulosic fibres. Where possible, the discussion will be restricted to bast fibres, since it is these that are currently receiving significant attention as replacements for synthetic fibres in industrial applications.

Table 3.1 provides a comparison of the mechanical and physical properties of a number of natural fibres with those of several synthetic fibre types. What is immediately apparent from the table is the *range* of properties observed in natural fibres. To understand better the origin of this variability, it is necessary to first understand the structure of natural fibres.

Table 3.1 Selected physical and mechanical properties of some synthetic and natural (plant) fibres				
Fibre type	Density (x 10^3 kg m^{-3})	Young's modulus (GPa)	Tensile strength (MPa)	Failure strain (%)
Synthetic fibres				
E-glass	2.56	76	2000	2.6
High strength carbon	1.75	230	3400	3.4
Kevlar™ (aramid)	1.45	130	3000	2.3
Boron	2.6	400	4000	1.0
Natural fibres				
Flax	1.4-1.5	50-70	500-900	1.3-3.3
Hemp	1.48	30-60	310-750	2-4
Jute	1.4	20-55	200-450	2-3
Sisal	1.45	9-22	80-840	3-14
Cotton	1.5	6-10	300-600	6-8
Sources: Hull and Clyne [9], and Ivens and co-workers [10]				

3.4.1 The Chemistry and Ultrastructure of Natural Fibres

3.4.1.1 Bast Fibres

Bast fibres from flax (*Linum usitatissimum*) and hemp (*Cannabis sativa*) are obtained from the bark, or *bast*, of the plant stem. **Figure 3.1** shows the location of these fibres in the stem of the flax plant. Flax and hemp fibres have similar characteristics and end uses and the plants have long been cultivated by mankind [11]. Both species grow well in temperate climates [11].

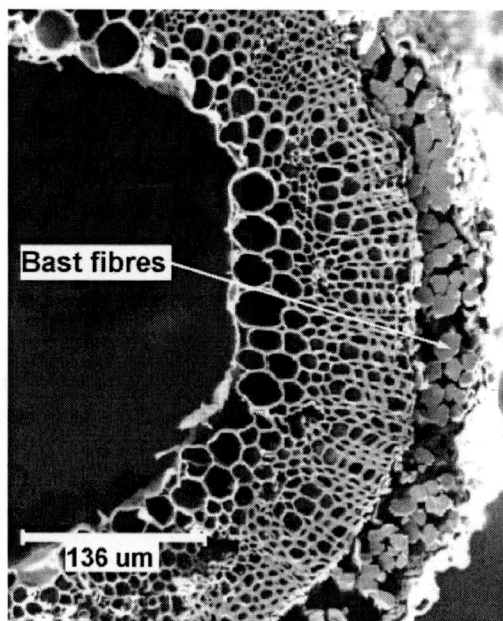

Figure 3.1 Bast fibres found in the outer portion of the stem

The bast fibres that are isolated from the plant stem are referred to as *technical* fibres or *fibre bundles*. It is these that are mainly used in textiles, or other industrial applications. The length of these fibres varies considerably. In flax, for instance, the length of the technical fibres range from 0.3 m to 0.6 m, whilst in hemp they can range from 0.9 m to 1.8 m [11]. The thickness of the technical fibres vary from around 50 to 500 μm in flax and from 500 μm to 5 mm in hemp.

The fibre bundles are, in turn, composed of many individual cells or *'ultimate'* fibres. These too vary in length and diameter. In flax, for example, cell length varies from 5 to 50 mm, averaging around 25 mm, whilst the diameter ranges from 15 to 35 μm, giving the fibres an average aspect ratio of 1200 [11]. Hemp cells range from 5 to 55 mm in length, averaging around 20 mm and are between 125 and 375 μm in diameter [11]. The average aspect ratio for hemp cells is 1000 [11].

3.4.1.2 The Chemistry of Bast Fibres

Bast fibres are, themselves, composite structures of considerable complexity and elegance. Since the properties of the fibres are heavily dependent upon their structural organisation, it is appropriate to consider, briefly, the chemical composition and ultrastructure of bast fibres. Bast fibres are *lignocellulosic* fibres that consist of three main polymers: *cellulose*, *lignin* and *matrix polysaccharides* (such as *pectins* and *hemicelluloses*) associated with cellulose and lignin in the cell wall. In addition to these, a number of non-structural components such as waxes, inorganic salts and nitrogenous substances are also present [7, 12]. A summary of the main chemical constituents of bast fibres, compiled from a number of sources, is presented in **Table 3.2**. As may be seen, there is a degree of variation in the composition, due, in the main to fibre variability and different analysis methods. Nevertheless, it is possible to distinguish general trends and it is possible to derive average, or typical, compositions. Typically, a bast fibre such as flax or hemp is composed of around 75% (by mass) cellulose, 20% polysaccharides and 5% lignin.

Cellulose

Cellulose may be described as a high molecular weight, long chain molecule, consisting of β-D anhydroglucopyranose units, bound with $\beta - (1 \rightarrow 4)$ glycosidic linkages, as represented schematically in **Figure 3.2** [7, 12, 17]. Every alternate unit ($C_6H_{10}O_5$) is rotated through 180 degrees [7]. Two anhydroglucopyranose molecules form the smallest repeating unit in the chain, known as the *cellobiose* unit. The number of anhydroglucopyranose units (the

Table 3.2 Summary of the main chemical constituents of flax, hemp and jute						
Fibre type	Cellulose (%)	Hemi-cellulose (%)	Pectin (%)	Lignin (%)	Extractives (%)	Reference
Flax	80-90	-	2-6*	-	3-4	[11]
	79	20.7	2.6	2.4	1.9	[13]
	71.2	18.6	1.1	2.2	6	[14]
	81	14	2	3	-	[15]
Hemp	85.7	-	10.2*	-	5.3	[11]
	83.4	20.1	1.0	4.1	0.9	[13]
	74	18	1	4	-	[15]
Jute	59.9	12.1	-	17.5	9.9	[16]
	70-83	-	23-28*	-	1.5-6.9	[11]
	69.8	23.3	0.2	14.9	0.6	[13]
	71.6	13.3	0.2	13.1	1.8	[14]
	72	13	<1	13	-	[15]
Includes: 'pectose bodies, lignin', 'pectose and gummy substances' and 'incrusting and pectin matter'. These include hemicellulose, pectin and lignin						

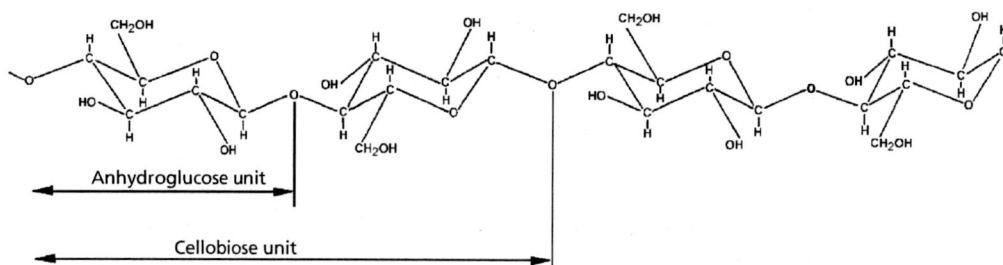

Figure 3.2 Schematic representation of part of a cellulose chain

Source: [7]

degree of polymerisation) in plant cellulose lies in the range 7,000-15,000 [17]. In retted (see Section 3.5) flax fibre, the degree of polymerisation is of the order of 2,500-3,000 depending upon the growing factors and retting conditions [12].

The structure of the anhydroglucopyranose unit is not flat, but bent in the form of a 'chair' (this is the lowest energy configuration [17]). Ribbon-like cellulose molecules pack together to form a crystal structure with neighbouring cellulose chains displaced by one quarter of a repeating unit. The existence of hydroxyl (-OH) groups facilitates intra- and inter-molecular hydrogen bonding, which bind adjacent molecules and imparts rigidity to the structure. Individual anhydroglucopyranose units are joined covalently, giving rise to a very strong molecular chain along its length [7] and partly explains the exceptional properties of plant fibre material. This packing arrangement is shown schematically in **Figure 3.3**.

Further aggregation of the cellulose molecules, gives rise to the crystalline 'backbone' of what may be regarded as the basic fibrous building element of all lignocellulosic fibres – the *microfibril*. The diameter of this unit is generally of the order of 10-30 nm. Nevertheless, even smaller units, termed *elementary fibrils* have been detected, having average diameters of between 2 and 4 nm [17]. As will be discussed further in Section 3.4.1.3, this cellulose 'core' is surrounded by *para-* and amorphous cellulose as well as other polymeric material [17]. In a longitudinal sense, the cellulose 'core' of the fibril is thought to consist of crystalline regions (the *crystallites*) interspersed with less ordered domains. The exact nature of this structure has not been fully elucidated and a number of models have been proposed [17, 18]. The length of the crystallites varies from around 50 to 100 nm, depending upon the origin of the cellulose [7, 18]. Between 70% and 80% of the cellulose present in bast fibres is crystalline [17]. It should be noted that a significant proportion of this non-crystalline cellulose corresponds to the surface chains [18].

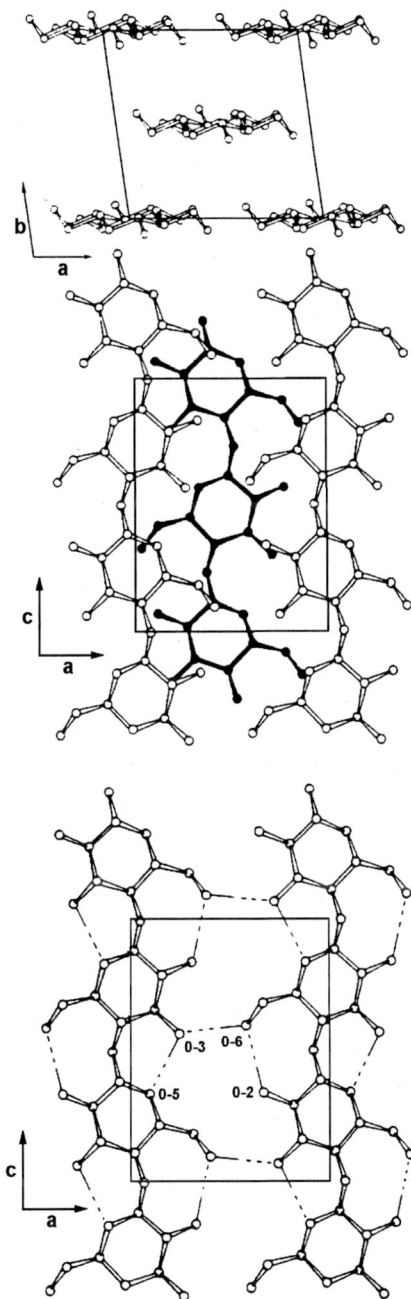

Figure 3.3 Schematic representation of the packing arrangement of cellulose chains, showing inter- and intra-molecular bonding

Source: [12]

Hemicelluloses

These are heterogeneous polysaccharides, composed of various monomeric units such as D-glucose, D-mannose, D-galactose, D-xylose, L-arabinose and small amounts of L-rhamnose in addition to D-glucuronic acid, 4-O-methyl-D-glucuronic acid and D-galacturonic acid [19]. Structurally, hemicelluloses are branched molecules having a low degree of crystallinity [7, 17]. Typically, the degree of polymerisation is only around 200 [19]. In wood, hemicelluloses are thought to account for some 20% to 30% of the dry weight [19] whereas in bast fibres this is generally less (see **Table 3.2**). Linkages probably exist between the polysaccharides an lignin, forming the so-called 'lignin-carbohydrate complex' [17]. It seems highly probable, therefore, that hemicelluloses function as 'interfacial coupling agents', linking the cellulose 'core' to the surrounding lignin.

Pectins

Pectins are present in plant tissue to varying degrees, being found predominantly in fruit peel and gums [17]. Unretted flax contains between 3% and 4% pectins [20]. In flax, pectins are to be found in the cells surrounding fibre bundles, especially those separating bast fibres from core tissue [21]. High concentrations of pectins occur in the primary cell wall and middle lamella of fibres [12, 17, 19, 21]. Here, in conjunction with hemicelluloses, their function is that of a cementing material [12]. Because of their importance as a binding material, removal of pectins during the retting process is of particular importance in the production of the technical fibre.

The main building block of pectin is a linear chain of $\alpha - (1 \rightarrow 4)$ linked D-galacturonic acid units [17, 19, 20, 22, 23]. The predominant galacturonan backbone of pectin is, however, modified by the insertion, at intervals, of α-L-rhamnopyranose units [23]. Other neutral sugars such as D-galactose and L-arabinose [19, 23] are present in extended side chains, whilst D-xylose; L-fucose; D-glucuronic acid are present in short side chains [23].

The uronan chain is ordered into blocks of unbranched, unesterified segments and heavily branched, esterified blocks [24]. The latter are frequently interrupted by rhamnose units, linked by $\alpha - (1 \rightarrow 2)$ and $\alpha - (1 \rightarrow 4)$ bonds to adjacent galacturonic acid units [17]. Many of the rhamnose units carry arabinan and galactan side chains. The rhamnose units introduce 'kinks' into the chain. The unbranched, unesterified blocks are rarely interrupted by rhamnose units. The chains are thus able to aggregate together along the unbranched lengths and form junctions, through bonding together non-covalently by co-ordinated calcium ions. This enables gel formation to take place. High ester pectic polysaccharides are able to form acid gels [24].

Lignin

Lignin is an amorphous, highly crosslinked aromatic polymer resulting from the dehydrogenative radical polymerisation of *p*-coumaryl, coniferyl and synapyl alcohols, which are the precursors and building blocks of all lignins [12]. By mass, lignin accounts for some 20 to 40% of the cell wall in wood [12, 17]. In flax this figure lies between 2% and 5% depending upon the degree of retting [12].

3.4.1.3 Structural Organisation of the Cell Wall

Broadly speaking, lignocellulosic fibres are analogous to synthetic fibre reinforced composites, in which cellulose forms the reinforcement with other polysaccharides and lignin, the matrix. The basic fibrous building element of the cell wall may be regarded the microfibril. The microfibril can itself be viewed as a composite, in which crystalline cellulose forms the 'reinforcement' core, surrounded by a 'matrix' of *para-* and amorphous cellulose, hemicellulose and pectins. These are, in turn, sheathed in lignin. Various authors have proposed several models of the association of the cell wall components. **Figure 3.4(a)**, shows a schematic representation of the cross-section of the microfibril. As may be noted, in this model a single crystalline 'core' is surrounded by a 'matrix' of other polysaccharides and lignin. **Figure 3.4(b)** depicts, not only the supramolecular structure of the microfibril, but also takes into account, smaller elementary fibrils. Nevertheless, in both these models, it may be observed that there is a close association between the cellulose 'core' and the surrounding hemicelluloses and lignin.

In detail, the structure of an individual lignocellulosic fibre consists of a *primary* and *secondary* cell wall, with the arrangement of the microfibrils within each of these varying as depicted schematically in **Figure 3.5**. The very thin primary wall, lying at the outside of the fibre, adjacent to the *middle lamella* (the interstitial region between adjacent cells), consists of a more or less random arrangement of microfibrils [7]. The secondary wall may be further sub-divided into S_1, S_2 and S_3 layers. The microfibrils within the outermost S_1 layer, which forms around 10% of the wall thickness, are arranged in two distinct spirals: one left handed and the other right handed ('Z' and 'S' helices). The microfibrils in this layer lie at angles of between 50° and 70° to the fibre long axis [7]. The angle formed between the microfibril axis and the fibre axis is referred to as the *microfibrillar angle*. The microfibrils in the middle, S_2, layer, which comprises some 85% of the total wall thickness, are aligned parallel to one another in a steeply inclined helix. In wood, the microfibrillar angle of the S_2 layer lies between 10° and 30° [7]. In bast fibres, the microfibrillar angle of the S_2 layer is generally low; in hemp for instance, this has been reported to lie between 2° and 2.3° (depending upon the measurement technique used) and in jute at 7.9° [25]. The mechanical properties of the fibre are closely linked to the

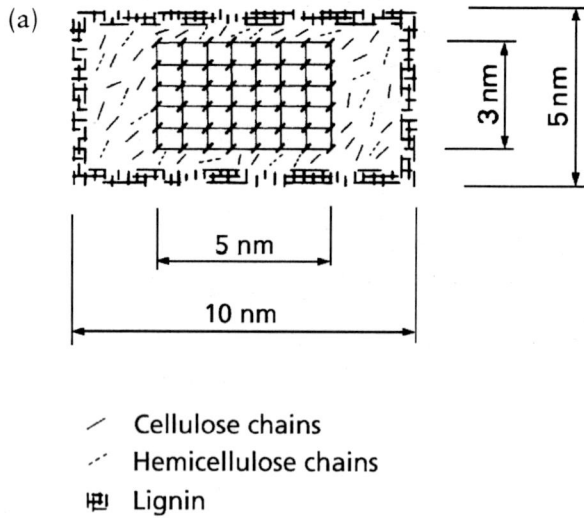

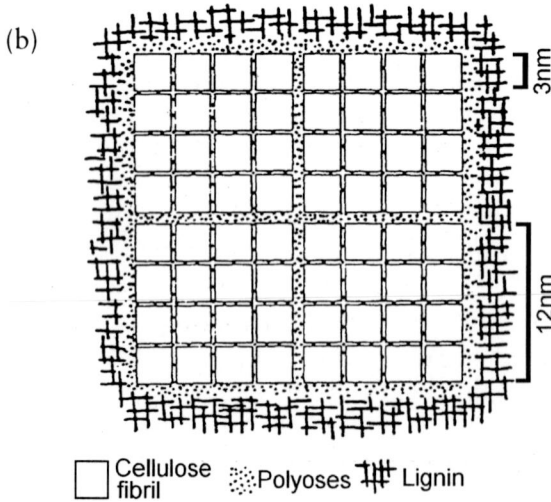

Figure 3.4 Schematic representation of the cross-section of an individual microfibril

Sources: (a) [7]; (b) [17]

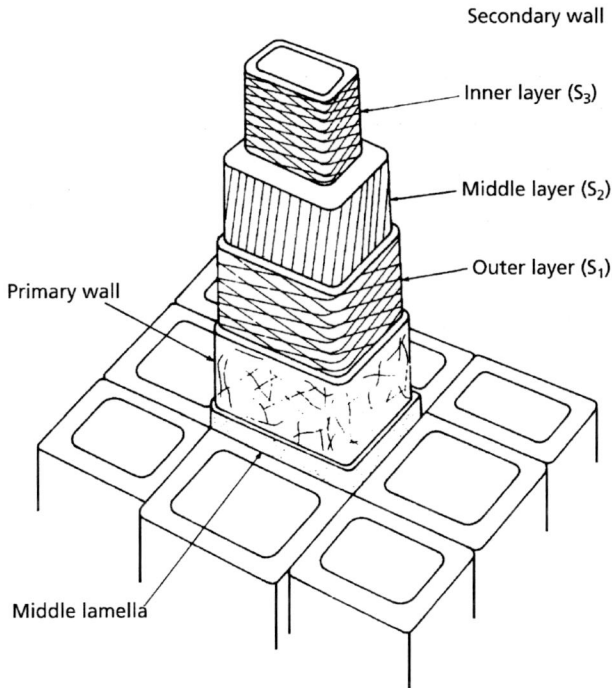

Figure 3.5 Schematic representation of the cell wall structure of a typical lignocellulosic fibre

Source: [7]

microfibrillar angle of the S_2 layer (see **Figure 3.6**); low angles being associated with higher strength and stiffness [26]. The S_3 layer in wood fibres accounts for only around 1% of the wall thickness [7].

3.4.2 The Influence of Fibre Ultrastructure Upon its Mechanical Properties

The structure of flax and other bast fibres is, in general terms, similar to that of other wood fibres in as much as the cell wall follows the same overall ultrastructural organisation of lignocellulosic fibres, outlined in Section 3.4.1.3 [28]. Whilst much work has been undertaken with a view to gaining a fundamental understanding of the mechanics of individual wood fibres in relation to their structure and chemical composition, comparatively little is to be found in the literature on similar studies conducted on bast fibres. This is most

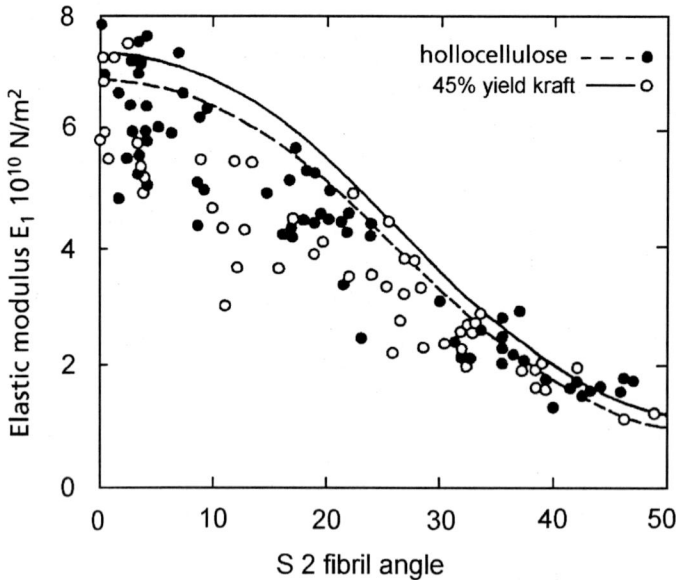

Figure 3.6 Variation of the fibre tensile (Young's) modulus with the microfibrillar angle of the S_2 layer

Source: [27]

likely due to the fact that the latter have, until recent times at least, been used almost exclusively in textile applications and that, compared with other natural fibres used in these applications (such as cotton and wool), only to a relatively limited extent.

3.4.2.1 Axial Tensile Properties

A significant volume of experimental work has been published over the years relating to the determination of the tensile properties of individual wood pulp fibres [29-31]. It has been clearly demonstrated that these properties are controlled by the ultrastructural organisation of the fibre, in particular the microfibrillar angle of the dominant S_2 layer. For example, it has been shown that the tensile strength and stiffness, as well as the mode of deformation and fracture of the fibres, are dependent upon the microfibril angle of the S_2 layer [26, 31, 32]. Whilst the S_2 layer dominates the properties of the fibre in tension, the S_1 layer is thought to be important in controlling fibre stability in compression, by limiting excessive lateral cell expansion [33]. The S_3 layer, on the other hand, is believed to resist hydrostatic pressure within the cell lumen [33], whilst the combined laminate structure is considered to be of importance in controlling trans- and intra-wall crack propagation [34].

Over the years, many models have been proposed to predict the elastic behaviour of the lignocellulosic fibre cell wall [26, 27, 35-40]. These models were based on orthotropic elasticity theory, in which the cell wall was considered to be a composite laminate structure composed of balanced plies [27, 38]. The models have been used to predict the Young's modulus and tensile strength of fibres as a function of the microfibril angle of the S_2 layer [27, 31]. Good agreement between theoretical and experimental values for stiffness and strength have been obtained, discrepancies being attributed to defects and inhomogeneities within the fibre [31, 41, 42]. Whilst most models have been constructed to predict the behaviour of wood fibres, some work has recently been undertaken with a view to modelling the behaviour of flax and other plant fibres [43].

3.4.2.2 The Toughness of Plant Fibres

One of the great attractions of wood as a structural material is its excellent work of fracture, which, on a specific basis is on a par with that of mild steel and higher than that of synthetic PMC [44]. This too has, in part at least, been attributed to the fibre ultrastructural organisation. The overall work of fracture of plant fibre material can be attributed to [45]:

- fracture of the cell wall material itself,

- plastic buckling of the cell wall into the lumen as a result of shear stresses,

- pull-out of intact cells.

Fracture toughness (K_{IC}) measurements conducted by Lucas and co-workers [45, 46] on a range of plant fibre material and wood species, indicated a figure which was identified as the work of fracture of the cell wall material itself. This figure, which was referred to as the 'intrinsic toughness' of the cell wall (independent of contributions from either plastic work or cell pull-out) was for a crack traversing a fibre perpendicular to its axis, in other words perpendicular to the axis of the microfibrils in the S_2 layer. The results of these investigations pointed to a figure of around 3.45 kJ m^{-2} for the intrinsic toughness of the cell wall. This value is in broad agreement with a figure of 1.65 kJ m^{-2} for the intrinsic toughness of cell wall material reported by Ashby and co-workers [47], obtained by a different method.

If the cell wall is considered to be an heterogeneous, orthotropic material itself, then a certain degree of variation in the intrinsic toughness, or cell wall work of fracture, would be expected. This would depend upon the path followed by a crack propagating through the cell wall. This is supported by the fact that Gibson and Ashby [48] report a figure for the intrinsic toughness of <0.35 kJ m^{-2} 'along the grain', whilst measurement of the fracture energy of solid wood conducted 'with the grain' indicates values of between 0.2 to 0.3 kJ m^{-2} [49]. In these tests it might be supposed that crack propagation would be either intra-wall, i.e., along the middle lamella or inter-wall, through the cell wall material, along the direction of the microfibrils in the S_2 layer.

Notwithstanding the above, the intrinsic toughness of the cell wall is substantially lower than the measured works of fracture of wood across the grain which fall in the region of 10-30 kJ m^{-2} [44]. The work of fracture attributed to the pull-out of individual cells is, however, believed to contribute little to the overall work of fracture (1.6 kJ m^{-2}) of the material [44]. The major contribution is thought to be due to a pseudo-plastic tensile buckling mechanism first proposed by Gordon and Jeronomidis [50, 51]. This was based on observations made by Page and co-workers [31] that some individual wood pulp fibres underwent unstable buckling when loaded in axial tension. Lucas and co-workers [45] were also able to estimate the amount of 'plastic work' occurring during fracture. They estimated that approximately 90% of the overall work of fracture could be attributed to plastic deformation, the balance being due to the intrinsic toughness of the cell wall and the pull-out of intact cells. This pseudo-plastic mechanism of energy absorption is due to shear failure occurring within the S_2 layer as the fibre buckles. In the process, considerable amounts of energy are absorbed irreversibly [44]. The effectiveness of this mechanism in enhancing energy absorption has been subsequently confirmed using synthetic analogues of the wood cell [52].

A picture emerges of the intricate chemical and ultrastructural organisation of natural fibres and the influence that this has upon the mechanical properties of the fibre. In essence plant fibres can be thought of as 'micro-composite structures' in their own right. Like their bigger synthetic cousins, these fibres are susceptible to damage through overloading and mishandling. There are a number of factors that can influence the realisable properties of these fibres. Perhaps the most important of these is the process of extracting the fibres from the stem. The following section describes the extraction and isolation processes and how this can influence the properties of the fibre.

3.5 The Influence of Fibre Extraction, Isolation and Processing upon the Properties of Bast Fibres

Flax and hemp, as well as other bast fibres such as jute have historically been extracted and isolated from the stem in a process that involves several operations. Firstly, the mature plant is either cut or pulled from the ground. Pulling is preferred to cutting as it preserves as much of the fibre length as possible [53]. Nevertheless at present, harvesting in the UK is by cutting. The straw is then *retted*. Retting may best be described as a form of controlled rotting. In this operation, the technical fibres are loosened from the surrounding stem tissue by enzymes secreted by micro-organisms. These degrade the non-cellulosic substances, mainly pectins, which are in abundance in the surrounding tissue. This process liberates the technical fibres [54-57]). Once retting is complete, the straw is dried, usually by spreading in the field and when it is dry it is *scutched*. Scutching, mechanically breaks the straw using a 'crimping' action. This fractures the woody core material known as *shive*, separating the loosened fibre from this extraneous material. The final operation, termed *hackling* involves drawing the fibres through a set of pins to align them and to remove any remaining woody material. Short or broken fibres known as *tow* are also separated from the long fibres in the hackling process [54].

The previous paragraph describes how fibre has traditionally been extracted from the retted straw. Fibre for textile applications is still processed in this manner, however, the process is slow and labour intensive. Whilst the quality of fibre produced in this way is generally excellent, it is too costly for many technical applications. In the past few years high throughput decortication methods for separating fibre from straw have been developed. The fibre produced in this way is generally shorter and coarser than that produced by traditional methods, however, it is suitable for many technical applications such as non-woven structures. The method is nonetheless mechanically intensive, leading to fibre damage (see Section 3.6) which can reduce the mechanical properties.

Fibre quality is very sensitive to the retting operation and whilst ultimately fibre quality is dependent upon the quality of the straw, poor retting can drastically lower the quality of otherwise good quality fibre [54]. In view of the influence of retting upon fibre quality, it is appropriate to briefly review the methods of retting. Traditionally, *dew* or *field* retting and *water* retting have been used [54]. Recently, however, attention has been directed towards retting procedures involving chemical or enzymic degradation of the binding materials, mainly pectins [58, 59].

3.5.1 Dew Retting

Dew retting involves laying pulled straw on the ground when the temperatures are above freezing and turning occasionally during the retting period. The retting period generally lasts for between 25 and 30 days in summer and between 50 and 100 days in winter [54]. The quality of fibre produced by this method is inferior to that of water retted fibre [54]. The principal micro-organisms involved in dew retting are fungi and although bacteria have been reported to be present they are often suppressed by the former [60, 61]. A major disadvantage of dew retting is that it is totally dependent upon weather conditions, which can sometimes lead to the loss of a whole crop. Further, the labour costs involved are high [60].

At the present time, the majority of retting carried out in the UK is dew retting. A more recent development involves applying a desiccant and allowing the straw to ret standing up. Amongst the advantages of this approach are that it is less susceptible to the vagaries of the weather and there is no requirement to 'turn' the straw on the ground. As retting by this process takes somewhat longer, there is also less likelihood of 'overshooting the mark' and over retting, leading to reduced fibre quality.

3.5.2 Water Retting

Although water retting used to be carried out in ditches, ponds and slow moving rivers, this practice has been largely discontinued in many countries due to pollution problems [54]. Most water retting nowadays is undertaken in purpose built retting tanks [54, 55, 61]. Pollution by rettery effluents, however, continues to be problematic and has led to the decline of flax retting in many countries [61].

In this process, the straw is firstly given a preliminary leaching for 2-24 hours [55, 61] after which the leach water is drained off, fresh water added and the ret commenced. Retting is completed in between 3 and 7 days at water temperatures of between 30 °C and 35 °C [61]. The retting action is carried out by enzymes secreted by a population of mainly anaerobic bacteria. [54, 55, 61-63].

3.5.3 Enzyme Retting

One of the drawbacks of retting as an industrial process, particularly dew retting, is the lack of control over the process. Furthermore, the length of time taken to complete the ret is comparatively long. In addition to this, the organic by-products of the process lead to pollution which can be problematic [54, 61]. The use of commercial enzyme preparations has been explored as an alternative method of retting to overcome these problems [57, 59, 61-63].

Enzymic retting is similar in concept to tank retting. However, rather than enzymes secreted by micro-organisms being responsible for the retting action, commercially prepared enzymes are used. To date, most studies on the enzymic retting of bast fibres have been carried out on a laboratory or semi-industrial scale. Nevertheless, it has been demonstrated that the retting time can be dramatically reduced. Majumdar and co-workers [63] for example, reported that the retting of jute can be completed in 48 hours, whilst the ret time of flax has been reduced to 20-24 hours [57, 62]. Vilppunen and co-workers [59] claim to have reduced the ret time to 3-6 hours for flax.

Sharma [57] found that there was no significant difference between the strength of enzyme retted fibre and that of water retted material, whilst Majumdar and co-workers [63], reported that the strength of enzyme retted jute was higher than that of conventionally retted fibre. However, Sharma and Van Sumere [61] stated that the strength of enzyme retted fibre is not as good as that of the best quality water retted fibre. This was attributed to the inability to control the enzyme action during retting; the enzyme concentration being higher, and the end point of the ret being reached faster, than in water retting. Owing to this, it is possible to 'overshoot' and over-ret the fibre, leading to a loss in fibre strength. Furthermore, after retting is complete, there is a tendency for the residual enzymes to continue to act and weaken the fibre. Methods have now been developed to overcome these problems [61]. Recently, excellent quality flax fibre has been produced on a semi-commercial scale utilising this retting method (Vilppunen and co-workers 1999 [64]).

3.5.4 Chemical Retting

Pectins may be degraded by chelating agents for calcium [23]. The use of chelating agents as a novel method for the retting of flax straw has been patented [58]. By chelating metal ions from the pectin complex, plant pectins and hemicelluloses can be extracted with an alkaline solution of chelating agents, leaving the fibre free from pectins and hemicelluloses [58]. On a semi-industrial scale, flax stems have been retted successfully using ethylene diamine tetra acetic acid (EDTA) sodium salt [57]. The quality and strength of the fibre were not significantly different from either water or enzymically retted fibre but chemical retting did remove a high percentage of hemicellulose and pectin [57].

Whilst a number of technologies have been or are being developed with a view to controlling the retting process and reducing the variability in subsequent fibre quality, the majority of industrial fibre from flax and hemp is produced by dew or field retting. This can lead to enormous variability in the quality of the resultant fibre. Dew retting is not only dependent upon the vagaries of the weather, but also the skill of the grower or agronomist in determining the correct degree of retting. Whilst spectroscopic techniques such as near infra red analysis (NIRA) have been developed, the majority of retted flax is assessed subjectively by human eye.

Once retting is complete, the straw is harvested and processed, either in the traditional manner described in the introduction to this section or by mechanical decortication. These processes, as well as subsequent operations can introduce damage to the fibres, leading to a reduction in their properties. For many technical applications this may have serious consequences. As composite reinforcement, it has been demonstrated that the occurrence of fibre damage in the form of small kink bands (similar to those found in synthetic polymer fibres) can lead to reduced fibre properties [65]. When such fibres are used as reinforcement in composite materials, it has been shown that the properties of the composites are also compromised by the presence of these defects [66, 67].

3.6 The Influence of Fibre Damage upon the Mechanical Properties of Natural Fibres

Sections 3.4 and 3.5 outlined the effect that fibre ultrastructure and extraction processes, respectively, have upon the properties of plant fibres. This section examines with how fibre damage, caused variously by extraction, processing and handling can affect the properties of the fibres.

3.6.1 Micro-Compressive Damage or 'Kink Bands' in Lignocellulosic Fibres

In solid wood, microscopic damage to individual cells caused by mechanical overloading in compression can dramatically affect its mechanical properties [68]. This damage takes the form of small creases in the cell wall known variously as slip planes, *kink bands* or *micro-compressions*. These features can be observed by polarised light microscopy and appear in the form of bright 'X' traversing the double cell wall as shown in **Figure 3.7**. The appearance of these features under polarised light is due to misalignment of the microfibrils in the cell wall [69]. Frequently, a line of micro-compressions forms, due to co-operative micro buckling of cell walls resulting in a crease in the structure of wood. Under certain circumstances these creases are visible to the naked eye and are readily

Figure 3.7 Micro-compressive damage in the cell wall of wood

Source: Dinwoodie [70]

observable under a microscope using low angle incident light [7]. In wood, these micro-compressive defects can arise as a result of growth or mechanical stresses within the tree [70]. Micro-compressions may also result from compressive stresses in the wood induced during harvesting, conversion or whilst in service [70].

3.6.1.1 The Effect of Fibre Damage on the Properties of Natural Fibres

Slightly lower tensile strengths and moduli have been noted in pre-compressed wood. However, the most significant effect observed in pre-compressed wood is the reduction in impact properties [68]. Failure is invariably noted to follow the line of the compression crease [68]. Micro-compressions formed during cyclic loading have been shown to propagate a line of compression creases in wood which can subsequently lead to crack formation and ultimately to failure [71].

Individual wood fibres carefully isolated from a block of pre-compressed wood, containing micro-compressive defects, have shown failure loads which are reduced by around 46%, as well as decreased stiffness [68]. Micro-compressions in wood pulp fibres, most likely caused by isolation and handling, have been shown to result in lowered fibre tensile strengths and moduli [31, 72]. It has also been shown that the elastic modulus of paper is reduced by the presence of fibre micro-compressive defects, crimps and curls [72, 73]. By drying fibres under tension, these kinks bands may sometimes be straightened out, resulting in improved tensile properties [41].

In addition to reduced tensile strength and stiffness, micro-compressions have been noted to act a critical defects in the fibre structure [31, 72] and fibre fracture has been successfully explained in a probabilistic manner [42].

Further, it has been shown that micro-compressions (as well as other features such as pit apertures, creases etc.) can affect the surface strain distribution in single wood pulp fibres [74].

3.6.1.2 Micro-Compressions in Bast Fibres

A number of features, similar to those seen in wood, have been noted in flax, hemp and jute fibres, and are referred to as nodes or dislocation marks [75, 76]. Morphologically, the appearance of these nodes or dislocation marks is similar to the kink bands observed in synthetic, highly oriented, polymeric fibres such as aramid fibre which result from compressive failure of the fibre. These features too can be observed, using a polarising microscope, as bright bands traversing the fibre in the same way as in individual micro-compressed wood fibres [65]. Such features have been shown to reduce both the tensile strength and modulus of flax and nettle ultimate fibres [65].

An examination of bast fibres using low angle incident light microscopy (**Figure 3.8**) at low magnification reveals creases (at X and Y), similar to those observed in compression damaged wood. At higher magnification under polarised light this damage takes the form of distinct kinks in the structure of the fibre (see **Figure 3.9**).

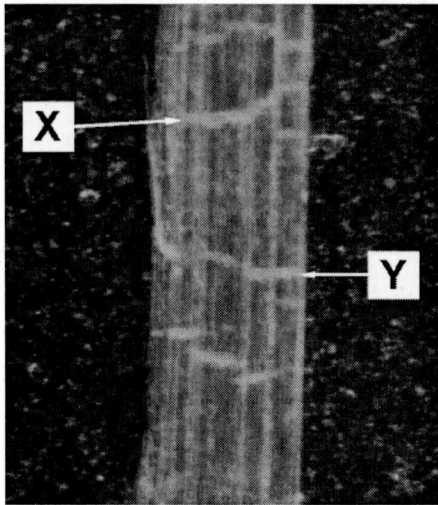

Figure 3.8 Water retted and scutched flax fibre bundle showing compression creases (X and Y) running across the fibre under low angle incident light (x12 magnification)

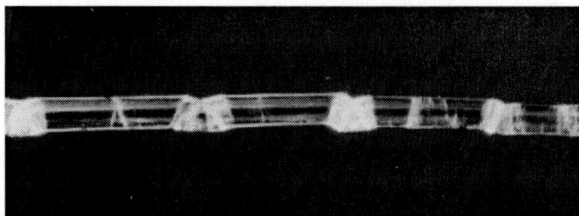

Figure 3.9 Section of a flax fibre ultimate under polarised light kink bands as bright bands traversing the fibre

3.7 Mechanical Properties of Natural Fibres

In the preceding sections, many of the factors, which have a bearing upon the mechanical properties of lignocellulosic fibres, have been introduced and discussed. An appreciation of why variability arises will have been gained and to illustrate the range of tensile properties, **Table 3.3** summarises the findings from a selection of published investigations into fibre mechanical properties. In addition to variability arising from the properties of the fibre itself, due regard must also be taken of the test method used to assess fibre properties. This aspect will be discussed in the following section (3.8).

3.7.1 Regenerated Cellulose Fibres

Before leaving the discussion on fibres and the factors influencing their mechanical properties, it is worthwhile considering the potential of fibres manufactured from a

Table 3.3 Mechanical properties of flax, hemp and jute fibre 'bundles' and 'ultimate' fibres			
Fibre type	Young's modulus (GPa)	Tensile strength (MPa)	Strain to failure (%)	Reference
Flax	-	814	-	[77]
	-	1500*	-	[78]
	103	690	-	[79]
	85	2000*	-	[80]
	50-70	500-900	1.3-3.3	[10]
	28	345-1035	2.7-3.2	[81]
	100	1100	2.4	[13]
	52	621	1.33	[65]
Hemp	-	690	-	[77]
	25	895	-	[80]
	30-60	310-750	2-3	[10]
	-	690	1.6	[13]
	57	-	-	[79]
Jute	-	455	-	[77]
	8	538	-	[80]
	10-78	-	-	[82]
	27.6	393-773	1.7-1.8	[81]
	13	550	-	[13]
* denotes 'ultimate' fibres			

renewable cellulose feedstock – regenerated cellulose fibres. Whilst, good agronomic practices and tightly controlled processing could undoubtedly lessen the degree of variability within the properties of natural fibres, it is unlikely that the degree of confidence in properties obtained with synthetic fibres will ever be achieved. In the same way as designers have had to learn to cope with variability in timber, methods will need to be developed to take into account variability in natural fibres. Using fibres in yarn form and forming these yarns into fibrous structures can, for example, help alleviate some of the problems associated with variable strength and other properties. To reduce the effects of variability found within naturally occurring plant fibres and yet at the same time retain the advantages of low density and biodegradability, regenerated cellulose fibres offer good potential.

Regenerated cellulose fibre is manufactured by dissolving high-grade cellulose pulp and passing the resultant solution through a spinneret to form filaments, thereby converting it, or 'regenerating' it, into almost pure cellulose. Regenerated cellulose is commonly known as 'viscose' or 'rayon'. Purified cellulose for rayon production usually comes from specially processed wood pulp. This purified cellulose is referred to as 'dissolving pulp' to distinguish it from lower grade pulps used for papermaking. Dissolving pulp is characterised by a high α-cellulose content and is relatively free from lignin and hemicelluloses.

Recent developments have lead to high strength, high modulus rayon fibres such as those produced by CORDENKA. CORDENKA 700 fibre, mainly used to reinforce tyres, hoses and for webbing, has a Young's modulus of ~20 GPa, a tensile strength of around 660 MPa and a failure strain of approximately 11% [83]. Another development is Lyocell, regenerated cellulose fibre obtained by an organic solvent spinning process. Lyocell is manufactured by dissolving cellulose an amine oxide solvent. The solution is filtered, extruded into an aqueous bath of dilute amine oxide, and coagulated into fibre form. The Young's modulus of Lyocell is around 15 GPa.

Reduced variability in the mechanical properties of the regenerated fibres was shown by Eichhorn and co-workers [83] in the lower number of sample replicates necessary to yield statistically significant results over that of flax and hemp fibre. Another significantly beneficial property of regenerated fibres is the fact that they are continuous and thus lend themselves readily to continuous processes.

The previous sections have discussed the various factors that can influence the properties of natural fibres namely, their chemistry and ultrastructure, the way in which the fibres are extracted from the plant and the effects of damage. The following section explores the testing of fibres and some of the factors, which can affect this.

3.8 Fibre Testing

A number of factors need to be taken into consideration when assessing the tensile mechanical properties of natural fibres. Many of these will contribute to variability in the observed properties. Amongst the factors that need to be considered are:

- Provenance of fibre (variety, agronomy, retting, processing)

- Fibre type (fibre 'ultimate', fibre bundle, or yarn)

- Testing conditions (temperature, relative humidity)

- Test protocol (measurement of fibre dimensions, fixing, load and strain measurement, type of instrument, number of specimen replicates)

All of these factors will, to a greater or lesser extent, affect the measured properties. The provenance of the fibre, the type of fibre to be tested and other influencing factors have been covered in Sections 3.4-3.6. A significant body of published work is available to the interested reader about test methodologies used to assess fibre properties. In terms of natural fibres, much of this work appears in pulp and paper literature [84, 32, 27], the textiles literature [65, 85] or the agricultural science literature [86]. In addition to the measurement of tensile properties (Young's modulus, tensile strength and failure strain), a number of workers have sought to elucidate specific effects, such as localised strain distributions in fibres [74] or fracture toughness [45].

Testing of fibres may be conducted in several ways. For all tests, however, it is vital to ensure constant and controlled conditions of temperature and relative humidity (RH), since properties are dependent upon these. In some instances, test methods will have been developed and standardised. For example, BS 5116 [87] provides a method for the determining the breaking tenacity of flat bundles of cotton fibres. Standard conditions for RH and temperature are set out in this document (RH 65% ± 2%; temperature 20 °C ± 2 °C). This (and other methods) have been developed for textile applications and the properties are expressed in terms of 'tenacity' and 'fineness'. Other standards have been developed for single textile fibres. These include ASTM D3822-01 [88] and BS EN ISO 5079:1996 [3]. However, for the technical application of bast fibres, such as in composite materials, properties expressed in true engineering units are essential.

To obtain accurate values for Young's modulus, tensile strength and failure strain, not only is it essential to accurately record load during testing, but also the extension and cross sectional area of the fibre. Additionally, the method used for gripping the fibre whilst being strained, must be given due attention. Methods for clamping the fibres during loading have included direct gripping in pin-vice mechanical grips [29] or gluing fibres between paper tabs [30]. An alternative method reported recently [89], allows for fibre realignment during loading by applying droplets of epoxy resin to the fibre ends. The droplets effectively form 'ball and socket' joints when the fibres are placed in a yoke on the tensile testing instrument. Accurate measurement of cross sectional area is essential if load readings are to be normalised to stress. Physical and optical methods have traditionally been used but confocal laser scanning microscopy offers the opportunity of providing accurate cross sectional measurements at the point of break.

Even taking all the precautions noted previously to minimise variation arising from the sampling and test procedures, tensile mechanical property data nevertheless remains highly variable. Snell and co-workers [89] estimated that to obtain a statistically significant mean value for flax fibre ultimates, some 1,200 specimens would be required!

Sections 3.4-3.6 introduced and discussed at some length, chemical, ultrastructural and processing factors that gave rise to variability in natural fibres. Section 4.8 outlined some of the pitfalls in the testing procedure for natural fibres. The discussion has focussed almost exclusively on the mechanical properties of natural fibres and the factors that account for the range in properties. This is principally due to the fact that adequate mechanical strength and/or stiffness of natural fibres is the main requirement for technical applications such as composite materials. Nevertheless, it must be remembered that the measurement of other physical and environmental properties are of interest for certain applications. Increasingly, the biodegradation profile of natural fibres will become important, especially when they are used in conjunction with the newly emerging 'breed' of biodegradable plastics and polymers produced from renewable resources to form true 'biocomposites'.

3.9 Biopolymers

3.9.1 Introduction

The vast majority of polymer 'plastics' and 'resins' are currently manufactured from fossil reserves and many are extremely environmentally stable, ensuring that decomposition can, in many instances, take centuries. It is probably fair to say that until recently, a major focus for 'resins' and 'plastics' polymer development has been to ensure that these materials are environmentally stable! With growing concern over the use of fossil reserves as feedstock and environmental concerns over the ultimate disposal of polymer products, the stage seems set for the widespread adoption of so-called 'green-', 'eco-' or 'biopolymers'.

New European and national directives based on the 'polluter pays' ethos (such as the Landfill of Waste directive), aimed at increasing the recycling of products and reducing the amount of material going to landfill have spurred on the development of 'sustainable biopolymers'. Mohanty and co-workers [90] define a 'sustainable' bio-product as one which is (i) derived from a renewable resource, (ii) has recycling capability and (iii) possesses 'triggered biodegradability'. Furthermore, the product should be commercially viable and environmentally acceptable. There are a number of 'biopolymers', which partially fulfil these criteria for sustainability. Few currently meet all criteria.

Renewable resources such as oils and starches from industrial crops as well as cellulose have all been utilised as feedstock for biopolymers. Oilseed rape (*Brassica napus spp*), soy bean (*Glycine max*)and linseed (*Linum usitatissimum (L)*) have formed precursors for polymer resins; cellulose, polylactides and starches have been used to manufacture 'plastics'. It should be remembered that even though a polymer is derived from a renewable resource it need not be biodegradable. Susceptibility to biodegradation is a function of the chemical structure and curing nature of the polymer resin or plastic and not necessarily the feedstock from which it was derived [90]. Indeed, a number of 'biodegradable' polymers have been developed from fossil reserves. The argument about how 'sustainable' a product is, is by no means clear cut. In addition to the 'feedstock' and 'disposal' issues, the energy required to manufacture the polymer in the first instance as well as to recycle it must be taken into account. Indeed, in this manner, it can be justifiably argued that the recycling of a fossil based plastic such as polypropylene (PP) can, in certain cases be a 'more sustainable' option since once manufactured, PP can be recycled several times without significant degradation in its properties [91]. Indeed, this line of argument has recently led to the development of 'all-PP' composites [91].

Broadly speaking, polymer 'plastics' or 'resins' can be defined as either thermoplastic or thermosetting. Thermoplastic polymers are those that flow readily under the effects of heat and stress. At normal temperatures they are solid. A key feature, differentiating thermoplastic from thermosetting polymers is that no chemical crosslinking occurs between the polymer chains during processing; the association between chains being the result of secondary (Van der Waal's) forces which are easily overcome. This enables thermoplastics to be easily recycled. Thermosetting polymers are initially liquid resins that harden into brittle solids at room temperature when cured. Curing involves chemical crosslinking, which is achieved by the application of heat and/or pressure and/or the addition of a catalyst to complete the process.

3.9.2 Biopolymer Types

Biopolymers too can be broadly classified as having either thermoplastic or thermosetting character (see **Table 3.4**). Whilst many of the thermoplastic biopolymers can be considered

Table 3.4 Some examples of thermoplastic and thermosetting biopolymers	
Thermoplastic polymers	Thermosetting polymers
Poly lactic acid	Cashew nut shell liquid
Polyhydroxyalkanoates	Epoxidised linseed
Chemically modified starch	Epoxidised soy

to be biodegradable, with many of the thermosetting resins, particularly those synthesised from vegetable oils like soy, linseed and rape, biodegradation is limited. Thermoplastic biopolymers are often used in short-term applications such as packaging or horticulture and generally have a short life cycle. With these types of product, rapid biodegradation or composting ability is desirable. Thermosetting resins on the other hand are often used to manufacture composite products with relatively long life spans, e.g., construction, infrastructure. In these cases, rapid biodegradation is most certainly not desirable.

As a result of the aforementioned environmental issues and legislative drivers, there has been a general increase in the interest being shown in so called 'green polymers' in recent years. It is beyond the scope of this chapter to give detailed descriptions of the various polymer classes and the interested reader is referred to a number of review articles and books on the subject that have appeared in the literature in recent years, such as Mohanty and co-workers [90], Nayak [92] and Scholz and Gross [93]. Rather, the remainder of this section will focus upon how variability in the properties of biopolymers may arise in the first instance and the types of tests that may be performed in order to assess these properties.

3.9.3 Properties

The choice of polymer for a particular application will depend very much upon the intended end-use. Long lasting, large and complex structures, favour the use of thermosetting polymers, whilst high production rates, of relatively small products may well be better served by, for example, injection moulded thermoplastics. The properties required will be very much dictated by the type of process and product attributes required. Broadly speaking, the properties can be categorised according to whether they are requirements for (i) processing, (ii) application or (iii) disposal.

Over the years, test methods have been developed to assess the properties of both 'traditional' polymer materials based upon fossil reserves as well as 'natural' materials like rubbers. Many of these test methods, particularly those for ascertaining 'processing' and 'application' properties can, with suitable care, be used to assess the properties of biopolymers too. On the other hand, the 'specialist' properties of biopolymers, e.g., biodegradation, have necessitated the development and introduction of new tests and standards.

3.9.3.1 Sources of Variability

In much the same way as the chemical composition of natural fibres varies, so too will that of the renewable feedstock used as precursors for biopolymers. Variability can arise, as a result of the agronomic conditions under which the crop is grown in the first place, as well as seasonal or climatic conditions. All can potentially combine to introduce variability into the feedstock, which without careful control could manifest itself as variability in the final polymer or product.

Standards have been developed to minimise variability arising from the differences in the test methods used to assess the particular property in question. Details of standard tests used on polymeric materials to assess processing, physical and mechanical properties can be found in publications such as the *Handbook of Plastics Test Methods* [94]. References to national standards organisations are given, e.g., British Standards Institution. These standards set out procedures for samples preparation (especially important for mechanical testing) and the test environment, which is particularly important for biological origin materials, which are often highly sensitive to variations in humidity and/or temperature. Brief mention of the types of test available and the range of process and applications properties, which can be assessed, are made in the following sections.

3.9.3.2 Process Properties

Process properties for thermoplastic and thermosetting polymers vary. With thermoplastic polymers, factors such as the processing temperature and melt viscosity are of importance. With thermosetting polymers, factors such as the initial viscosity of the liquid resin, gelation time and cure temperatures are of primary importance when assessing the processability of the polymer.

3.9.3.3 Application Properties

For applications, physical and mechanical properties are of prime consideration. Ultimately the physical and mechanical properties are, to a greater or lesser extent, dependent upon the chemical constitution of the polymer [94]. As such, it is of importance to be able to assess parameters such as the degree of crystallinity or the molecular weight distribution.

When discussing the physical properties of polymers, density is perhaps the most familiar, but also electrical properties, thermal properties, environmental resistance (see Section 3.9.3.3), fibre performance, optical properties, appearance (colour) and surface roughness are important factors are far as the final product is concerned.

The mechanical properties of a material generally dictate whether it can be used for a structural application or not. Mechanical properties can be assessed either in the short or long-term. Short-term mechanical tests; include hardness, tensile, flexural or impact tests. Polymeric materials exhibit time dependent behaviour and thus the choice of loading rate can significantly affect the measurements of properties obtained. Assessment of the dynamic stress-strain properties of a polymeric materials can be undertaken and yield data concerning the appropriate modulus of a material [94]. Such tests can produce fundamental insights into the molecular transformations occurring in polymers. Time dependent behaviour of polymers should also be considered. Creep and relaxation are of importance when a polymer is subjected to a continuous load or extension, and fatigue measurements are important when a polymer is subjected to long-term cyclic loading.

3.9.3.3 Disposal Properties

One of the great attractions of biopolymers is the possibility of being able to dispose of a product or article manufactured from these materials through 'natural' decomposition processes. As discussed previously, the term 'biopolymer' does not necessarily imply that the material is, therefore, biodegradable. However, for emerging biopolymers, the ability to biodegrade is an important feature.

Whilst it may be desirable for a product or material to biodegrade at the end of its useful working life, it may not be so desirable for the said material or product to biodegrade during its working life! The superior environmental stability of many synthetic polymers manufactured from fossil reserves has ensured that they have replaced many natural alternatives. A prime example is the almost complete substitution of ropes and cordage made from natural fibre, such hemp, flax or sisal, with that of synthetic fibres like polypropylene or Nylon. In this example, the susceptibility of these natural polymers to biodeterioration and other degradation processes during service can present a real danger. This same argument can be applied to wood and other 'natural' products used in any form of structural application. Indeed, a great deal of attention has been paid to understanding and controlling the degradation processes in these natural polymers. Good design using naturally durable species, preservative treatments or protective coatings can slow down the degradation process. A great deal of scientific interest has been shown in potential of modifying not only wood but also other lignocellulosic fibres by physical and chemical means, with a view to improving their durability [95].

During the life span of a product, a number of degradation processes may come into play. For a product to fulfil its designed role, it must be ensured that the degradation profile of the product is suited to its intended application. For example, a product that is to be used in exterior applications must be sufficiently robust to ensure that it can survive the ravages of often large changes in humidity and temperature, to intermittent wetting by liquid water, or to other chemicals, to ultra violet (UV) radiation and to biological attack. If the materials itself does not possess adequate resistance to the various degradation processes that come into play, then it is often necessary to protect the product or material in question, to increase its resistance. This is a commonly used strategy to 'tailor' the degradation process and is most usually seen in the painted surfaces of wood and metals especially steel.

Clearly, it is impossible to entirely separate the in-service and post-service degradation profiles of biopolymers. It is, of course, entirely feasible to change the environment in which the product is located in, from say a warm, dry interior environment, whilst in service, to a moist exterior environment in which biological agents, which may attack the product, are present, post service. Further, the designed service life of the product will dictate the applicability of a particular material to that application. Perhaps one of the most pertinent examples, is food packaging. Generally speaking, in terms of longevity, or in service life, the packaging material may only be required for a matter of days. Nevertheless, many foodstuffs are packaged in materials that may take decades or more to biodegrade!

When considering the biodegradation process, we are generally concerned with the action of colonising micro-organisms that, effectively, de-structure the polymer through the action of enzymes or by other means. Nevertheless, it must be remembered that de-structuring of the polymer can take place as a result of other processes such as direct chemical attack, or cleavage of the molecular bonds by UV radiation.

A number of test methods have been developed to assess the rate at which a polymer biodegrades under controlled conditions. ISO 14855 (evaluation of the ultimate aerobic biodegradability and disintegration of plastics under controlled composting conditions-method by analysis of evolved carbon dioxide) [4], for example, assesses the 'compostability' of the polymer by monitoring the amount of carbon dioxide evolved in the biodegradation process. ASTM (American Society for Testing and Materials) has produced a number of standards pertaining to the evaluation of the biodegradability of plastic materials. ASTM D5338-98e1 [96], for example provides for a standard test method for determining aerobic biodegradation of plastic materials under controlled composting conditions.

ASTM D5988-03 [97], provides for a standard test method for determining the aerobic biodegradation in soil of plastic materials or residual plastic materials after composting, whilst ASTM D6400-99e1 [98] provides for a standard specification for compostable plastics.

As the role of biopolymers becomes more prevalent, it seems likely that further standards will be introduced, or existing ones revised to keep apace of new developments.

3.10 Conclusions

Quantification of the range of properties of natural origin based fibres and polymers is essential if these materials are to be used in an industrial context. It is important that the designer, engineer or scientist can be confident in the properties of the material with which he or she is working with. Most 'natural' materials tend to be inherently variable, and a significant portion of the text has been given over to a discussion of how and where this variability arises. This is particularly so of natural plant fibres, which are currently receiving a significant amount of attention as potential replacements for synthetic fibres in load bearing composite applications.

References

1. *Our Common Future: the Bruntland Report*, World Commission on Environment and Development Report, Oxford University Press, Oxford, UK 1987.

2. R. Elias, M. Hughes, A. Norton and S. Turunen, *Renewable Feedstock for Sustainable Materials – BioProducts – Their Importance to Wales: A Scoping Study*, A report prepared by the Centre for Advanced and Renewable materials (CARM), The University of Wales, Bangor for the Welsh Development Agency, 2002. www.carmtechnology.com.

3. BS EN ISO 5079, *Textiles – Fibres – Determination of Breaking Force and Elongation at Break of Individual Fibres*, 1996.

4. ISO 14855, *Determination of the Ultimate Aerobic Biodegradability and Disintegration of Plastic Materials under Controlled Composting Conditions – Method by Analysis of Evolved Carbon Dioxide*, 1999.

5.   D6400-99e1, *Standard Specification for Compostable Plastics*, 1999.

6.   *British Timber Statistics 2001*, Forestry Commission, Edinburgh, UK, 2002. http://www.forestry.gov.uk

7.   H.E. Desch and J.M. Dinwoodie, *Timber: Structure, Properties, Conversion, and Use, Seventh Edition*, Food Products Press, New York, NY, USA, 1996.

8.   G.A. Sanderson in *Proceedings of the Sixth European Panel Products Symposium*, 2002, Llandudno, Wales, UK, p.94.

9.   D. Hull and T.W. Clyne, *An Introduction to Composite Materials*, Cambridge University Press, Cambridge, UK, 1996.

10. J. Ivens, H. Bos and I. Verpoest in *Renewable Bioproducts: Industrial Outlets and Research for the 21st Century*, 1997, Wageningen, The Netherlands.

11. L. Weindling, *Long Vegetable Fibers: Manila, Sisal, Jute, Flax and Related Fibers of Commerce*, Columbia University Press, New York, NY, USA, 1947.

12. Focher in *The Biology and Processing of Flax*, Eds., H.S.S. Sharma and C.F. Van Sumere, M Publications, Belfast, UK, 1992, p.11-32.

13. A.K. Bledzki, S. Reihmane and J. Gassan, *Journal of Applied Polymer Science*, 1996, **59**, 8, 1329.

14. J. Gassan and A.K. Bledzki, *Proceedings of the 54th Annual SPE Technical Conference, ANTEC '96*, Indianapolis, IN, USA, 2553.

15. D. Robson, J. Hague, G. Newman, G. Jeronomidis and M.P. Ansell, *Survey of Natural Materials for Use in Structural Composites as Reinforcement and Matrices*. The BioComposites Centre, University of Wales, Bangor, UK, 1993.

16. H. Lilholt and A.B. Bjerre in *Proceedings of the 18th Risø International Symposium on Materials Science: Polymeric Composites – Expanding the Limits*, Roskilde, Denmark, 1997, 411-423.

17. D. Fengel and G. Wegener, *Wood: Chemistry, Ultrastructure, Reactions*, Walter de Gruyter, Berlin, Germany, 1984.

18. H. Chanzy in *Cellulose Sources and Exploitation: Industrial Utilisation, Biotechnology and Physico-Chemical Properties*, Eds., J.F. Kennedy, G.O. Phillips and P.A. Williams, Ellis Horwood, New York, NY, USA, 1990, p.3-12.

19. E. Sjöström, *Wood Chemistry: Fundamentals and Applications*, Academic Press, San Diego, CA, USA, 1993.

20. Z.I. Kerestz, *The Pectic Substances*, Interscience Publishers, New York, NY, USA, 1951.

21. D.E. Akin, G.R. Gamble, W.H. Morrison III, L.L. Rigsby and R.B. Dodd, *Journal of the Science of Food and Agriculture*, 1996, **72**, 2 155.

22. A.M. Stephen in *The Polysaccharides, Volume 2*, Ed., G.O. Aspinall, Academic Press, New York, NY, USA, 1983, p.154-166.

23. G.O. Aspinall in *The Biochemistry of Plants: A Comprehensive Treatise, Volume 3, Carbohydrates: Structure and Function*, Ed., J. Preiss, Academic Press, New York, NY, USA, 1980, p.480-486.

24. M.C. Jarvis, *Plant, Cell and Environment*, 1984, **7**, 153.

25. R.D. Preston, *The Physical Biology of Plant Cell Walls*, Chapman and Hall, London, UK, 1974.

26. R.E. Mark, *Cell Wall Mechanics of Tracheids*, Yale University Press, New Haven, 1967.

27. D.H. Page, F. El-Hosseiny, K. Winkler and A.P.S. Lancaster, *Tappi*, 1977, **60**, 4, 114.

28. J-C. Roland, M. Mosiniak and D. Roland, *Acta Botanica Gallica*, 1995, **142**, 5, 463.

29. B.A. Jayne, *Forest Products Journal*, 1960, June, 316-322.

30. J.M. Dinwoodie, *Nature*, 1965, **205**, 763.

31. D.H. Page, F. El-Hosseiny and K. Winkler, *Nature*, 1971, **229**, 252.

32. D.H. Page, F. El-Hosseiny, K. Winkler and R. Bain, *Pulp and Paper Magazine of Canada*, 1972, **73**, 8, 72.

33. R.E. Booker and J. Sell, *Holz als Roh und Werkstoff*, 1998, **56**, 1.

34. R.E. Booker in *Recent Advances in Wood Anatomy*, Eds., L.A. Donaldson, A.P. Singh, B.G. Butterfield and J. Whitehouse, NZ Forest Research Institute Ltd., Rotorua, New Zealand, 1995, p.273-282.

35. I.D. Cave, *Wood Science and Technology*, 1968, **2**, 268-278.

36. I.D. Cave, *Wood Science and Technology*, 1969, **3**, 40-48.

37. R.C. Tang and N.N. Hsu, *Wood and Fiber Science*, 1973, 5, 2, 139.

38. L. Salmén and A. de Ruvo, *Wood and Fiber Science*, 1985, **17**, 3, 336.

39. J.J. Harrington, R. Booker and R.J. Astley, *Holz als Roh und Werkstoff*, 1998, **56**, 37.

40. R.J. Astley, K.A. Stol and J.J. Harrington, *Holz als Roh und Werkstoff*, 1998, 56, 43.

41. C.Y. Kim, D.H. Page, F. El-Hosseiny and A.P.S. Lancaster, *Journal of Applied Polymer Science*, 1975, **19**, 1549.

42. D.H. Page and F. El-Hosseiny, *Svensk Papperstidning – Nordisk Cellulosa*, 1976, **14**, 471.

43. G.C. Davies and D.M. Bruce, *Journal of Materials Science*, 1997, **32**, 20, 5425.

44. G. Jeronimidis, *Proceedings of the Royal Society of London Series B*, 1980, **208**, 447-460.

45. P.W. Lucas, H.T.W. Tan and P.Y. Cheng, *Philosophical Transactions of the Royal Society of London Series B*, 1997, **352**, 1351.

46. P.W. Lucas, B.W. Darvell, K.D. Lee, T.D.B. Yuen and M.F. Choong, *Philosophical Transactions of the Royal Society of London Series B*, 1995, **348**, 363.

47. M.F. Ashby, K.E. Easterling, R. Harrysson and S.K. Maiti, *Proceedings of the Royal Society of London Series B*, 1985, **398**, 261.

48. L.J. Gibson and M.F. Ashby, *Cellular Solids: Structure and Properties*, Pergamon Press, Oxford, UK, 1988.

49. S.E. Stanzl-Tschegg, D.M. Tan and E.K. Tschegg, *Wood Science and Technology*, 1995, **29**, 31.

50. J.E. Gordon and G. Jeronimidis, *Nature*, 1974, **252**, 116.

51. G. Jeronimidis in *Wood Structure in Biological and Technical Research*, Eds., P. Baas, A.J. Bolton and D. Catling, Leiden Botanical Series No.3, Leiden University Press, Leiden, The Netherlands, 1976, p.253-265.

52. J.E. Gordon and G.J. Jeronomidis, *Philosophical Transactions of the Royal Society of London Series A*, 1980, **294**, 545.

53. A.G. Searle and J.W. Tuck, *The King's Flax and the Queen's Linen*, The Larks Press, Dereham, UK, 1999.

54. J.M. Dempsey, *Fiber Crops*, University Presses of Florida, Gainsville, FL, USA, 1975.

55 A. Chesson, *Journal of Applied Bacteriology*, 1978, **45**, 219.

56. A. Chesson, *Journal of Applied Bacteriology*, 1980, **48**, 1.

57. H.S.S. Sharma, *International Biodeterioration*, 1987, **23**, 329.

58. H.S.S. Sharma, inventor; Lanberg Industrial Research Association, assignee; GB 2186002, 1987.

59. P. Vilppunen, T. Solin and K. Harmaa, *Proceedings of the 6th EC Conference: Biomass for Energy, Industry and Environment*, Athens, Greece, 1992, p.1357.

60. H.S.S. Sharma, P.C. Mercer and A.E. Brown, *International Biodeterioration*, 1989, **25**, 327.

61. H.S.S. Sharma and C.F. Van Sumere, *The Genetic Engineer and Biotechnologist*, 1992, **12**, 19.

62. H.S.S. Sharma, *International Biodeterioration*, 1987, **23**, 181.

63. S. Majumdar, A.B. Kundu, S. Dey and B.L. Ghosh, *International Biodeterioration*, 1990, **27**, 223.

64. P. Vilppunen, K. Oksman, Mäentausta, E. Keskitalo and J. Sohlo, *Proceedings of the 5th International Conference on Woodfiber-Plastic Composites*, Madison, WI, USA, 1999, p.309.

65. G.C. Davies and D.M. Bruce, *Textile Research Journal*, 1998, **68**, 9, 623.

66. M. Hughes, L. Mott, J. Hague and C.A.S. Hill in *Proceedings of the 5th International Conference on Woodfiber-Plastic Composites*, Madison, WI, USA, 1999,

67. M. Hughes, G. Sèbe, J. Hague, C. Hill, M. Spear and L. Mott, *Composite Interfaces*, 2000, 7, 1, 13.

68. J.M. Dinwoodie, *Wood Science and Technology*, 1978, **12**, 271.

69. J.M. Dinwoodie, *Journal of the Institute of Wood Science*, 1968, **21**, 37.

70. J.M. Dinwoodie in *Wood Structure in Biological and Technical Research*, Eds., P. Baas, A.J. Bolton and D. Catling, Leiden Botanical Series No.3, Leiden University Press, Leiden, The Netherlands, 1976, p.238.

71. N. Imayama, *Holz als Roh und Werkstoff*, 1994, **52**, 49.

72. D.H. Page and R.S. Seth, *Tappi*, 1980, **63**, 10, 99.

73. D.H. Page, R.S. Seth and J.H. De Grace, *Tappi*, 1979, **62**, 9, 99.

74. L. Mott, S.M. Shaler and L.H. Groom, *Wood and Fiber Science*, 1996, **28**, 4, 429.

75. D. Catling and J. Grayson, *Identification of Vegetable Fibres*, Chapman and Hall, London, UK, 1982.

76. M.M.M. Rahaman, *Indian Journal of Agricultural Sciences*, 1979, **49**, 6, 483.

77. W.J. Brown, *Fabric Reinforced Plastics*, Cleaver-Hume Press Ltd., London, UK, 1947.

78. H.L. Bos, M.J.A. Van den Oever and O.C.J.J. Peters in *Proceedings of the 4th International Conference on Deformation and Fracture of Composites*, Manchester, UK, 1997.

79. P. McMullen, *Composites*, 1984, **15**, 3, 222.

80. A.J. Bolton, *Materials Technology*, 1994, **9**, 1/2, 12.

81. M.K. Sridhar, G. Basavarajappa, S.G. Kasturi and N. Balasubramanian, *Indian Journal of Textile Research*, 1982, **7**, 87.

82. H. Wells, D.H. Bowden, I. Macphail and P.K. Pal in *Proceedings of the 35th Annual Technical Conference of the Reinforced Plastics/Composites Institute and The Society of the Plastics Industry, Inc.*, New Orleans, LA, USA, 1980, Section 1-F, p.1-7.

83. S.J. Eichhorn, J. Sirichaisit and R.J. Young, *Journal of Materials Science*, 2001, **36**, 3129.

84. P.P. Kärenlampi, H.T. Suur-Hamari, M.J. Alava and K.J. Niskanen, *Tappi*, 1996, **79**, 5, 203.

85. A. Lahiri, *Indian Journal of Textile Research*, 1987, **12**, 149.

86. D. Sager and P. Pütz, *International Agrophysics*, 1994, **8**, 681.

87. BS 5116, *Method of Test for Determination of Breaking Tenacity of Flat Bundles of Cotton Fibres*, 1974.

88. ASTM D3822-01, *Standard Test Method for Tensile Properties of Single Textile Fibers*, 2001.

89. R. Snell, J. Hague and L. Groom in *Proceedings of the Fourth International Conference on Woodfiber-Plastic Composites*, Madison, Wisconsin, USA, 1997, p.5.

90. A.K. Mohanty, M. Misra and L.T. Drzal, *Journal of Polymers and the Environment*, 2002, **10**, 1/2, 19.

91. T. Peijs, *Materials Today*, 2003, **6**, 4, 30.

92. P.L. Nayak, *Journal of Macromolecular Science C*, 1999, **39**, 3, 481.

93. *Polymers from Renewable Resources*, Eds., C. Scholz and R.A. Gross, ACS Symposium Series No.764, American Chemical Society, Washington, DC, USA, 2000.

94. *Handbook of Plastics Test Methods*, 2nd Edition, Ed., R.P. Brown, George Godwin Limited in association with The Plastics and Rubber Institute, London, UK, 1981.

95. C.A.S. Hill, H.P.S. Abdul Khalil and M.D. Hale, *Industrial Crops and Products*, 2000, **8**, 1, 53.

96. D5338-98e1, *Standard Test Method for Determining Aerobic Biodegradation of Plastic Materials under Controlled Composting Conditions*, 1998.

97. ASTM D5988-03, *Standard Test Method for Determining Aerobic Biodegradation in Soil of Plastic Materials or Residual Plastic Materials after Composting*, 2003.

98. ASTM D6400-99el, *Standard Specification for Compostable Materials*, 1999.

4 Natural Fibres as Fillers/Reinforcements in Thermoplastics

Anand R. Sanadi

4.1 Introduction

Recent interest in reducing the environmental impact of polymers is leading to the development of new polymers or composites that can reduce the stress upon the environment. In the light of petroleum shortages and pressures for decreasing the dependence on petroleum products, there is an increasing interest in maximising the use of renewable materials. The use of agricultural materials as a source of raw materials to the industry not only provides a renewable source, but could also generate a non-food source of economic development for farming and rural areas.

Several billion pounds of fillers and reinforcements are used annually in the plastics industry. The use of these additives in plastics is likely to grow with the introduction of improved compounding technology, and new coupling agents that permit the use of high filler/reinforcement content [1]. The authors suggested fibre loadings of up to 75 parts per hundred could be common in the future: this could have a tremendous impact in lowering the usage of petroleum-based plastics. It would be particularly beneficial, both in terms of the environment and also in socio-economic terms, if a significant amount of the fillers were obtained from a renewable agricultural source. Ideally, of course, an agro-/bio-based renewable polymer reinforced with agro-based fibres would make the most environmental sense.

There are a wide variety of agro-fibres and wastes that are available. Our work has primarily concentrated on fibres that are easily available and are presently an important source for clothing, rope, sacking and rugs in Asia and Africa. Fibres such as kenaf are now being grown in the United States and the quantity and the types of fibres grown are likely to increase with new uses being developed for natural fibres. Recent work on pure cellulose pulp have some distinct advantages, although the cost of these fibres can be significantly more than typical plant-based natural fibres.

4.1.1 Agro-Fibres and Their Use in Thermoplastics

There are a wide variety of natural lignocellulosic fibres available in the world. The properties of these fibres can differ considerably. The quality and type of natural fibres affect the reinforcing efficiency of the composites. Natural lignocellulosic fibres are themselves composites that consist of a framework of crystalline cellulose in a matrix consisting mainly of lignin and polysaccharides. Several other components such as inorganic salts, waxes, etc., are also present in smaller quantities. The properties of the fibres are thus dependent on the characteristics, amount and orientation of the reinforcing element, which is the crystalline cellulose. The degree of polymerisation of the crystalline cellulose is an important factor that influences fibre strength and modulus. The fibre history includes the species, growing conditions, climatic and soil conditions that can also influence the properties of the fibre. An interesting example is that a species of flax fibres that grow in the summer months in Iceland have very good Young's moduli and strengths. This has been attributed to an unusually high degree of polymerisation of the crystalline cellulose, possibly due to the long hours of daylight in the region [2].

Natural plant-based fibres can be characterised into three major categories, bast (the outer fibres of a seed like plant – examples are kenaf, hemp, jute, etc.), leaf and seed fibres. In general, the bast fibres strands tend to be long and have long been used for making ropes, e.g., jute, hemp, flax, kenaf, etc. The fibres occur in bundles and these are composed of many cells, that are known as 'ultimate fibres' [3]. Leaf fibres, such as sisal, abaca (leaf fibre like sisal), etc., tend to be coarser than their bast counterparts, and tend to be used in carpets, etc. Seed fibres such as coconut fibres (coir) and cotton are attached to the seed of plants. Their characteristics can vary from coarse coir, a fibre with a high lignin content, to cotton that has a comparatively high percentage of cellulose.

Extraction of the fibres is thus an important process that determines the properties of the fibres. For example, kenaf fibres are extracted from the bast of the plant *Hibiscus cannabinus*, and filament lengths longer than 1 m are common. These filaments consist of discrete individual fibres, generally 2 mm to 6 mm long, which are themselves composites of predominantly cellulose, lignin and hemicelluloses. The properties of the fibres are difficult to measure and we have made no attempt to measure the properties of kenaf. Earlier work on a similar natural bast filament, sunhemp (*Crotalaria juncea*) suggested that the filament properties ranged widely. The tensile strengths of the filaments or strands of sunhemp varied from about 325 MPa to 450 MPa, while the tensile modulus ranged from 27 MPa to 48 MPa [4]. Information on the strength of various fibres is available, although it is sometimes unclear whether the properties are that of the strands or the ultimate fibre [3, 5]. However one must remember that filament and individual fibre properties can vary depending on the source, age, separating techniques, and the history of the fibre.

Separation of the fibres from the original plant source is an important step to ensure the high quality of fibres. Recent work indicates defects impact the final strength properties of the fibres. There is little one can do about the inherent defects in the fibres. However, major defects are introduced during the fibre separation procedure. Recent work, in a laboratory scale operation by Bos and co-workers [6] and Vilppunen and co-workers [7] using controlled separation of bast fibres by enzymes indicates a significant reduction in the amount of defects present in the fibres. Strengths of flax fibres, separated to give short fibres with high aspect ratios, could well be over 1.5 GPa. The real challenge is to have a commercially viable technique to separate these fibres.

The primary advantages of using annual growth lignocellulosic fibres as fillers/ reinforcements in plastics are:

- **Property Advantages:**

  Low densities

  Not abrasive (relatively)

  High filling levels possible resulting in high stiffness properties

  High specific properties

  Easily recyclable

Unlike brittle fibres, the fibres will not fracture when processed over sharp curvatures.

- **Environmental and Socio/Economic Advantages:**

  Biodegradable

  Wide variety of fibres available throughout the world

  Generates rural jobs

  Non-food agricultural/farm based economy

  Low cost

Material cost savings due to the incorporation of the relatively low cost agro-fibres and the higher filling levels possible, coupled with the advantage of being non-abrasive to the mixing and melding equipment are benefits that are not likely to be ignored by the plastics industry for use in the automotive, building, appliance and other applications.

Early work on lignocellulosic fibres in thermoplastics has concentrated on wood (including wood bark) based flour or fibres and significant advances have been made by a number of researchers [8-16]. A study on the use of annual growth lignocellulosic fibres indicates that these fibres have the potential of being used as reinforcing fillers in thermoplastics

[17]. The use of annual growth agricultural crop fibres such as kenaf has resulted in significant property advantages when compared to typical wood based fillers/fibres such as wood flour, wood fibres and recycled newspaper. Compatibilised composites of polypropylene (PP)/kenaf have mechanical properties comparable with those of commercial PP composites [17].

The primary drawback of the use of agro-fibres is the lower processing temperature, due to the possibility of lignocellulosic degradation and/or the possibility of volatile emissions that could affect composite properties. Generally, the processing temperatures are thus limited to about 200 °C, although it is possible to use higher temperatures for short periods. This limits the type of thermoplastics that can be used with agro-fibres, to commodity thermoplastics such as polyethylene (PE), PP, polyvinyl chloride (PVC) and polystyrene (PS). However, it is important to note that these lower priced plastics constitute about 70% of thermoplastics consumed by the plastics industry, and subsequently the use of fillers/reinforcement presently used in these plastics far outweigh the use in other more expensive plastics.

However, recent work by Caulfield and co-workers [18], suggests that purified cellulose pulp fibres can be effective reinforcements in Nylon 6 and Nylon 6,6 where processing temperatures can exceed 220 °C. Another advantage of using these pulp fibres is that the problem of odour (which can be a deterrent in end use applications) when some natural fibres are processed with plastics is reduced dramatically. However, the cost of the pulp fibres is more than agro-fibres due to the pulping process involved in their production.

The second drawback is the high moisture absorption of the natural fibres. Moisture absorption can result in swelling of the fibres and concerns about the dimensional stability of the agro-fibre composites cannot be ignored. The absorption of moisture by the fibres is minimised in the composite due to encapsulation by the polymer. It is difficult to entirely eliminate the absorption of moisture without using expensive surface barriers on the composite surface. If necessary, the moisture absorption of the fibres can be dramatically reduced through the acetylation of some of the hydroxyl groups present [19] in the fibre, but with some increase in the cost of the fibre. Good matrix-matrix bonding can also decrease the rate and amount of water absorbed by the composite.

It is important to keep these limitations in perspective, when developing end use applications. We believe that by understanding the limitations and benefits of these composites, these renewable fibres are not likely to be ignored by the plastics/composites industry for use in the automotive, building, appliance and other applications. This is now apparent since the use of natural fibres has dramatically increased in recent times, for both automotive, building and other applications.

4.2 Processing Considerations and Techniques

The limiting processing temperatures, when using lignocellulosic materials with thermoplastics, is important in determining processing techniques. In general, high processing temperatures that reduce melt viscosity and facilitates good mixing cannot be used (except for short periods) and other routes are needed to facilitate mixing of the fibres and matrix in agro-matrix thermoplastics.

A review by Milewski [20] on short fibre composite technology covers a variety of reasons why composite properties fall short of their true reinforcing potential. The major factors that govern the properties of short fibre composites are fibre dispersion, fibre length distribution, fibre orientation and fibre-matrix adhesion. Mixing the polar and hydrophilic fibres with non-polar and hydrophobic matrix can result in difficulties associated with the dispersion of fibres in the matrix. Clumping and agglomeration must be avoided to produce efficient composites. The efficiency of the composite also depends on the amount of stress transferred from the matrix to the fibres. This can be maximised by improving the interaction and adhesion between the two phases, and also by maximising the length of the fibres retained in the final composite [21]. Using long filaments during the compounding stage can result in higher matrix length distribution. However, long fibres sometimes increase the amount of clumping resulting in areas concentrated with fibres and areas with excessive matrix; this ultimately reduces the composite efficiency. Uniform matrix dispersion cannot be compromised, and a careful selection of processing techniques, initial matrix lengths, process conditions and processing aids are needed to obtain efficient composites. Several types of compounding equipment, both batch and continuous, have been used for blending lignocellulosic fibres and plastics.

The ultimate matrix lengths present in the composite depend on the type of compounding and moulding equipment used. Several factors contribute to the matrix attrition (breaking down of the long chain polymers to shorter chain lengths) such as the shearing forces generated in the compounding equipment, residence time, temperature and viscosity of blends. An excellent study on the affect of processing and mastication, of several types of short fibres in thermoplastics was conducted by Czarnecki and White [22]. They concluded that the extent of breakage was most severe and rapid for glass fibres, less extensive for Kevlar (aramid) fibres, and the least for cellulose fibres. The level of matrix attrition depends on the type of compounding and moulding equipment used, level of loading, temperatures and viscosity of the blend.

The properties of the agro-based thermoplastic composites are thus very process dependent. Yam and co-workers [14] at Michigan State University, studied the effect of twin-screw blending of wood fibres and high density polyethylene (HDPE) and concluded that the level of fibre attrition depended on the screw configuration and the processing temperature.

Average fibre lengths decreased from about 1.26 mm prior to compounding to about 0.49 mm after extrusion. Modification of the screw configuration reduced fibre attrition to an average length of about 0.78 mm. Fibre weight percentages of up to 60% were used. The tensile strength of the pure HDPE was higher than that of the wood fibre-HDPE, irrespective of the level of fibre filling. This was explained as a lack of dispersion with fibres clumping in bundles and poor fibre-matrix bonding. Use of stearic acid in HDPE/ wood fibres improved fibre dispersion and improved wetting between the fibre and matrix [9] and resulted in significant improvement in mechanical properties. Work by Raj and Kokta [23] indicate the importance of using surface modifiers to improve fibre dispersion in cellulose fibres/PP composites. Use of a small amount of stearic acid during the blending of cellulose fibres in polypropylene decreased both the size and number of fibre aggregates formed during blending in an internal mixer (Brabender roll mill).

Another technique that is gaining acceptance is high intensity compounding using a turbine 'Z' blade mixer (thermokinetic mixer). Woodhams and co-workers [9] and Myers and co-workers [12], found the technique effective in dispersing lignocellulosic fibres in thermoplastics. Addition of dispersion aids/coupling agents further improved the efficiency of mixing. The high shearing action developed in the mixer decrease the lengths of fibres in the final composite. However the improved fibre dispersion resulted in improved composite properties. A recent study of evaluating fibre lengths of a jute-PP composite blended in a thermokinetic mixer and then injection moulded was conducted by dissolving the PP in xylene and the length of fibres measured using image analysis [24]. The ultimate fibres lengths varied from about 0.10 mm to about 0.72 mm with an average of 0.34 ± 0.13 mm. Work using a thermokinetic mixer to blend kenaf in PP [17] has confirmed the usefulness of the compounding technique in effectively dispersing natural fibres in the thermoplastic matrix. An added advantage is that no pre-drying of the fibres is needed before the blending stage in the mixer.

When using Nylon 6 (PA-6) with cellulose pulp fibre, a different compounding method for producing cellulose reinforced engineering thermoplastic composites was used [18]. The pulp fibres need to be pelletised to reduce the bulk volume of the fibres. Careful attention to the compounding sequence is required to produce high quality composites. The cellulose pellets are delivered to the melted polymer stream by a side-stuffer/crammer, which is calibrated to deliver cellulose pellets at a rate equal to half that of the polymer feed-rate. This results in a composite blend of 33% cellulose and 67% PA-6. As cellulose is added into the polymer, the melt viscosity will begin to rise and the side-stuffer/crammer feed-rate needs to be increased slightly to counteract the increased extrusion melt pressure in the extruder. At steady-state conditions, the cellulose feed-rate no longer requires adjustment.

A new technique by Sanadi and Caulfield [25, 26] has led to very high filling rates for kenaf fiber-PP composites. Kenaf fibres were coated with 1 to 4% by weight with glycerine to attain such high fibre loadings in composites. The short fibres, maleated polypropylene (MAPP; a coupling agent), and PP, the latter two in pellet form, were compounded in a one litre high intensity kinetic mixer where the only source of heat is generated through the kinetic energy of rotating blades. The fibres and polymers were inserted into the blender and compounded at 5000 rpm, which resulted in a blade tip speed of about 30 m/s. The compounded mass was then automatically discharged at 200 °C. The precise time to discharge depended on the amount of fibre and polymer, and their ratios. On average, the time of blending varied from 2 to 3 minutes with higher fibre content blends taking longer to reach 200 °C. The compounded mass was then immediately compressed using a hot press. We have been able to achieve a fibre weight loading of up to 90%, although the major part of our testing has been conducted on composites consisting of 85% by weight of fibres.

4.3 Properties

4.3.1 Mechanical Properties: Effects of Coupling and Fibre Content and Type

The interphase characteristics influence the mechanical and physical behaviour of natural fibre/PP composites. Details of the physical state such as whether the composite is above or below the glass transition temperature (T_g) and the nature and degree of crystallinity affect the stress transfer efficiency. Grafted polymers such as MAPP have been used to improve the mechanical properties of natural fibre composites [27]. Property enhancement can be significant and a two-fold increase (approximately) in tensile strengths of 50% by weight of kenaf fibre filled PP has been reported with the addition of an appropriate MAPP [28].

The molecular weight and amount of grafted maleic anhydride of MAPP are important parameters in determining the effectiveness of stress transfer. Gatenholm and co-workers [29] have indicated covalent bonding of the grafted moiety to the hydroxyl group on the fibre surface through infra red and electron spectroscopy for chemical analysis (ESCA) analysis. Felix and Gatenholm [30] have suggested that brush like interfaces exist while using the MAPP. This brush like interphase for lignocellulosic-PP composites was postulated partly based on nucleation densities calculated from model systems observed under a polarising microscope. These experiments were performed using a fibre placed in a PP melt. For observing coupling agent effects, the MAPP was first covalently bonded to the fibres using a solvent. The fibres were then dried and then placed in a PP melt to be observed under a polarising microscope.

In the previously mentioned work, to observe transcrystallinity in coupled systems, model systems were used where the MAPP was first reacted to the fibre using a solvent. The fibres are then placed between films of the polymer before proceeding with the hot stage microscopy. In this work, a new 'solventless' technique was used to obtain thin films of fibres in polymer which were then placed in a polarising microscope with a hot stage [31]. The fibres, PP and the MAPP (if used) were first blended in the thermokinetic mixer as previously described. The granulated blend was then diluted to about 1% by adding pure PP. Samples were injection moulded at higher than typical pressures. This higher pressure resulted in 'flashing' in the moulded specimen. The fibres in the 'flashed' film were well dispersed so individual fibres could be clearly observed. We did try using a 20% 'flashed' layer, but it was not possible to observe individual fibres. Both coupled and uncoupled blends were prepared. Since the MAPP reacts with the fibres during the compounding process, the fibres dispersed in the diluted 'flashed' samples already have the reacted surface layer. If necessary, the films were again pressed to thinner films so that the transcrystallinity could be observed. Samples were heated to 220 °C to remove the polymer morphological history, and crystallisation was conducted under isothermal conditions at 130 °C. As soon as the transcrystalline zones were formed, the samples were quenched and photographed.

Using the 'solventless' technique for the formation of thin films we observed transcrystalline zones for both the uncoupled and coupled systems. It was observed that the crystallisation rate was higher for the coupled system, although the rates of crystallisation were not studied. The zone nucleated at several points on the fibre surface, and then grew outwards radially [31]. Neighbouring crystals impinged on each other to limit the range of lateral growth.

The stress transfer efficiency of the composite is dependent on the properties of the interphase zone, and this includes the physical or chemical interactions between two different *interfaces*. The first is the interface between the transcrystalline zone and the fibres, and the second is the interface between the outer edges of transcrystalline zone and the neighbouring spherulites or (when two fibres are close together due to high fibre volume fractions) adjacent transcrystalline zones. For the uncoupled systems, the interaction between the fibres and the transcrystalline zone is limited to dispersive forces, topographical effects and normal shrinkage stresses. In case of the coupled system there are the additional factors of covalent and acid-base interactions. The interface strength between the transcrystalline zone and neighbouring spherulites or the transcrystalline zone from neighbouring fibres depends on molecules bridging between the crystals. Long molecular chains are important to maximise the effectiveness of the 'physical crosslinking'. The presence of low molecular weight chains will likely reduce this interface strength.

Interface failure can occur either at the fibre surface interface or the outer crystalline zone interface. Earlier work suggests that tensile and flexural strengths of kenaf-PP composites increase with a small amount of MAPP added to the blend [28]. This suggests that the addition of MAPP increased the transcrystalline-transcrystalline zone interface strength. Addition of more than 3% (by composite weight) of MAPP reduces the strength of the composite [31] and the failure shifts to other regions. This is possibly due to the presence of larger amounts of lower molecular weight polymer reducing the strength of the physical cross-links between crystals and also between the transcrystalline surface and the transcrystalline zone. Secondly, the transcrystalline zone of the coupled composites has more defects that could lead to failure at low strains. The molecular weight of the MAPP, and the amount of MA are important factors in determining stress-transfer efficiency [31, 32].

Figure 4.1 shows typical stress strain curves of pure PP, uncoupled 50% by weight of kenaf-PP, and coupled systems with increasing amounts of kenaf in the composite. The strength of the uncoupled 50% by weight composite was about the same as that of the PP matrix. Composites made with the use of MAPP resulted in higher strengths and failure strains than the uncoupled 50% composites. The decrease in the failure strain with increasing amounts of kenaf for the coupled systems is apparent. The non-linearity of the curves is mainly due to plastic deformation of the matrix. The tensile energy absorption, the integrated area under the stress-strain curve up to failure behaves in roughly the same manner as the tensile failure strain.

Figures 4.2, 4.3 and **4.4** show the effect of the amount of MAPP on composite properties of 50% by weight (about 39% by volume) of kenaf fibres in PP. A small amount of the MAPP (0.5% by weight) improved the flexural and tensile strength, tensile energy absorption, failure strain and un-notched Izod impact strength. The anhydride groups present in the MAPP can covalently bond to the hydroxyl groups on the transcrystalline surface [29]. Any maleic anhydride (MA) that is in acid form can interact with the transcrystalline surface through acid-base interactions. The improved interaction and adhesion between the fibres and matrix, either through covalent bonding or acid-base interaction (such as H-bonding) or a combination of both, leads to better matrix to transcrystalline stress transfer. There was little difference in the properties obtained between the 2% and 3% (by weight) of the MAPP. The drop in tensile modulus with the addition of the MAPP is probably due to molecular morphology of the polymer near the fibre surface or in the bulk of the plastic phase.

Transcrystallisation and changes in the apparent modulus of the bulk matrix can result in changes in the contribution of the matrix to the composite modulus and will be discussed in Section 5.3.5. There is little change in the notched impact strength with the addition of the MAPP, while the improvement in un-notched impact strength is significant. In the notched test, the predominant mechanism of energy absorption is through crack propagation as the notch is already present in the sample.

Addition of the coupling agent has little effect in the amount of energy absorbed during crack propagation. On the other hand, in the un-notched test, energy absorption is through

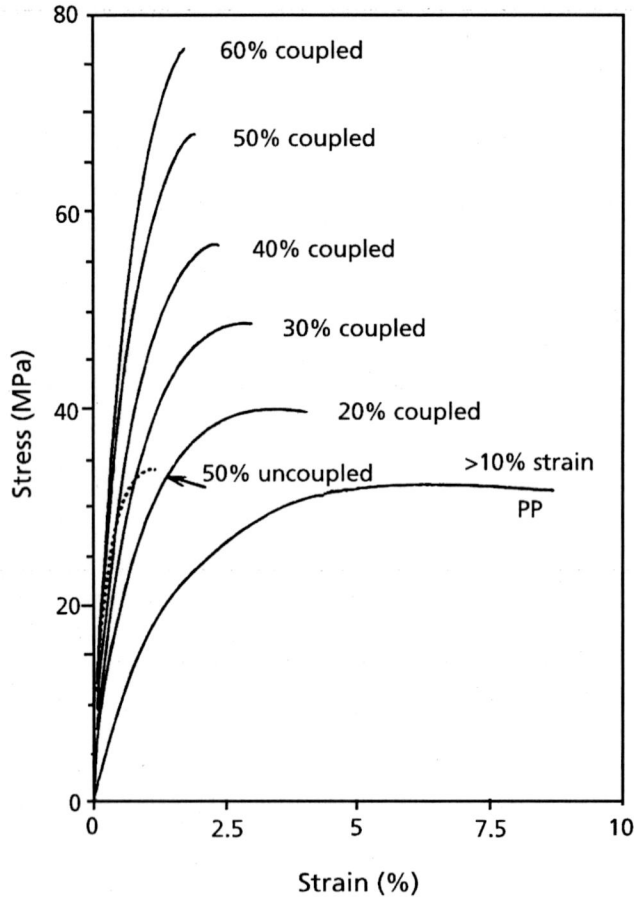

Figure 4.1 Stress-strain curves of PP and kenaf-filled PP. The numbers at the end of the curve indicate weight% of kenaf [28]

Reproduced with permission from A.R. Sanadi, D.F. Caulfield, R.E. Jacobson and R.M. Rowell, Industrial and Engineering Chemistry Research, 1995, Copyright, 1995, 34, 1889, Figure 7, American Chemical Society

a combination of crack initiation and propagation. Cracks are initiated at places of high stress concentrations such as the fibre ends, defects, or at the interface region where the adhesion between the two phases is very poor. The use of the additives increases the energy needed to initiate cracks in the system and thereby results in improved un-notched impact strength values with the addition of the coupling agent.

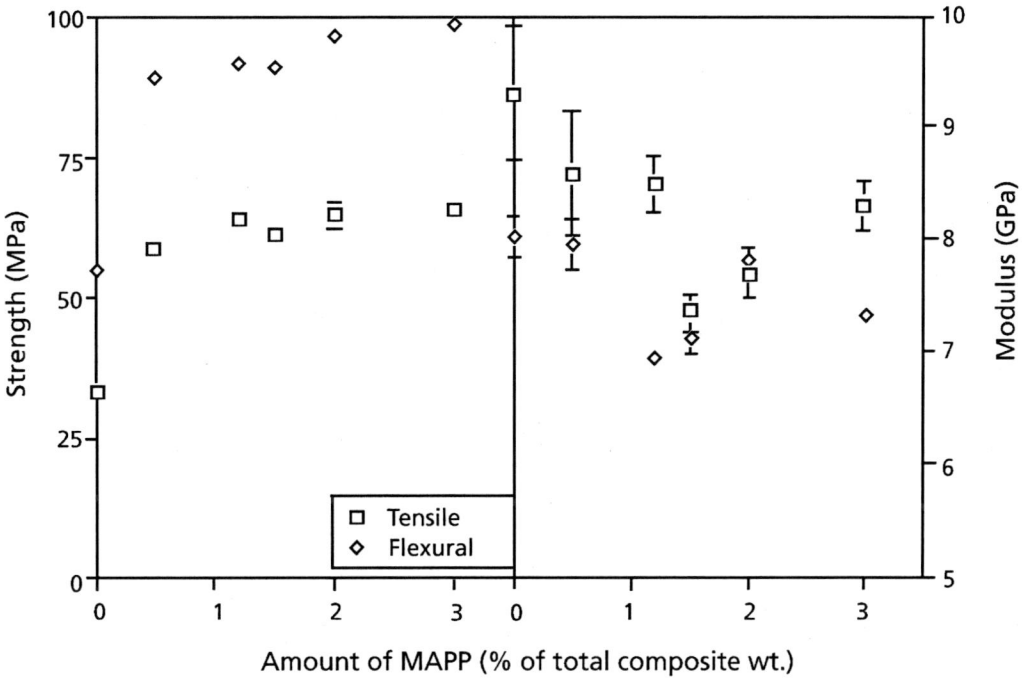

Figure 4.2 Effect of the amount of coupling agent (MAPP) on the strength and modulus (both tensile and flexural) properties of 50% by weight of kenaf in PP [28]

Reproduced with permission from A.R. Sanadi, D.F. Caulfield, R.E. Jacobson and R.M. Rowell, Industrial &Engineering Chemistry Research, 1995, Copyright, 1995, 34, 1889, Figure 1, American Chemical Society

Use of MAPP increases the failure strain and the tensile energy absorption. Thermodynamic segregation of the MAPP towards the interphase can result in covalent bonding to the -OH groups on the fibre surface. Entanglement between the PP and MAPP molecules results in improved interphase properties and the strain to failure of the composite. There is a plateau after which further addition of coupling agent results in no further increase in ultimate failure strain. The average molecular weight of the MAPP is about 20,000 da and the amount of entanglements between the PP molecules and MAPP is limited, and molecules flow past one another at a critical strain. Any further increase in the amount of coupling agent does not increase the failure strain past the critical amount. However, a minimum amount of entanglements are necessary through the addition of about 1.5% by weight (**Figure 4.2**) for the critical strain to be reached.

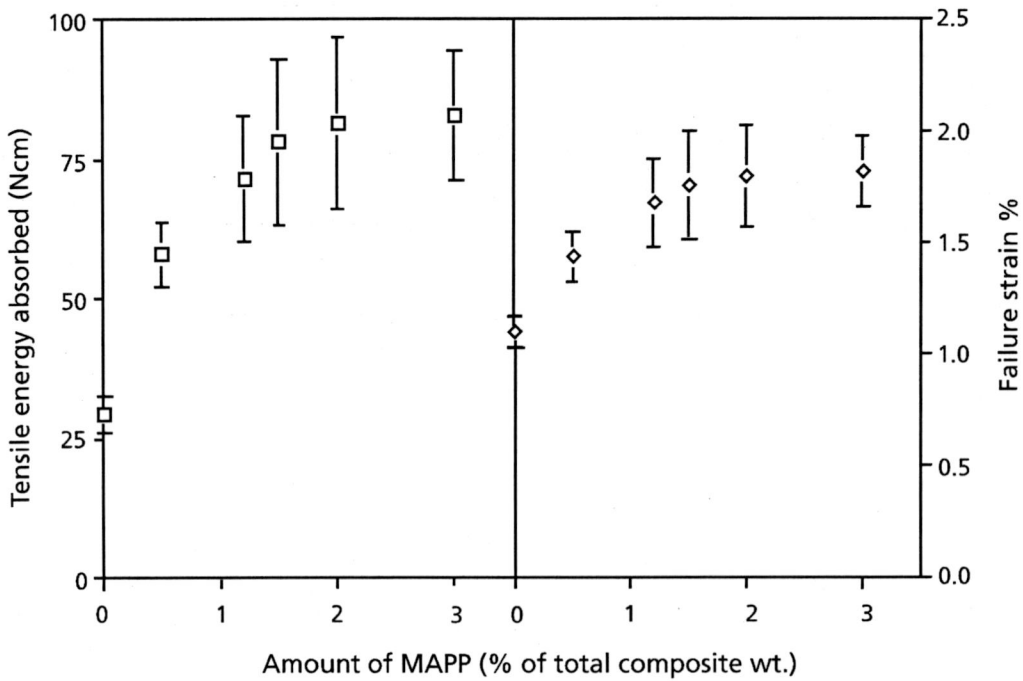

Figure 4.3 Effect of the amount of MAPP on the tensile energy absorption and ultimate tensile failure strain of 50% by weight of kenaf in PP [28]

Reproduced with permission from A.R. Sanadi, D.F. Caulfield, R.E. Jacobson and R.M. Rowell, Industrial & Engineering Chemistry Research, 1995, 34, 1889, Figure 2, Copyright, 1995, American Chemical Society

There is little difference in the tensile strength of uncoupled composites compared with the unfilled PP, irrespective of the amount of fibre present (**Figure 4.5**). This suggests that there is little stress transfer from the matrix to the fibres due to incompatibilities between the different surface properties of the polar fibres and non-polar PP. The tensile strengths of the coupled systems increase with the amount of fibre present and strengths of up to 74 MPa were achieved with the higher fibre loading of 60% by weight or about 49% by volume. As in case of tensile strength, the flexural strength of the uncoupled composites were approximately equal for all fibre loading levels, although there was a small improvement when compared to the unfilled PP. The high shear mixing using the thermokinetic mixer causes a great deal of fibre attrition. The strengths obtained in our

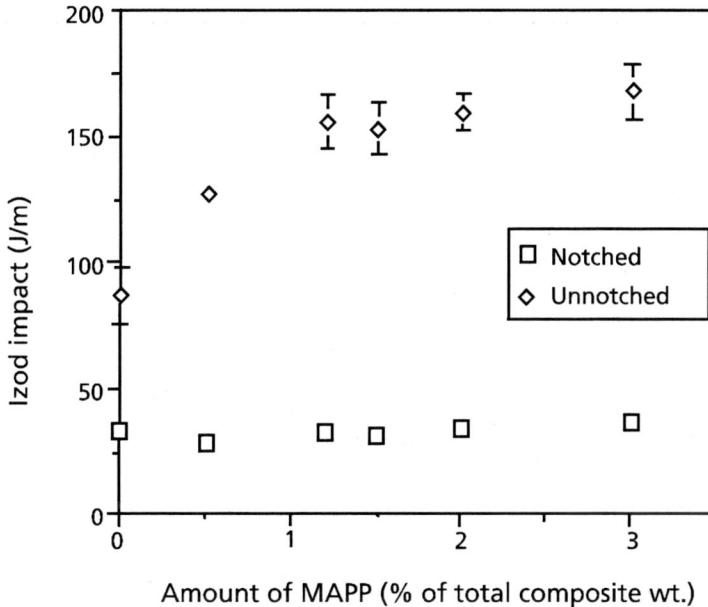

Figure 4.4 Effect of the amount of MAPP on the notched and un-notched Izod impact strength of 50% by weight of kenaf in PP [28]

Reproduced with permission from A.R. Sanadi, D.F. Caulfield, R.E. Jacobson and R.M. Rowell, Industrial Engineering & Chemistry Research, 1995, Copyright, 1995, 34, 1889, Figure 3, American Chemical Society

composites were thus limited by the short fibre lengths. Higher strengths are likely if alternate processing techniques are developed that reduce the amount of fibre attrition while at the same time achieving good fibre dispersion.

The tensile modulus of the kenaf composites showed significant improvements with the addition of the fibres (**Figure 4.6**). The uncoupled composites showed some very interesting behaviour, with moduli higher than the coupled systems at identical fibre loading. The possibility of a high stiffness transcrystalline zone forming around the fibre in the unmodified systems could lead to the high moduli observed, although the possibility of different fibre orientations contributing to the higher modulus cannot be ruled out. However, studies of jute-PP composites on fibre orientation suggest that the difference in moduli cannot be explained exclusively by the difference in fibre orientations. Several studies suggest that the morphology of the polymer chains is affected by the presence of

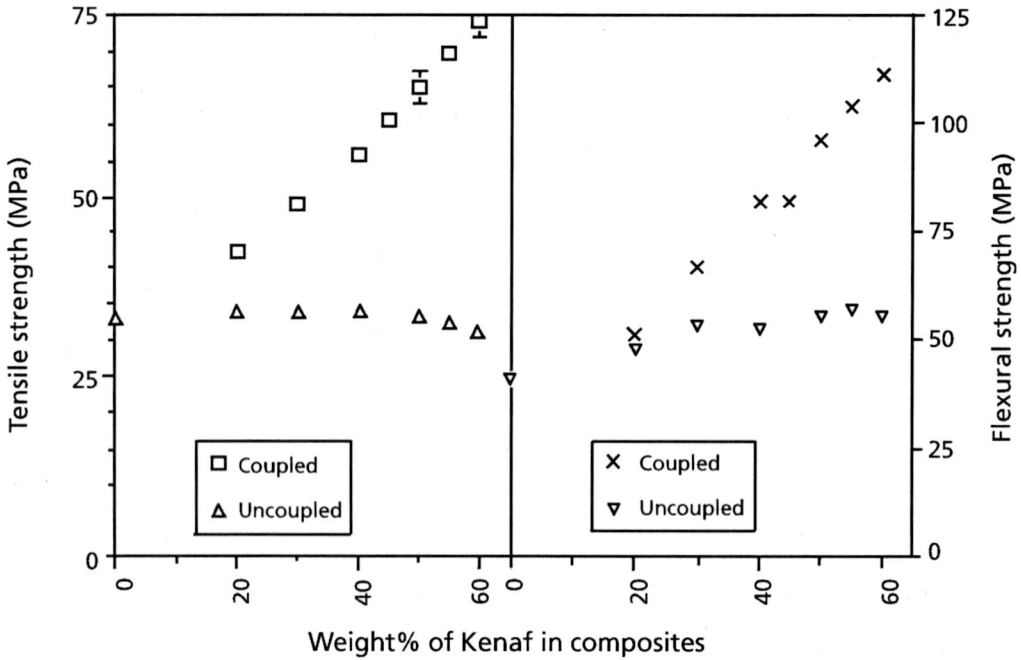

Figure 4.5 Effect of the amount of kenaf *versus* the tensile and flexural strengths of kenaf-PP composites. All coupled composites had 2% by weight of MAPP [28]

Reproduced with permission from A.R. Sanadi, D.F. Caulfield, R.E. Jacobson and R.M. Rowell, Industrial Engineering & Chemistry Research, 1995, 34, 1889, Figure 4, Copyright, 1995, American Chemical Society

filler particles. Addition of a rigid filler/fibre restricts the mobility of the polymer molecules to flow freely past one another and thus causes premature failure. Addition of MAPP followed a similar trend to that of the uncoupled system, although the drop in failure strain with increasing fibre amounts was not as severe.

Figure 4.7 shows the effect of fibre content and MAPP coupling agent on the maximum load values for notched and un-notched impact strength. Second order polynomials fitted to the data describe the general trend of behaviour. In both notched and un-notched impact tests, the addition of MAPP did increase the strengths. In case of the un-notched test, there was a decrease when no coupling agent was used, indicating potential lack of interface efficiency, which leads to crack initiation.

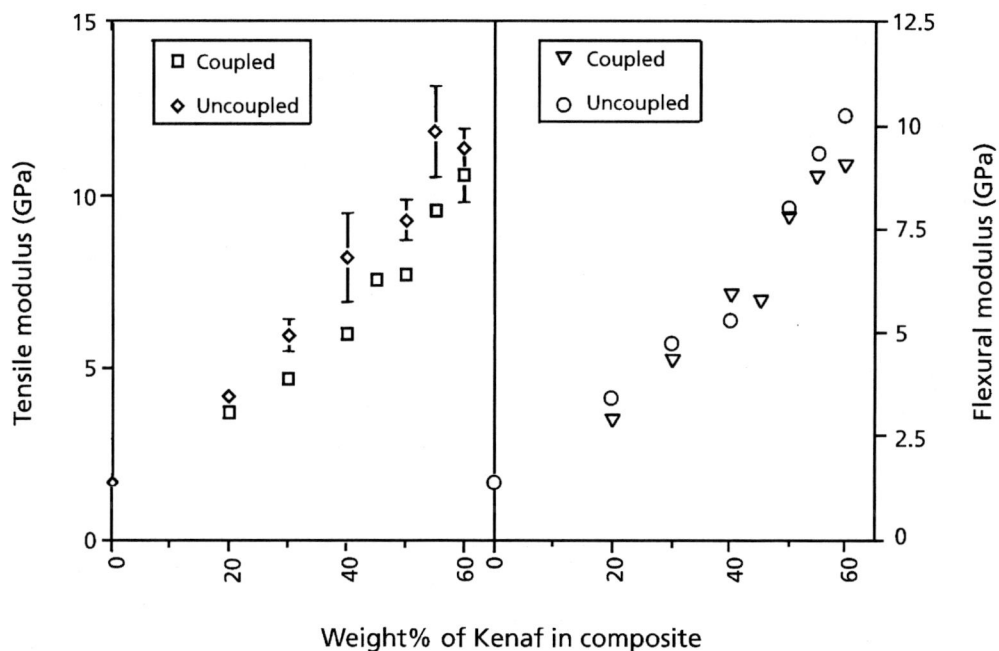

Figure 4.6 Effect of the amount of kenaf *versus* the tensile and flexural moduli of kenaf-PP composites. All coupled composites had 2% by weight of MAPP [28]

Reproduced with permission from A.R. Sanadi, D.F. Caulfield, R.E. Jacobson and R.M. Rowell, Industrial Engineering & Chemistry Research, 1995, 34, 1889, Figure 5, Copyright, 1995, American Chemical Society

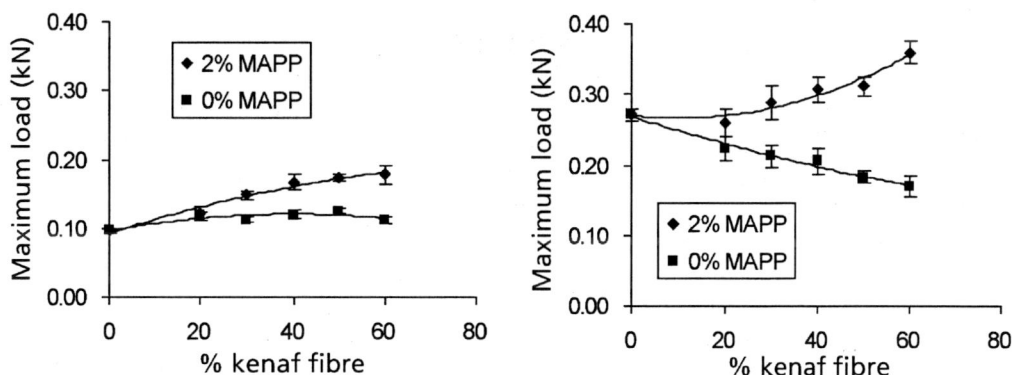

Figure 4.7 Fibre content and coupling agent (MAPP) effects on maximum load in notched (left) and un-notched (right) impact tests, from Clemons and Sanadi, in preparation [39]

4.3.2 Effect of Fibre and Polymer

Table 4.1 shows some typical data of commercially available injection moulded PP composites and the comparison with typical kenaf-PP composites. Data on the talc, mica, calcium carbonate and glass composites were compiled from Resins and Compounds, in the *Modern Plastics Encyclopaedia* (1993) [30] and Thermoplastic Moulding Compounds, in *Material Design* (1994) [31]. The properties of kenaf-based fibre composites have properties superior to typical wood (newspaper) fibre-PP composites. The specific tensile and flexural moduli of a 50% by weight of kenaf-PP composite compares favourably with the stiffest of the systems shown, that of glass-PP and mica PP. The properties of 40% by weight of kenaf-PP are far superior to fillers such as calcium carbonate, newspaper fibres and talc.

The two main disadvantages of using kenaf-PP rather than glass-PP are the lower impact strength and the higher water absorption. The lower notched impact strength can be improved by using impact modified PP copolymers and the use of flexible maleated copolymers, albeit with some loss in tensile strength and modulus. Care needs to be taken when using these fibres in applications where water absorption and the dimensional stability of the composites are of critical importance. Judicious use of these fibres will make it possible for agro-based fibres to be used in the plastics industry for the manufacture of low-cost, high-volume composites using commodity plastics.

An interesting point to note is the higher fibre volume fractions of the agro-based composites compared to the inorganic filled systems. This can result in significant material cost savings as the agricultural fibres are cheaper than the pure PP resin, and far less expensive than glass fibres. Environmental and energy savings by using an agriculturally grown fibre instead of the high energy utilising, glass fibres or mined inorganic fillers are benefits that can not be ignored, although a thorough study needs to be conducted to evaluate the benefits.

Table 4.2 shows data on some mechanical properties of, coupled, 40% by weight fibres filled with PP. There is quite a bit of difference between some of the fibres. As previously mentioned, the composite properties are reflected by the fibre properties: degree of polymerisation, cellulose and lignin content, micro-fibril angle, defects, and other factors contributes to the fibre properties. Of course, the level of stress transfer due to the matrix-matrix interface cannot be ruled out. The fibres have different waxes and other ingredients that could enhance or decrease the stress transfer efficiency. For example, coir has low cellulose content and that is reflected by the lower strength and modulus of the composite.

Table 4.2 shows the some of the properties of various fibres (40% by weight) in PP, with the same amount of coupling agent [34].

Table 4.1 Comparison of natural fibre composites with commercially available filler/reinforcements

Filler/reinforcement in PP	ASTM Standard	None	Kenaf	Kenaf-PP impact copolymer	Talc	CaCO$_3$	Glass	Mica
Filler % by weight		0	50	50	40	40	40	40
Filler % by volume (estimated)		0	39	39	18	18	19	18
Tensile modulus, GPa	D638 [35]	1.7	8.3	7.5	4	3.5	9	7.6
Specific tensile modulus, GPa		1.9	7.8	7.0	3.1	2.8	7.3	6.0
Tensile strength, MPa	D638	33	65	53	35	25	110	39
Specific tensile strength, MPa		37	61	50	28	20	89	31
Elongation at break, %	D638	>>10	2.2	2.5			2.5	2.3
Flexural strength, MPa	D790 [36]	41	98		63	48	131	62
Specific flexural strength, MPa		46	92		50	38	107	49
Flexural modulus, GPa	D790	1.4	7.3		4.3	3.1	6.2	6.9
Specific flexural modulus, GPa		1.6	6.8		3.4	2.5	5.0	5.5
Notched izod impact - J/m	D256A [37]	24	32	74	32	32	107	27
Specific gravity		0.9	1.07	1.07	1.27	1.25	1.23	1.26
Water absorption, % - 24 h	D570 [38]	0.02	1.05	1.3	0.02	0.02	0.06	0.03
Mould (linear) shrinkage cm/cm		0.028	0.003	0.004	0.01	0.01	0.004	

Data on mineral filled systems from various sources, Modern Plastics Encyclopaedia (1993) [3] and Machine Design: Materials Guide Issue (1994) [31]

Caulfield and other workers have looked at using cellulose pulp fibres [18, 43] in higher temperature melting polymers such as PA 6. **Table 4.3** compares the effect of the pulp fibre as compared to PA-6, and also with that of wollastanite and glass fibres. Although the cost of the pulp fibres is higher than most natural fibres, an advantage is that that odour problems are reduced since the fibres are almost pure cellulose. These pulp fibre composites may have potential for use in the automotive industry.

Table 4.2 Properties of various fibres (40% by weight) in PP with the same amount of coupling agent

Fibre type	Tensile strength (MPa)	Tensile modulus (GPa)	Notched Izod-strength (J/m)	Un-notched Izod strength (J/m)
Kenaf	55.8	6.4	28.3	157.3
Coir	45.8	3.5	18.4	100.1
Henequen	55.4	5.1	50.1	156.3
Milkweed	53.6	5.7	19.2	131.6
Ixtle	49	4.5	50.1	156.3
Jute	64	6.24	35.6	189.3
Newsprint	53	4.4	21	190
Pineapple	63.7	5.55	42.2	169.6
Abaca pulp	64.2	5.7	32.6	227.9

Table 4.3 Mechanical properties of Rayonier Terracel (cellulose fibre) reinforced PP and Terracel reinforced PA-6 composites compared to PP, PA-6 and other composites [18]

	Tensile strength (MPa)	Tensile modulus (GPa)	Notched Izod (J/m)	Unnotched Izod (J/m)
PA-6	60.2	2.75	24	746.3
PA-6 + 33% Wollastonite	62.7	6.51	25.8	173.9
PA-6 + 33% Glass fibre	111.2	8.02	45.9	406.1
PA-6 + 33% Terracel 10J	86.5	5.71	25.3	318.3

Our work on PLA filled (30% by weight in the plastic) with biodegradable fillers/fibres (ground corn cob and kenaf fibres) indicates that there is improvement in mechanical properties. **Table 4.4** shows data on some of the initial work that the USDA Forest Service has done. The strength, modulus and impact strength of the 30% kenaf fibre filled PLA was in the same range or higher than the filled system reported by Cargill. However, in order to maximise the amount of fibres in the PLA, to improve strength and at the same time reduce the cost of the composite, it is important to ensure good dispersion, while maintaining fibre lengths. Cargill data of filled PLA [40] indicate that there is an improvement of composite strength and modulus, but no improvement in impact strength. Furthermore, talc is not biodegradable and the density of talc (2.4 g/cm^3) is much more than natural fibres (1.4 g/cm^3).

	Modulus (GPa)	Strength (MPa)	Failure strain (%)	Izod notched (J/m)	Izod unnotched (J/m)
PLA*	3.5	53	4.1	16	—
Talc filled-PLA*	6.9	58	2.0	16	—
Toughened PLA*	2.1	25	40	208	—
Corn cob filled PLA	5.8	48	0.9	24	64
Kenaf filled PLA	7	65	1	35	112

Table 4.4 Data using biodegradable agro-based fillers and fibres filled with polylactic acid (PLA) compared with properties of PLA

* *Cargill data from Plastics Technology, 1995, 41, 17 [42]*

4.3.3 High Fibre-Filled Composites

This new technique of highly filled composites shows some interesting properties [25, 26]. Studies on 85% kenaf-thermoplastic composite indicates that properties such as flexural modulus and strength are superior to most types of wood particle, low and medium density hardboards. These results indicate the stiffness of the new materials (85% by weight of kenaf) is very high, and could even compete with oriented strand board in some applications, (**Table 4.5**). The data for the 85% composites were obtained from three different boards, with four specimens from each of the boards. *Note that the 85% composites were compression moulded and therefore the fibres were randomly*

Table 4.5 Comparison of flexural properties of commercial available formaldehyde-based wood composites with data on ~85% filled Kenaf-PP composites. Standard deviation given for the kenaf composites [26]				
	Strength range (MPa)		Modulus range (GPa)	
	Low	High	Low	High
High density fibreboards (commercial)	38	69	4.48	7.58
Medium density fibreboards (commercial)	13.1	41.4	2.24	4.83
85% kenaf-PP (compression moulded-random alignment)	75 ± 9		6.8 ± 0.5	
60% kenaf-PP injection moulded (significant fibre alignment)	110		10.1	

oriented, while that of the 60% injection moulded specimen is likely to have significant fibre alignment.

The lignin content in kenaf fibre is reported to be between 15 to 19% and the hemicellulose content around 22 and 23%. The crystalline cellulose content of kenaf fibres can vary between 44 and 57% in the natural fibre. Thus for an 85% kenaf filled PP/MAPP, the amount of lignin and hemicellulose content in the composite works out to be approximately 34%. If one uses 55% for the amount of cellulose content in kenaf fibre, then the 85% by weight composite will have about 47% crystalline cellulose (the reinforcing element), 15% PP/MAPP, and about 34% lignin and hemicellulose in the blend. The lignin and hemicellulose is plasticised to allow molecular chain mobility. Thus, if one looks at this point of view, it is not inconceivable to have highly filled kenaf composites, since the reinforcing element is far below the limits imposed by packing of fibres in a composite.

4.3.4 Dynamic Mechanical Properties, Temperature and Creep Behaviour

Figure 4.8 shows the change in storage modulus with temperature of kenaf-PP composites and pure PP. Two consistent variations of the storage modulus (E′) with temperature can be observed for all the systems. There is a sharp drop in E′ from about –10 °C to about +10 °C, and the other is a reduction in the rate of drop in E′ with temperature after

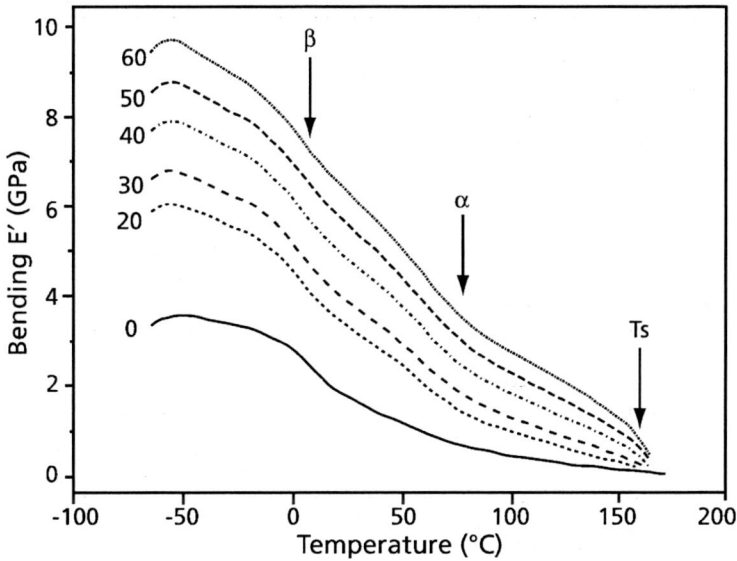

Figure 4.8 Bending dynamic mechanical spectra of PP and kenaf-PP composites. All composites were coupled and the numbers at the beginning of the curves indicate amount of kenaf (wt.%) in the composite [43]

Reproduced with permission from A.R. Sanadi, D.F. Caulfield, N.M. Stark and C. Clemens, Proceedings of the 5th International Conference on Wood-Fiber-Plastics, 1999, 67, Figure 1. Copyright 1999, Forest Products Society

around 75 °C – both shown by arrows. The first change between –10 °C to about +15 °C is the relaxation associated with the amorphous phase (β relaxation). In this case, the glassy state of the amorphous phase goes through its T_g and there is a sharp drop in E′. At about 15 °C, E′ continues to fall and the slope is similar to that before the β relaxation started. After about 70 to 80 °C, depending on the system, the reduction in E′ is less severe than before until the softening temperature (T_s) and then the melt region is reached. It is interesting to note that the dynamic storage modulus of PP at about 10 °C is about the same that of a 60% kenaf-PP composite at around 100 °C.

Figure 4.9 shows the loss modulus of 20% to 60% by weight kenaf-PP composites and that of PP. Comparing the coupled composite β relaxations with that of pure PP, **Figure 4.5**, it is clear that there is an increase in the molecular mobility of the amorphous phase with the addition of fibres and coupling agent. Several factors need to be considered to explain these phenomena. The presence of low molecular weight MAPP could reduce the T_g by

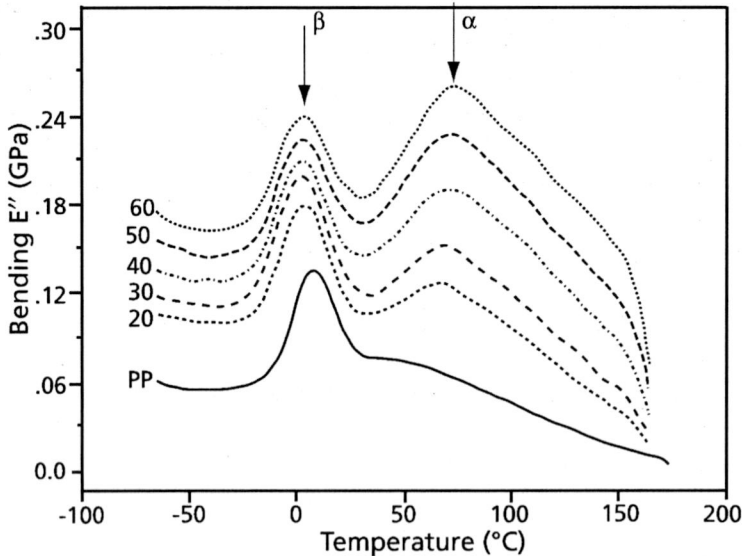

Figure 4.9 Bending dynamic mechanical loss modulus spectra of PP and kenaf-PP composites. All composites were coupled and the numbers at the beginning of the curves indicate amount of kenaf (wt.%) in the composite [33]

Reproduced with permission from A.R. Sanadi and D.F. Caulfield, Composite Interfaces, 2000, 7, 1, 31, Figure 5. Copyright 2000, VSP

enhancing the ability of the polymer to undergo a glass-rubber transition [42]. Good interaction (covalent or acid-base) between the MAPP and the lignocellulosic surface restricts the mobility of the molecules and increases the β peak temperature. Acid-base interactions between the anhydride groups of the MAPP can restrict the mobility of the molecules. *An important factor that has been neglected previously* is that during blending, some of the surface matrix material of the kenaf fibres themselves are sheared off from the fibre and blend in with the matrix. This sheared material could have lignin and cellulose, and oils and other surface extractives that are inherently present in the fibre. This can be clearly seen as the dark brown tinge of the polymer matrix after blending. In other words, the matrix material is a blend of PP and several constituents that are contributed by the fibre.

The T_α relaxation in semi-crystalline polymers is due to the presence of amorphous chains in the crystalline phase (defects) also called 'rigid' amorphous molecules. An empirical relationship was observed! The peak amplitudes or intensities of the 20, 30, 40 and

50 and 60% fibre blend (all weight percent) were proportional to their respective *fibre volume fractions* (V_f) estimated from specific gravity data (**Table 4.6**). The α peak amplitude or intensity is the E′ increase from the onset of the α transition to the peak of the transition. The similarity in the ratios of the intensities to the V_f is remarkable. This proportionality should also hold true between the intensities and the fibre surface area. It is thus possible that the defects in the crystals that cause the T_α transition for kenaf-PP composites are predominantly near the fibre–matrix interface and exist in the transcrystalline zone.

The dynamic mechanical analysis spectra of the 85% filled composites (**Figure 4.10**) show some interesting features when compared to a 60% filled kenaf composite that was injection moulded [26]. It should be noted that the 60% composites were injection moulded, which results in a significant amount of fibre alignment. This is as opposed to the compression moulded 85% composites where the fibres were randomly distributed. At low temperatures, the 60% composites have a higher E′, which suggests a brittle material at these temperatures. However, after about 10 °C, the 85% composites have a higher modulus. The difference in storage moduli becomes even more pronounced as the temperature increases. The softening temperature of the 85% filled composite is also higher than the 60% composite. The lower modulus at low temperature and the higher modulus at higher temperature is an advantage of the 85% composite, since the composite is not as brittle as the 60% at lower temperatures but maintains its integrity better at higher temperatures.

Figure 4.11 shows notched impact performance above and below T_g for unfilled PP and PP-kenaf fibre composites with 2% MAPP using an instrumented Izod impact test [33]. There was little change in the deflection of maximum load (DML) at −25 °C; a temperature that is below the T_g for the composite (see **Figure 4.9**). The DML was lower at −25 °C as compared

Table 4.6 Intensity of the α relaxation for the coupled kenaf composites [31]			
Fibre weight fraction (%)	Estimated fibre volume fraction - V_f (%)	Intensity (MPa)	Intensity/V_f
60	47.79	76.52	1.60
50	37.89	60.87	1.61
40	28.92	45.22	1.56
30	20.73	33.91	1.64
20	13.24	20.87	1.58

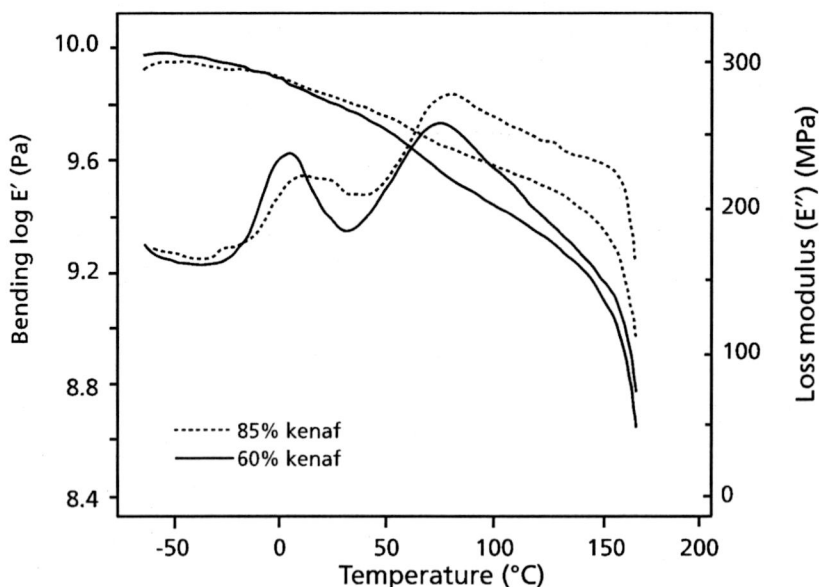

Figure 4.10 Storage and loss modulus of 85% kenaf-PP as compared to 60% injection moulded kenaf-PP composite [26]

Reproduced with permission from A.R. Sanadi, J.F. Hunt, D.F. Caulfield, G. Kovacsvolgyi and B. Destree, Proceedings of the Sixth International Conference on Woodfiber-Plastic Composites, Forest Products Society, Madison, WI, USA, Figure 2. Copyright 2002

to that above T_g (greater than about 0 °C). The DML peaks at about 35% kenaf weight. The lower DML at higher fibre fractions can be explained by molecular mobility restriction of the polymer due to the high amount of transcrystalline zone present in the system.

In case of the energy-to-maximum load (EML), the load increased irrespective of the temperature. This is possibly due to the crack bridging and pull-out energy mechanisms.

It is obvious that the energies increase with temperature, particularly above the T_g of the polymer, **Figure 4.12**. However, it is clear that the low temperature impact energy absorbed at −50 °C of the kenaf composites was about three times that of the PP and the wood flour/PP composites. This is likely to be due to the higher aspect ratios of the fibres, giving rise to energy absorbing mechanisms due to crack bridging and pull-out.

The creep behaviour of natural thermoplastic-thermoplastic is definitely a consideration when it comes to load bearing applications. The linear macromolecules of a thermoplastic allow flow during load, at temperatures above the T_g. However, improving the adhesion

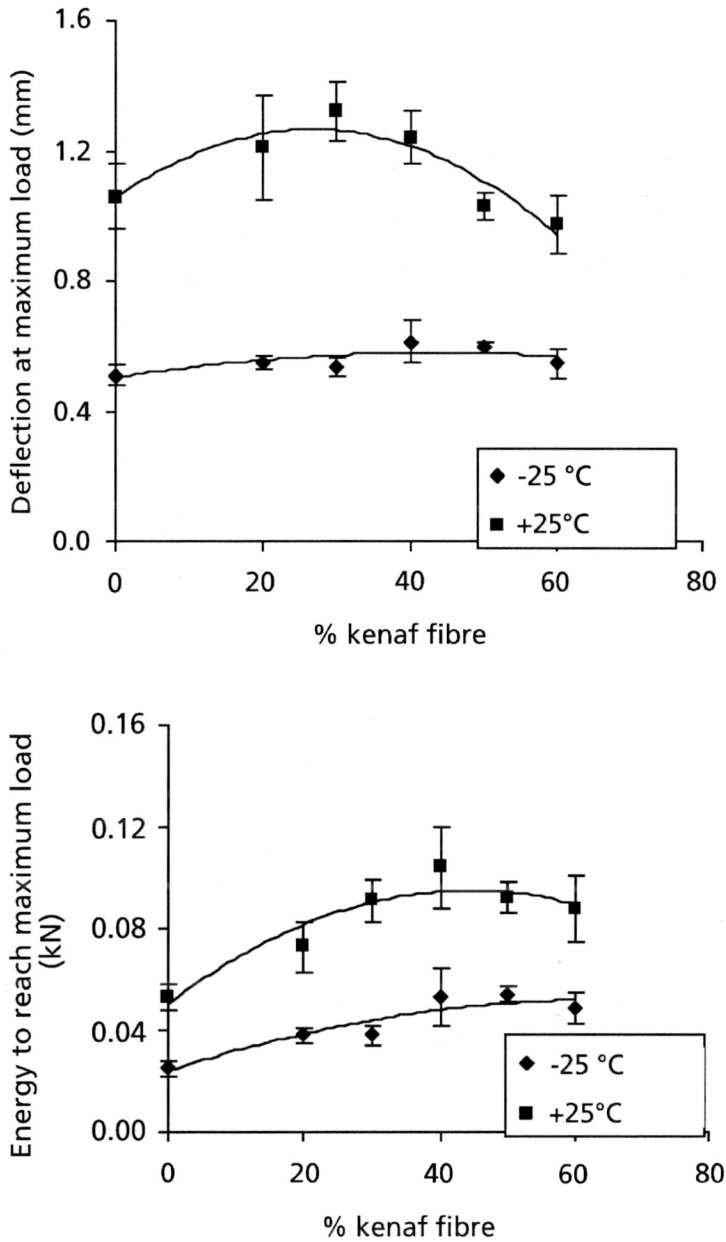

Figure 4.11 Deflection at maximum load (top) and energy to reach maximum load (bottom) for notched impact tests above and below T_g of PP for kenaf composites, Clemons and Sanadi, in preparation [39]

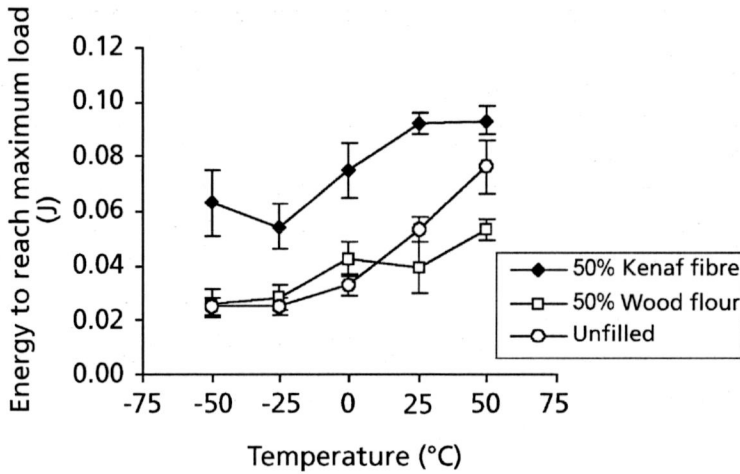

Figure 4.12 Notched impact energies to maximum load of PP and wood flour/PP and kenaf-PP composites, at various temperatures, using an instrumented impact tester, Clemons and Sanadi, in preparation [39]

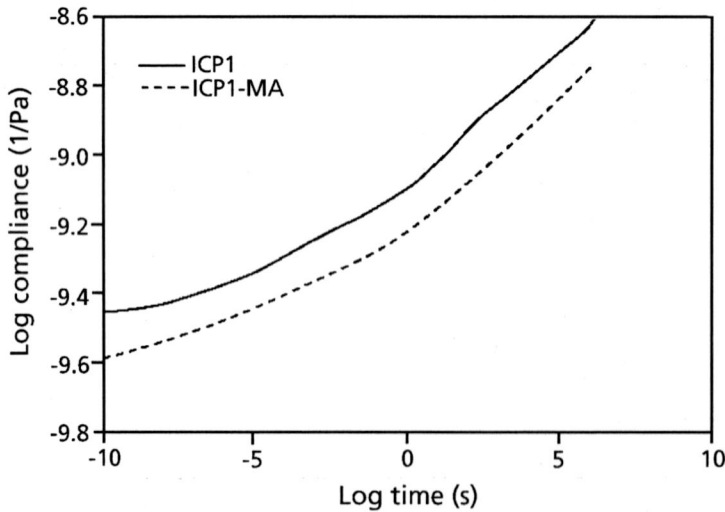

Figure 4.13 Dynamic creep spectra of low molecular weight, impact PP co-polymer ICP1 (Amoco 3541, ethylene-propylene low MW, impact copolymer) and ICP1-MA (ICP2 coupled with maleated PP) blends [45]

Reproduced from D. Feng, D.F. Caulfield and A.R. Sanadi, Polymer Composites, 2001, 22, 4, 506, Figure 12 and the presence of high molecular weight polymers may reduce the impact of the creep. The dynamic creep curves of a low molecular weight impact copolymer, ICP1, are shown in **Figure 4.13**. The ICP-MA denotes composites that have been compatibilised. The molecular weights of the copolymers have a strong influence on the creep behaviour of the blends. For the low molecular weight impact copolymer (ICP1), the coupling agent improves the creep behaviour significantly (**Figure 4.13**). The difference between the coupled and uncoupled blends of the high molecular weight copolymer (ICP2) was evident (**Figure 4.14**), but much less than the lower molecular weight copolymers. The creep behaviour of the ICP/kenaf blends, or for any composite, is dependent on several factors including crystallinity, adhesion between the fibre and polymer matrix, and the polymer chain entanglements. The high molecular weight copolymer has a higher entanglement density than the lower molecular weight polymers, and the effect of the coupling agent on the creep of the high molecular weight copolymers is much less pronounced [45].

4.3.5 Water Absorption

Water absorption in lignocellulosic composites can be significant and the rate of absorption is affected by the amount of coupling agent. **Figure 4.15** shows data on sisal-PP composites [46], after the samples were immersed in boiling water for two hours.

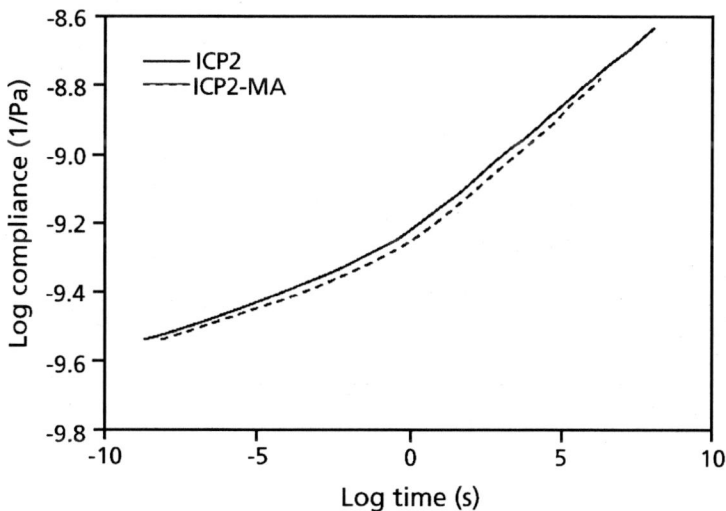

Figure 4.14 Dynamic creep spectra of high molecular weight, impact PP co-polymer ICP2 (uncoupled) and ICP2-MA (coupled) blends [45]

Reproduced from D. Feng, D.F. Caulfield and A.R. Sanadi, Polymer Composites, 2001, 22, 4, 506, Figure 13

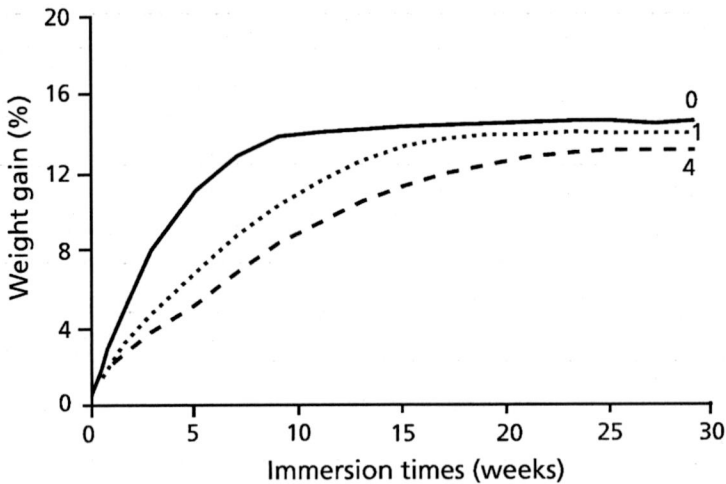

Figure 4.15 Weight gain in weeks for a 50% loading sisal-PP composites. Numbers at the end of curves indicate % of coupling agent used [46]

Reproduced with permission from R.A. Young, A.R. Sanadi and S. Prabawa, Proceedings of the 4th International Conference on Woodfiber-Plastic Composites, 1997, 94, Figure 8. Copyright 1997, Forest Products Society

Data are presented in **Table 4.7** where properties of the dry samples (control) were compared to samples that were boiled in distilled water for two hours [47]. The tensile modulus was reduced after the samples go through the two-hour boiling cycle and this decrease was reduced when using a larger amount of coupling agent, i.e., 4% by weight of the composite. The tensile strength decreased after the water exposure for all the samples, but no clear distinction was observed between the coupled and uncoupled samples. The unnotched impact strengths for the all samples increased by about 15% after the boil test, and no real distinction was observed between the coupled and uncoupled systems. In the un-notched Izod test, the increase was dramatic in all cases. The increase for the uncoupled system was about 18%, while that for the 4% by weight coupled systems was about 40%. Moisture plasticises the fibres, which increases the energy absorption contribution to the composite toughness. Furthermore, pull-out energies due to increased diameter of the fibre, due to water absorption, and matrix plasticisation may also be responsible for the increased toughness.

4.3.6 Recycling and Reprocessing

Agro-based fibres are less brittle and softer than glass fibres and are likely to result in composites that are easier to recycle than mineral-based fibres. Although no post-consumer

Table 4.7 Effect of water on 50% sisal filled PP composite properties [47]					
Properties	Coupling agent (%)	Control		2 hour boil	
		Average	SD	Average	SD
Tensile modulus (GPa)	0	5.5	0.6	4.6	0.1
	1	5.9	0.5	5.1	0.1
	2	6	0.4	5.1	0.3
	4	5.9	0.7	5.8	0.1
Tensile strength (MPa)	0	37.1	1.2	35.3	0.2
	1	64.8	1.5	60.8	0.6
	2	65.7	2	61.1	1.8
	4	67.2	0.6	64.2	0.9
Izod impact notched (J/m)	0	68.2	1.1	81.4	5.9
	1	63.6	3.8	75.2	3
	2	66.6	3	78.2	5.2
	4	62.7	6.4	72.2	5.7
Izod impact un-notched (J/m)	0	130.4	16.7	153.6	32.3
	1	212.9	25.9	243.7	28
	2	210	24	255.7	27.5
	4	203.8	26.5	285.4	34.7
SD: Standard deviation					

based recycling studies have been done on agro-based fibres a short study on the effect of reprocessing has been conducted at the Forest Products Laboratory and the University of Wisconsin-Madison [48]. Experimental details are:

Short kenaf filaments were compounded with PP and MAPP using the thermokinetic mixer described earlier in the text. The blend ratio was 50% kenaf to 49% PP to 1% MAPP, based on dry weight of material. The mixer was operated at 5200 rpm. A total of 2.25 kg (15 batches of 150 g each) of material was blended for the experiment.

All the compounded material was then granulated, dried at 105 °C for four hours and then moulded at 190 °C using an injection moulder. Specimens were randomly selected to evaluate the tensile, flexural, and impact properties and five samples were used for each test: this first set of data was the control or virgin data and is denoted by '0' in **Figures 4.16, 4.17** and **4.18**. All the remaining non-tested specimens were once again granulated, and the injection moulded. Once again five specimens were randomly selected for mechanical property evaluation: this set was labelled as the first recycle data point. This procedure of injection melding and granulation was repeated for a total of nine recycle data points. **Figures 4.16, 4.17** and **4.18** show that the repeated grinding and melting does cause a deterioration of composite properties. However, the unnotched impact strength appears to be little affected by reprocessing. The loss in properties is a combination of repeated fibre attrition and oxidative degradation of the PP through chain scission.

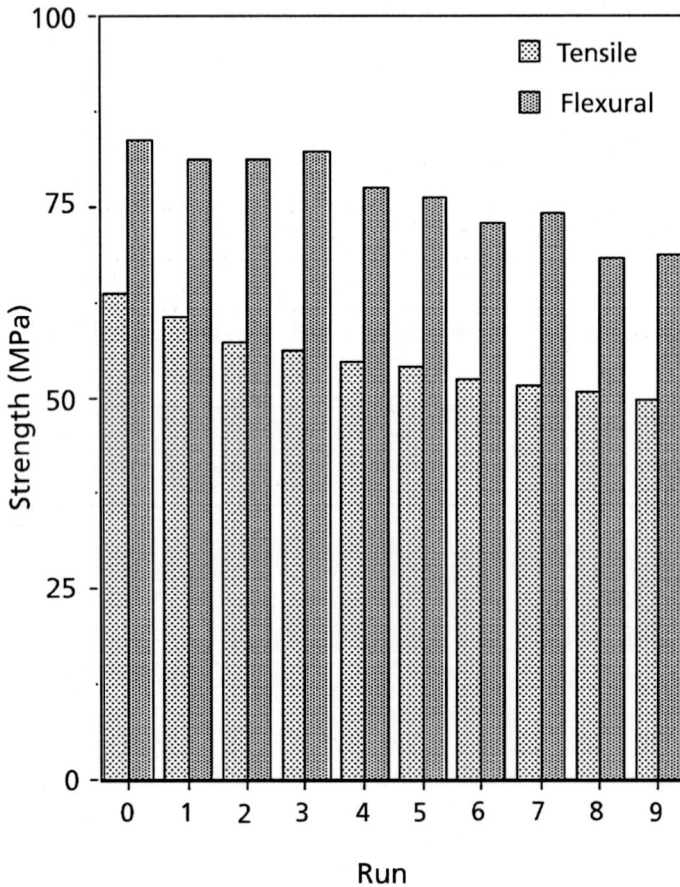

Figure 4.16 Effect of repeated reprocessing on the tensile and flexural strengths of kenaf-PP composites [48]

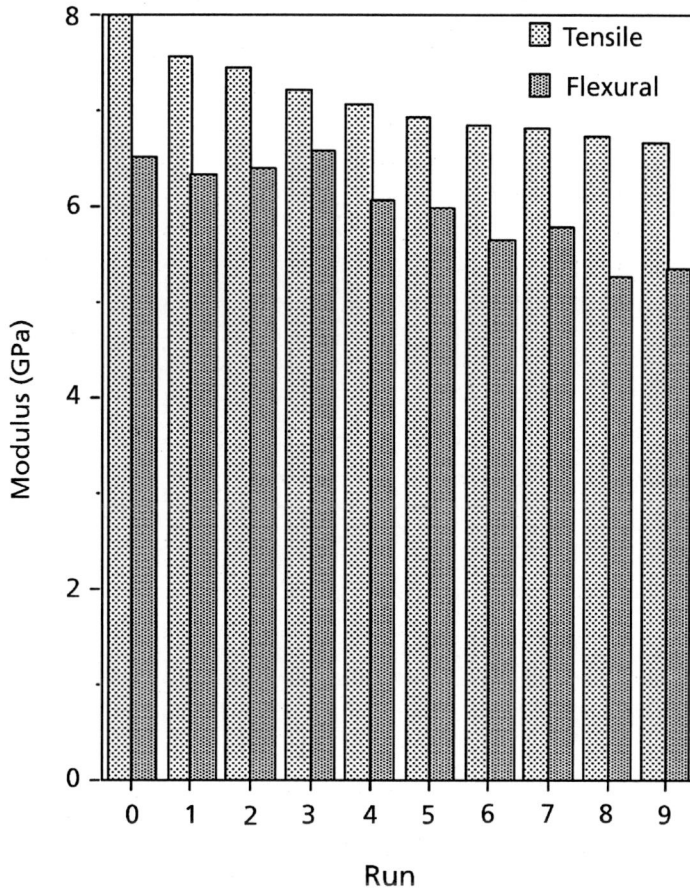

Figure 4.17 Effect of repeated reprocessing on the tensile and flexural moduli of kenaf-PP composites [48]

It is important to point out that these regrinds/reprocessed samples have not undergone any environmental effects such as UV, moisture, load cycling, temperature and other effects. These environmental factors must be accounted for in any real design consideration. This will be discussed briefly in the next section.

4.3.7 Accelerated Environmental Tests

Accelerated weathering of natural fibre filled HDPE was performed by Falk, Lundin and Felton [49, 50]. An Atlas Electric Devices xenon-arc weatherometer was used for the exposure. The irradiance was maintained at 0.35 W/m^2 at 340 nm.

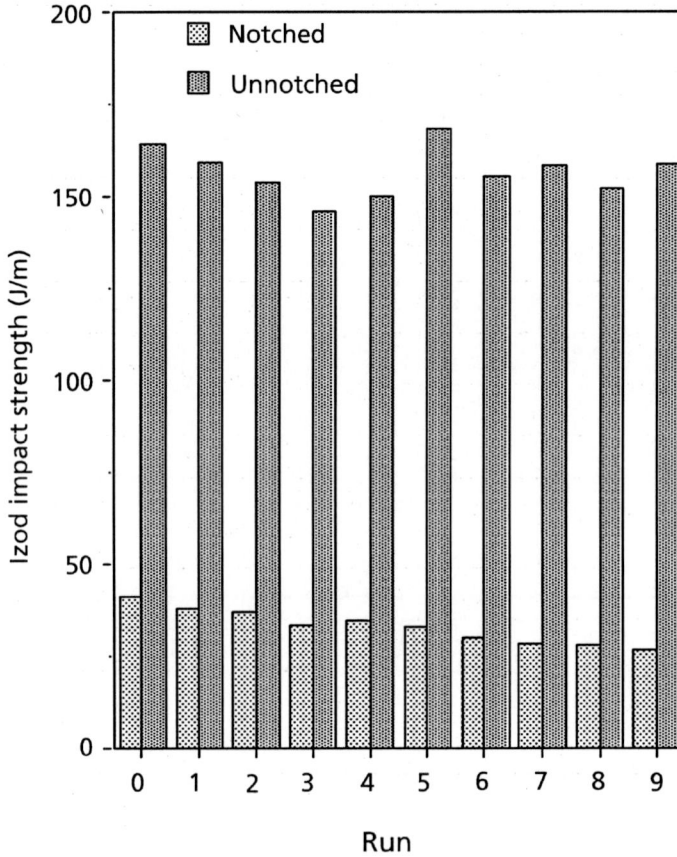

Figure 4.18 Effect of repeated reprocessing on the notched and un-notched Izod impact strength of kenaf/PP composites [48]

HDPE, 50% by weight wood flour-HDPE and 50% by weight of kenaf-HDPE were tested. All composites included a hindered amine UV stabiliser, an antioxidant and a compatibiliser (maleic anhydride grafted PE). **Tables 4.8** and **4.9** show the change in properties with exposure time in hours. It must be noted it is difficult to correlate real life exposure to accelerated tests. However, the study was performed to give an indication of the degradation composite properties with exposure to water and UV light, and thereby the importance of carefully designing these materials for outdoor exposure applications.

References

1. *Handbook of Fillers for Plastics*, Eds., H.S. Katz and J.V. Milewski, Van Nostrand Reinhold, New York, NY, USA, 1987.

Table 4.8 Effect of elastic bending modulus on hours of exposure using a xenon-arc weatherometer that exposes samples to UV and water

	Average modulus of elasticity (GPa) at different exposure times (h)				
	0	500	1000	1500	2000
HDPE	0.817	0.773	0.751	0.744	0.768
HDPE-kenaf (50%)	5.948	4.577	3.596	3.471	3.427
HDPE-wood (50%)	3.951	3.329	2.798	2.650	2.631
Data from Lundin, Falk and Felton [49]					

Table 4.9 Effect of bending strength on hours of exposure using a weatherometer

	Average bending strength (MPa) at different exposure times (h)				
	0	500	1000	1500	2000
HDPE	16.5	16.8	16.0	16.0	16.5
HDPE-kenaf (50%)	48.2	41.3	37.6	37.0	36.5
HDPE-wood (50%)	36.9	32.3	30.8	29.8	29.7
Data from Lundin, Falk and Felton [49]					

2. R. Koslowski, *Personal Communication at the Fifth International Conference on Frontiers of Polymers and Advanced Materials*, Poznan, Poland, 1999.

3. T. Rials and M. Walcott in *Paper and Composites from Agro-Based Resources*, Eds., R. Rowell, R.A. Young and J.K. Rowell, CRC Press, Boca Raton, FL, USA, 1997, 63.

4. A.R. Sanadi, S.V. Prasad and P.K. Rohatgi, *Journal of Materials Science*, 1985, 21, 4299.

5. A.K. Bledski, S. Reihmane and J. Gassan, *Journal of Applied Polymer Science*, 1996, 59, 1329.

6. H.L. Bos and Martien J.A. van der Oever in *Proceedings of the 5th International Conference on Woodfiber-Plastic Composites*, Forest Products Society, Madison, WI, USA, 1999, 79.

7. P. Vilppunen, K. Oksmann, O. Maentausta, E. Keskitalo and J. Sohlo in *Proceedings of the 5th International Conference on Woodfiber-Plastic Composites,* Forest Products Society, Madison, WI, USA, 1999, 309.

8. H.E. Hendrickson and C.N. McCain, inventors; Weyerhaeuser Co., assignee; US 3,577,368, 1971.

9. R.T. Woodhams, G. Thomas and D.K. Rodgers, *Polymer Engineering and Science,* 1984, **24**, 1166.

10. C. Klason and J. Kubat in *Composite Systems from Natural and Synthetic Polymers,* Eds., L. Salmen, A. de Ruvo, J.C. Seferis and E.B. Stark, Elsevier Science, Amsterdam, The Netherlands, 1986.

11. C. Klason and J. Kubat in *Polymer Composites,* Ed., B. Sadlacek, Walter de Gruyter & Co., Berlin, Germany, 1986, 153.

12. G.E. Myers, C.M. Clemons, J.J. Balatinecz and R.T. Woodhams, *Proceedings of SPE Antec '92,* Detroit, MI, 1992, Volume 1, 602.

13. B.V. Kokta, R.G. Raj and C. Daneault, *Polymer-Plastics Technology and Engineering,* 1989, **28**, 247.

14. K.L. Yam, B.K. Gogoi, C.C. Lai and S.E. Selke, *Polymer Engineering and Science,* 1990, **30**, 693.

15. P. Bataille, L. Ricard and S. Sappieha, *Polymer Composites,* 1989, **10**, 103.

16. A.R. Sanadi, R.A. Young, C. Clemons and R.M. Rowell, *Journal of Reinforced Plastics and Composites,* 1994, **13**, 54.

17. A.R. Sanadi, D.F. Caulfield and R.M. Rowell, *Plastics Engineering,* 1994, 50, 27.

18. D.F. Caulfield, R. Jacobson, K. Sears and J. Underwood, *Proceedings of On Global Outlook for Natural Fiber Reinforcements,* Orlando, FL, USA, 2001.

19. R.M. Rowell, A.M. Tillman and R. Simonson, *Journal of Wood Chemistry and Technology,* 1986, **6**, 427.

20. J.V. Milewski, *Polymer Composites,* 1992, **13**, 223.

21. D.M. Bigg, D.F. Hiscock, J.R. Preston and E.J. Bradbury, *Journal of Thermoplastic Composite Materials,* 1988, **1**, 146.

22. L. Czarnecki and J.L. White, *Journal of Applied Polymer Science,* 1980, **25**, 1217.

23. R.G. Raj and B.V. Kokta, *Journal of Applied Polymer Science,* 1989, **38**, 1987.

24. A. Karmarker, *Internal report,* USDA Forest Products Laboratory, 1994.

25. A.R. Sanadi and D.F. Caulfield, in *Third International Symposium on Natural Polymers and Composites (ISNAPOL 2000),* San Pedro, Brazil, 2000.

26. A.R. Sanadi, J.F. Hunt, D.F. Caulfield, G. Kovacsvolgyi and B. Destree in *Proceedings of the 6th International Woodfiber-Plastics Conference,* Forest Products Society, Madison, WI, USA, 2001, 121.

27. H. Dalvag, C. Klason and H-E. Stromvall, *International Journal of Polymeric Materials,* 1985, **11**, 9.

28. A.R. Sanadi, D.F. Caulfield, R.E. Jacobson and R.M. Rowell, *Industrial & Engineering Chemistry Research,* 1995, **34**, 1889.

29. P. Gatenholm, J. Felix, C. Klason and J. Kubat in *Contemporary Topics in Polymer Science,* Volume 7, Eds., J.C. Salamone and J. Riffle, Plenum Press, New York, NY, USA, 1992.

30. *Modern Plastics Encyclopedia,* McGraw Hill, Hightstown, NJ, USA, 1993.

31. *Machine Design,* Materials Guide Issue, 1994.

32. J.H. Felix and P. Gatenholm, *Journal of Applied Polymer Science,* 1993, **50**, 699.

33. A.R. Sanadi and D.F. Caulfield, *Composite Interfaces,* 2000, **7**, 31.

34. M.H.B. Snijder and H.L. Bos, *Proceedings of the 5th International Woodfiber-Plastics Conference,* Forest Products Society, Madison, MI, USA, 1999, 123.

35. ASTM D638-03, *Standard Test Method for Tensile Properties of Plastics,* 2003.

36. ASTM D790-03, *Standard Test Methods for Flexural Properties of Unreinforced and Reinforced Plastics and Electrical Insulating Materials,* 2003.

37. ASTM D256-04, *Standard Test Methods for Determining the Izod Pendalum Impact Resistance of Plastics,* 2004.

38. ASTM D570-98, *Standard Test Method for Water Absorption of Plastics,* 1998.

39. C. Clemons and A.R. Sanadi, in preparation, 2002.

40. R. Jacobson, K. Walz, A.R. Sanadi and D.F. Caulfield, *Effect of Fiber Type on Composite Properties*, Internal USDA Forest Service, Forest Products Report, 1996.

41. K.D. Sears, R.E. Jacobson, D.F. Caulfield and J. Underwood, inventors; The United States of America as represented by the Secretary of Agriculture, assignee; WO 00/21743, 2000.

42. M.H. Naitov, *Plastics Technology*, 1995, **41**, 15.

43. A.R. Sanadi, D.F. Caulfield, N.M. Stark and C. Clemans, Proceedings of the 5th International Conference on Wood-Fiber-Plastics, Forest Products Society, Madison, WI, USA, 1999, 67.

44. P. Jarvela, L. Shucai and P. Jarvela, *Journal of Applied Polymer Sciences*, 1996, **62**, 813.

45. D. Feng, D.F. Caulfield and A.R. Sanadi, *Polymer Composites*, 2001, **22**, 506.

46. R.A. Young, A.R. Sanadi and S. Prabawa, *Proceedings of the 4th International Conference on Woodfiber-Plastic Composites*, Forest Products Society, Madison, WI, USA, 1997, 94.

47. S. Prabawa, *Lignocellulosic Fiber Plastic Composites Using Sisal Fibers from Indonesia*, Department of Forestry, University of Wisconsin-Madison, 1995. [Masters Thesis]

48. K. Walz, A.R. Sanadi and R. Jacobson, *Effect of Reprocessing/Recycling on the Mechanical Properties of Kenaf-PP Composites*, Internal Data, University of Wisconsin-Madison and Forest Products Laboratory.

49. T. Lundin, R.H. Falk and C. Felton in *Proceedings of the 6th International Conference on Woodfiber-Plastic Composites*, 2001, Forest Products Society, Madison, WI, USA, 2001, 87.

50. R.H. Falk, T. Lundin and C. Felton in, *Proceedings of the 2nd Annual Conference on Durability and Disaster Mitigation in Wood-Frame Housing*, Forest Products Society, Madison, WI, USA, 2000, 175.

5 Manufacturing Technologies for Biopolymers

Nick Tucker

5.1 Introduction

In this chapter, the processes used to manufacture articles from biopolymeric materials will be discussed. Some of these methods are specific to biopolymers, for example, the Shimada press, used to produce lignin bound wood fibre extrusions mostly used for fuel briquettes. Some are fundamental raw material production processes such as fibre spinning, and some, such as the Davis-Standard 'woodtruder' are normally used with fossil polymers, but having a clear potential for use with biopolymers.

The methods under discussion are all based on established polymer manufacturing techniques, but the control and application of these methods must be varied to cope with certain factors associated with exploiting the advantages of biopolymers. For a biomass origin polymer to be used to full advantage, it must be biodegradable at the end of its useful life. This means that enough of its molecular structure must retain enough similarity to a biological structure to enable bacteria and fungi to digest it. The conditions under which biological structures can survive are limited in terms of temperature (degradation starts with the denaturing of the tertiary structure (the way a protein molecule is 'coiled') of proteins at about 45 °C: to consider the effect of this, think about what happens to the white of an egg during cooking. In terms of the kitchen, the process continues through caramelisation to incineration resulting in an inedible (non-biodegradable) materials that is largely carbon. The temperatures reached during polymer processing exceed the range found in the kitchen by up to 200 °C, therefore some care may need to be taken not to entirely destroy these structures. However some manufacturing processes exploit the thermal alteration of biopolymers, for example the manufacture of packing peanuts (chippings formerly made from expanded polystyrene used for packing) from starch-based polymers (*cf.* popcorn) by melting moist amylose starch under pressure. The pressure is then released, and the water boils, foaming the structure into that familiar to us as, for example, in prawn crackers.

In solidification to produce thermosetting polymers, the chemical 'curing' that occurs will also affect the biodegradability of the finished article. Since the crosslinking reaction leads to an inherently more stable chemical structure (showing for example, better high temperature performance, increased stiffness, less water absorption), it can follow that degradation by microorganisms will also be reduced.

The chapter also considers the use of admixtures. Commercial polymer formulations are rarely consist of only the simple polymer – ingredients are added to improve processing or properties of the finished article. For biopolymers, currently expensive due to novelty and low production volumes, the use of fillers and extenders to reduce raw material cost is of particular interest.

5.2 Manufacturing Methods

The manufacturing routes using biopolymers show some fundamental similarities (see **Figure 5.1** and **Figure 5.2**) with the principal differences depending on whether a thermosetting or thermoplastic biopolymer is being processed.

Thermoplastic materials are melted, and then moulded into the shape of the finished article. The melt is then frozen and the manufacture of the article is complete. This process can be repeated, but not indefinitely, since even though the raw material is only melted (that is, only a physical change of state happens), damage can still be done to the long polymer molecules that make up the polymer melt. Subjecting the melt to energetic mixing (the high shear conditions experience when pumping or stirring, can break the molecular chains. A similar process of chain breakage is part of the process of degrading

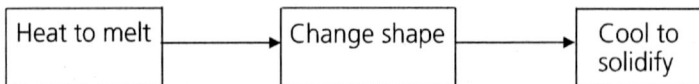

Figure 5.1 Thermoplastic manufacturing cycle

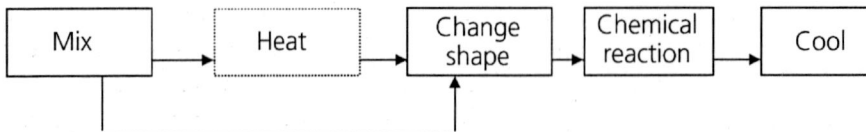

Figure 5.2 Thermoset manufacturing cycle

the engine oil in a motor-car. The net effect of this to the manufacturer is an unquantified decline in the mechanical properties of the solid polymer, as well as an unquantified change in the nature of the flow properties (rheology) of the polymer melt. In terms of molecular chain length, the chains are on average shorter, and variation of these molecular lengths has become larger. The former leads to a reduction in mechanical properties (tensile strength and stiffness), and the latter to some unpredictable changes in flow properties. This results in difficulties in controlling the manufacturing process, unless steps are taken to monitor the change in material properties and to control the amount of recycled material added to the manufacturing process. Due to the temperature sensitive nature of biopolymers there is usually a strict limitation in the practical amount of in-process recycling. For example: Novamont quote a maximum scrap-recycling rate of between 10% and 20% for Y101U grade Mater-Bi [1].

It should be understood that this process of degradation is not limited to biopolymers. Degradation during processing presents a particular problem in the use of recycled thermoplastic polymers (leaving aside the unpleasant nature of sorting through rubbish, and the financial and energy costs of such operations) because the provenance and composition of the materials is unknown. Biopolymers have the potential to avoid this problem, by being recovered via composting: composting neatly sidesteps the need to separate waste streams into very specific polymer types to avoid problems with compatibility.

The route for manufacturing thermoset articles is significantly different; thermosets are not reacted to form polymers until they have been formed into their final shape. The unreacted thermosets still consist of short chain molecules and hence are much less viscous than thermoplastic polymer melts. This opens up a number of thermoset specific production routes. It is possible to inject the low viscosity unreacted liquid thermosets through fibre reinforcement to produce large long-fibre composite articles, e.g., boat hulls.

5.2.1 Spinning and Fibre Production

Bio-origin biodegradable natural fibres are available either directly from nature or by spinning from liquefied precursors (usually cellulosic). The production of fibre directly from crop materials such as flax, hemp, and kenaf, will not be examined here, except to note that the current state-of-the-art in the production of these materials, are rooted in technologies that date from the dawn of civilisation. The production route is typically:

- Retting - an initial stage of fungal attack on the harvested crop as it lies in the field – the degree of separation is heavily dependent on the weather.

- Ginning - the separation of fibres from seeds.

- Scutching - decorticating by bending or crushing the stalks.

- Hackling - to orientate the fibres.

And then conversion into a fabric by either:

- Needling - to produce non-woven felts.

- Spinning – converting the fibres into yarn, and hence to a woven or laid fabric.

The materials produced tend to be either very high cost, such as linen, or very low cost such as soil stabilising fabrics. It is likely that this process route can be better optimised for the production of a technical grade of reinforcing fibres for sustainable polymer matrix composites.

Spinning of fibres from liquid precursors is a significant and mature use of biopolymers; the most well known spun fibre is rayon. The use of cellulose as a raw material for artificial fibres dates from the late 1800s, and the main route of production is the viscose process. This process is based on treating the cellulose in alkaline solution and reacting it to form cellulose xanthate. The xanthate is dissolved in dilute caustic soda sodium hydroxide solution. The solution is then forced through a spinneret into an acidic (H_2SO_4 and sodium sulfate) 'regenerating' bath. In this bath, the xanthate is regenerated to cellulose and a controlled degree of water removal occurs. In the next stage the fibre is stretched to align the cellulose molecules, and to maximise mechanical properties. Viscose rayon is the material that created the market for artificial fabrics. However, figures quoted by Woodings [2] suggests that against a background of increasing use of fibre non-wovens, the use of these established biodegradable fibres is declining, in favour of cheaper non-biodegradables that are perceived as being easier to process and more versatile. Since the advent of commercial availability of fibres made from fossil origin polymers in the 1940s their market share has risen to about half of the 45 million tonnes world annual consumption of fibres.

One reason for the high cost of viscose rayon is that the traditional routes for rayon manufacture use considerable quantities of water. Turbak [3] says that the viscose process consumes 420-750 litres of water per kilogramme of rayon produced, with an additional 8-10 times that amount being required for plant supplementary service facilities. However, new methods of rayon manufacture are being devised such as the N-methylmorpholine N oxide (NMMO), (Lyocell) process. This process is a wet spinning process that is faster than viscose production, and it is claimed by Courtaulds that the process is completely closed loop (all component materials are recycled). Neste Oy in Finland are developing the so-called CC process, which uses ammonia and urea to make cellulose carbamate. The carbamate is soluble in caustic soda solution, and is spun from this solution. The Celsol process uses an enzyme-based method of activating cellulose to make it more reactive and directly alkali soluble.

More recent spun biopolymer fibres are discussed by Woodings [2], it should be noted that formulations of these materials are also used as feedstocks for other manufacturing methods.

Monsanto's Biopol™ (originally developed by ICI Zeneca in the UK) who planned a 10,000 tonnes per year plant in 1987) was based on a random copolymer of 3-hydroxybutyrate and 3-hydroxyvalerate made by bacterial fermentation. Monsanto have since ceased production of this material due to difficulties in competing on price with established commodity polymers. Polycaprolactone has been used in blend with other plastics to make biodegradable films since the late 1970s. Bayer's BAK™, a polyester amide polymer is based on hexamethylene diamene, butane diol and adipic acid. Butane diol is also the basis for Bionolle™, Showa Polymer's biodegradable synthetic. Eastman's EASTAR BIO™ is a copolyester based on terephthalic acid and ethylene glycol. Dupont's Biomax™ is said to be based on three proprietary aliphatic polyesters and to cost only slightly more than conventional polyethylene terephthalate (PET).

These fibres are used either in clothing, medical applications such as sutures, or nappies or sanitary protection. It is anticipated that these biopolymer fabrics will also be used as reinforcements for sustainable composite articles.

5.2.2 Extrusion and Compounding

The extrusion process originated as a biopolymer processing method, being industrialised by pasta makers in the late 1800s. Extrusion is now the pre-eminent polymer processing method, it is used to mix the components (called compounding) of the pellets that are used by other thermoplastic manufacturing processes as well as producing sheets, films and profiles, and continuous geometries, e.g., pipes. It has therefore been extensively studied and described [4].

The material is loaded into a hopper and then passes through the extruder throat into the extruder barrel. Here, the plastic is conveyed by the turning screw, which transports the material forward towards the front end and die of the machine (see **Figure 5.1**). The mechanical energy put into moving the material against the screw and barrel produces frictional heat; this is sufficient to melt the plastic. By the time the material has moved into the final third of the barrel all the material has melted. In a typical single screw

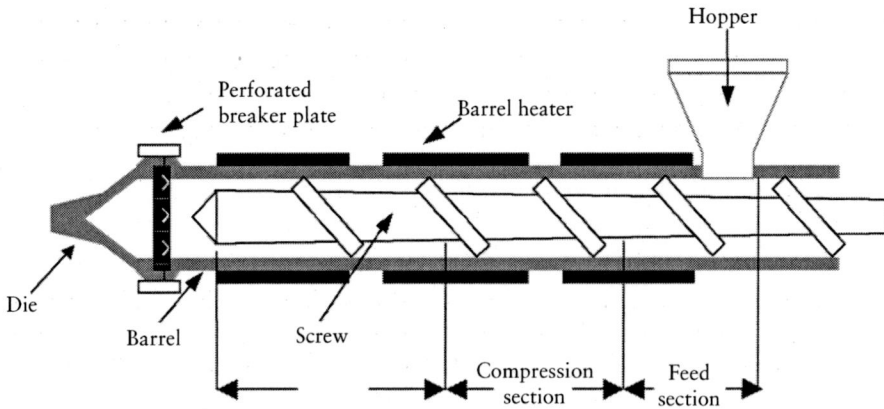

Figure 5.3 Illustration of a single-screw plasticating extruder

Adapted from McCrum [4]

extruder this mechanical activity (shear) produces around 90% of the energy required to melt the plastic, the remainder is supplied by heater bands placed on the outside of the barrel. These serve mainly to start the process and insulate the screw from excessive heat loss. When the plastic reaches the die, the movement of the screw and the material behind it forces it out. The shape of the die dictates the shape of the extrudate. The extrudate can be pulled away as required to form the product, and the molten material is usually cooled by immersion in water. The screw geometry is specific to the type of material being processed, but the tapered shape of the screw is typical to most geometries. The feedstock is therefore subjected to continual and increasing compression, with the barrel only running completely full at the metering section. The effect of this is that volatile components of the melt have some opportunity to escape, this process may be assisted by putting extraction ports along the compression section; typical volatiles could be moisture, or waxy components of the mix, either inherent or added for example, to improve flow characteristics.

The most established use of the extruder to produce low environmental impact polymers (LEIP) products is the manufacture of wood fibre or flour polymer composites. The driver for this seems to be a perception that there will be a shortage of timber, in particular, continuous geometries for the building industry. For a full discussion of the use of wood flour and fibre loaded polymers the reader is referred to Chapter 7. In processing terms, Bledzki, Reihmane and Gassan [5], reviewed the use of thermoplastics reinforced with wood fillers, the conclusions to this work were that the major problems associated with the use of wood fillers are high levels of moisture absorption and the fact that untreated wood fibres are inherently difficult to couple to the polymer matrix, hence composites with a low tensile strength result. Schut [6], says that wood does not adhere well to certain polymers, as it is a polar material. The cohesion of molecules within a material is related to the polarity of a material, and also to dipole forces that are present in some materials [7]. This results in material having poor strength and stiffness as the load will not be shared between the multiple components of the material.

Equally, some polymers are chemically non-polar [8], and it may be necessary to modify the plastic to develop strong bonds with the polar fibres. The coupling of the fibre to the matrix can be achieved through the use of a compatibiliser such as maleic anhydride. Dicumyl peroxide or chlorosulfonated polyethylene are alternatives [9].

The fossil origin polymers that are usually the material of choice for the matrix at the time of writing, are chosen for reasons of cost rather than any intrinsic difficulty in processing LEIP polymers. The main technical difficulty with extrusion of wood fibre-polymer mixtures is the removal of the moisture from the wood. Excessive moisture in raw materials can lead to problems such as surface defects in the extrudate and strand breakage, although the use of water as a foaming agent in wood polymer composites has been reported as an academic exercise. The industrial user is therefore faced with the prospect of drying the fibre content before compounding, or using a technology able to cope with the high moisture levels in the raw materials.

The lowest capital cost route to wood fibre polymer composite production is a conventional 24-30:1 single-screw single-stage extruder. The disadvantage of this method is that extensive pre-drying of the wood fibre is required, and the relatively high shear stresses and wide distribution residence time in the extruder can lead to thermal damage of the fibres.

A single screw extruder designed to cope with moist wood fibre feedstock is the Shimada press [10]. The Shimada press is a short barrel extruder that takes hard or softwood at a moisture content of 8 wt%, an average particle size of 2-6 mm and a bulk density of approximately 200 kg m^3. After extrusion the moisture content is reduced to 4 wt% and the bulk density has increased to approximately 1400 kg m^3 (manufacturers figures).

This method of extrusion is designed to produce a very competent material bound together by the lignin present in the feedstock. Strand breakage is not an issue as the product is parted into short lengths for use as fuel briquettes, but it is the author's opinion that with further development it may be suitable for article manufacture.

If twin-screw technology is used, a 28:1 counter rotating conical screw machine is preferred. An example of this is the Cincinnati company's Fiberex process [11], where the manufacturer claims up 90% fibre loadings are achievable. The screws taper from a maximum at the feed end to a minimum at the die end. This geometry gives a low heat and shear history to the extrudate, particularly suited to the fibrous wood material and biopolymers under discussion. The moisture level of the feedstocks should be <1%, and devolatilisation ports are used between the compression zone and the metering zone of the extruder.

The Standard-Davis Woodtruder process [12], adds more sophistication to the process by using two extruders. The larger of the two is a 40:1 co-rotating twin-screw that takes the wood fibre, and dries it using vents to atmosphere before the polymer content is added through a side port via a 10:1 single-screw extruder, where further devolatilisation takes place via a vacuum port. The manufacturers claim 80% wood fibre loading.

5.2.3 Injection Moulding

Injection moulding is used for high production volume complex shapes [13].

The process is one of the most important methods for converting thermoplastics from pellet or powder form to a wide range of articles such as forks, beakers, or computer cabinets. The high pressure used in injecting polymer melts, and the high precision of manufacture of the moulds, limit this process to high production volumes. However, the rapidity and ease of manufacture of complex shapes associated with this method ensure its popularity.

The process involves heating the powder or pellets by extrusion, until it melts. The melt is then injected by the forward movement of the screw into a cooled mould and held under pressure until it returns to the solid state.

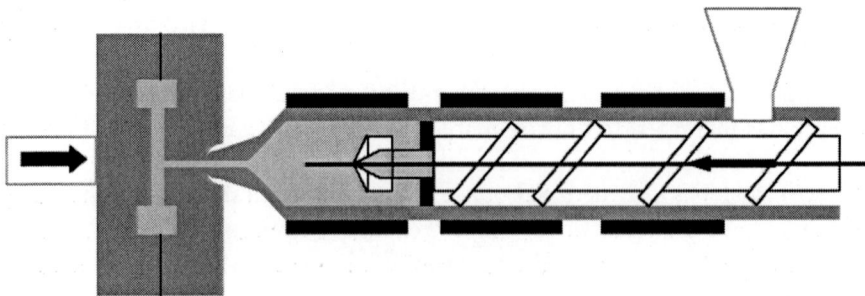

Figure 5.4 Reciprocating screw injection moulding machine

Adapted from: Crawford (1999) [14]

The cycle of operation is as follows:

- As the screw rotates, material is conveyed along the barrel by an Archimedean action, and is heated in a uniform fashion by a combination of conduction from the barrel and mechanical shearing action (note the similarities with extrusion). The first section of the screw is called the feed section. Usually melting begins about half-way down the length of the screw. The depth of the screw flights decreases at this point to begin compressing the melt. This is the start of the compression section. Eventually the material has completely melted and reaches the point where minimum flight depth occurs. This is the beginning of the metering section. The metering section has screw flights of constant depth. The melt is sheared to give a melt of uniform composition and temperature. This section is acting as a constant volume-metering pump. The metering section discharges material into the barrel cavity in front of the screw. At this stage in the operation the screw is being withdrawn down the barrel giving an increasing volume of homogenous plastic melt in front of it.

- The screw now stops rotating, and is driven forward, expelling the polymer melt into the mould cavity. Pressure is held at injection levels until the polymer melt begins to freeze – this reduces the effects of shrinkage upon solidification. A non-return valve prevents melt from being driven back up the barrel. Pressures in the barrel of 60 MPa are not uncommon at this point.

The cycle is then repeated.

5.2.4 Thermoset Injection Moulding

As with thermoplastics, the process is used to make high production volume, complex shapes. Care must be taken (by control of residence times and temperatures) to avoid premature cure of the resin within the injector.

5.2.5 Film Blowing

Is another extension of the extrusion process, used to manufacture of films for thin sheet materials [15]. A tubular extrusion die is used, with the addition of a high volume air supply leading into the centre of the annulus. The extruded tube can then be inflated to produce thin polymer film. As the bubble of film freezes off, it is gathered together and collected on rollers. Periodic weighing of the rolls of product stock enables the operator to control the thickness of the film. The critical value for film blowing is the blow-up ratio: the ratio of the final bubble diameter to the die diameter.

5.2.6 Calendering and Coating

This process is used to produce sheet stock production of polymers and polymer-coated products [16]. Polymer is extruded into an arrangement of rollers (sometimes called 'bowls') somewhat resembling in form and function, a pair of old fashioned mangles. The arrangement provides a staged reduction in film thickness. Product thickness is greater than that produced by film blowing. Calendering also allows the polymer to be surface coated onto a substrate. An example of this is the application of a polymer coating to paper to produce a waterproof feedstock for the manufacture of packaging material. The rollers can also be textured to impart surface relief to the product.

5.2.7 Blow Moulding

Blow moulding is used to make hollow plastic parts such as fuel tanks and bottles [17]. Initially it was used almost exclusively for the production of small plastic containers such as bleach bottles. However advances in blow moulding technology has enabled the production of articles such as surfboards, cool boxes, toys and automotive parts, for example: spoilers and hoses. There now exist two distinct processes: extrusion blow moulding and injection blow moulding.

- **Extrusion blow moulding**

Extrusion blow moulding uses an extruder to form a tubular extrudate called a parison. Before this polymer can completely solidify it is trapped between the mould plates and sealed. A blow pin is inserted which forces air into the parison and inflates it against the sides of the mould. After the material cools the mould opens and the moulding is ejected. One disadvantage of extrusion blow moulding is that flash is produced on both the neck and base of the container. This is necessary since the parison must be longer than the mould to enable neck formation and for the base to seal. Generally some of this scrap is recycled back into the system. Articles such as bleach bottles and fuel tanks are made by this method.

- **Injection blow moulding**

Injection blow moulding, as the name implies, uses injection moulding to produce the parison. By first producing an injection moulded parison, this process offers advantages on extrusion blow moulding, for example the use of an injection moulded parison allows the formation of precision, gas-tight seals, e.g., carbonated beverage bottles, and no flash is formed during the blow moulding operation.

5.2.8 Thermoforming

Thermoforming is a process that shapes a heated thermoplastic sheet by the application of vacuum or pressure [18].

It is one of the least expensive polymer processes as the equipment required is relatively simple in both design and operation. The material is placed in a clamp frame to hold it securely on all edges. The material is then heated to soften it, usually by the action of convection and radiant heat from electrical heating elements. Heating the top, bottom or sometimes both sides of the film surface can do this. Once the sheet/film is sufficiently softened, air pressure is applied from underneath to stretch the sheet, the mould tool is then moved up under the stretched sheet. This part of the process ensures that the sheet stretches evenly, and does not catch and tear during forming. A vacuum is applied which facilitates the drawing of the material towards the mould by atmospheric pressure. Male or female tools can be used and sometimes matched moulds or plug assists are used to help the forming process. A variety of products can be produced ranging from plastic cups and signs to complex refrigerator liners. Thermoforming tools are simple in both design and operation. Most production tools are made of aluminium and have a number of small holes machined in them to provide the vacuum or allow the passage of air through the tool as required in the operation of the process.

5.2.9 Compression Moulding

Compression moulding is used for the manufacture of simple geometries, at medium production volumes. A hydraulic press is used to close a mould or tool to form the shape of the manufactured article. This method is mainly used for thermoset polymers and rubbers (the tool is heated to initiated the chemical crosslinking reaction) but is gaining considerable popularity as a method of manufacturing thermoplastic composite articles - the process is known as GMT or glass mat thermoplastics. Note that in this case the glass and plastic are heated before being placed in the cold mould cavity.

5.2.10 Pultrusion

Pultrusion is used to make continuous constant geometry composite shapes [19]. Thermoset resin impregnated bundles of fibre are pulled through a heated die, curing the material and producing the rigid profile. Rowell [20] describes the use of jute rovings and woven cloth being pultruded with a phenolic matrix into structural components such as doorframes and ceiling panels.

Pultrusion can also be used with thermoplastic matrices. Commingled tows of thermoplastic fibres are blended with the reinforcing fibres. The fibres are then heated in an oven before being pulled through a chilled consolidation die. Van de Velde [21] describes this process being applied to flax fibres.

5.2.11 RTM (Resin Transfer Moulding) and RIM (Reaction Injection Moulding)

These processes are related, with RTM being used for low to medium volume composite mouldings and SRIM for high volume composite and polymer mouldings. Both methods inject thermosetting resin systems into closed moulds. A closed mould is a mould with a cavity - similar in principal to those used in thermal injection moulding. The injection pressures are much lower than the reciprocating screw injection methods described earlier, and the methods have the advantage of isolating the operator from the unreacted thermoset polymer systems.

RTM was developed for the production of high precision (wall thickness, dimensional tolerance and material uniformity) radomes – the fairings used to cover radar scanners in aeroplanes [22]. The first attempts to make radomes were by single-sided contact moulding techniques in the 1940s with RTM becoming the established technique in the following decade. RTM is, therefore, a useful method for the production of closed mould composite articles [22]. RTM machines usually inject long pot life (slow reacting) resin at low pressures (less than 10 MPa) and slow speeds, producing high quality mouldings. This resin behaviour is typical of low functionality, bio-origin, epoxy resins, and these materials may be particularly suited to this method. The low injection pressures mean that tooling can be composite, and consequently the whole process has a low capital cost.

Reaction injection moulding (RIM) [23] is a faster more sophisticated version of RTM. RIM resin systems have two components that are mixed and then co-react, usually with the addition of heat, to form a solid cured resin. RIM technology was developed for the polyurethane (PU) industry. It is therefore likely that very fast cure time PU or epoxy systems of biological origin can be developed. The fast reaction times (approximately 30 seconds for polyurethanes) necessitate the two components being kept separate until just before injection. Injection pressures are in the order of 15 MPa. Low injection pressures (*cf.* Reciprocating screw injection moulding) allow lower tooling costs. Further developments of RIM include reinforced reaction injection moulding (RRIM). In RRIM short (approximately 10 mm) fibres are mixed with the resin to form a short fibre reinforced component. RIM technology can be used to inject over preplaced reinforcement (the so called 'preform'). This technique is known as structural reaction injection moulding (SRIM).

Disadvantages of RIM are the high capital cost of plant (equivalent to thermal injection moulders) and higher cost of tooling - higher than RTM, but not as high as thermal injection moulding

5.3 Processing Conditions

Cogswell [24] suggests that a typical thermoplastic process will last from 3-5 min from the addition of the granules to the machine to the emergence of a solid product at the other end. This suggests that due to the distribution of residence times inherent in the process, about 5% of the material will be exposed to the high shear and temperature conditions within the processing unit for a duration in the order of 10 min, and smaller quantities in the upper range of the time distribution for much longer than that. This assumes that no reclaim (sometimes called in-process recyclate) is added to the melt. Process scale is also an important consideration: for small screw diameter extruders (in the region of 25 mm diameter) the melt temperature may well be maintained within 5 °C of the set barrel temperature and 10 °C of the polymer melt temperature (this difference presumably being because the element of heat imparted to the melt by the mechanical work done on it by the screw elements (typically 90%) is not under the direct control of the operator). In larger scale operations the mean melt temperature may well be in the region of 30 °C above the melting point and in localised hot spots 20 °C above that.

Cogswell [24] also asserts that the conditions in intermittent operation processes such as injection moulding are least damaging to polymer melts, and most problematic in continuous processes such as extrusion, particularly in processes where the extrudate is stretched, such as film blowing. The limiting factors for processing conditions for biopolymers are the same as for fossil origin materials: degradation at the upper limits of temperature and shear, and lack of homogeneity at the lower limits. However, these limits are somewhat more tightly drawn at the upper limits for biopolymers. The results of exceeding these upper limits are degradation of the polymer, resulting in moulding defects such as weld lines, discolouration or a strong odour in the final product.

5.4 Additives or Admixtures

In addressing the problems mentioned in the previous section, when processing a fossil origin polymer melt, the operator has the twin advantages of not be bound by strictures of environmental impact, and having access to a wide range of processing aids and property improvers. These additives are usually added in small quantities, and with the exception of the biocides, are unlikely to have any effect on the gross biodegradation of the polymer.

However, the selection of additives should be undertaken with care to avoid compromising the biodegradability certification of the finished article.

Harper [25] lists the following as important:

- Plasticisers

- Fillers

- Flame retardants

- Lubricants

- Colorants

- Blowing (foaming) agents

- Crosslinkers

- Biocides and antimicrobials

5.4.1 Plasticisers

Plasticisers increase the flexibility of the polymer product and also incidentally reduce the viscosity of the polymer melt, a role also played by lubricants. In terms of biological origin materials, epoxidised soyabean oil is used as a plasticiser and heat stabiliser in polyvinyl chloride (PVC) production. Epoxidised linseed oil and tall oil are also used as PVC plasticisers. It may be that these materials will also find application in biopolymers. Recent work by Brooks and co-workers [26], and others describe the use of liquid CO_2 (sometimes called supercritical CO_2) injected into the melt. At atmospheric pressure, liquid CO_2 will boil at −60 °C. However, if the CO_2 is kept under high enough pressure, the CO_2 can be heated up to polymer processing temperatures whilst still remaining liquid. Under these conditions, the CO_2 exhibits properties of both a liquid and a gas, and is described as supercritical. If supercritical CO_2 is mixed into the polymer melt, it insinuates into the interstices between the long chain polymer molecules, and provides internal lubrication resulting in a lowering of the apparent viscosity of the melt, and lowering of melt temperature. This requires the addition of a mixing stage to the end of the extruder or moulder barrel, but significant reductions in melt viscosity and melt temperature are reported. It should however be noted that the plasticising effect is transient. The CO_2 diffuses out of the solid product over a period of hours after solidification, and after this aging process the product does not show a long-term increase in flexibility.

5.4.2 Fillers

Fillers are the most commonly used additives, see also Chapter 7. Fine mineral powder fillers are added as nucleating agents in small (~1%) quantities to limit the size of crystalline structures. This is of limited application with biopolymers as the complexity of the molecules in most biopolymers limits the degree of crystalline structure formation. The DIN CERTCO Certification scheme: products made of compostable materials, 3rd revision, Berlin 2001 (available at: http://www.ibaw.org/eng/downloads/BAW_certification_engl.doc). lists a number of mineral fillers as certified compostable up to levels of 49%.

5.4.3 Flame Retardants

The chemicals used to reduce the flammability of polymers are often chlorinated and brominated compounds (sometimes with antimony-based synergists) used as flame quenchers, or phosphorus compounds used to improve char strength. High char strength is desirable because it means that burnt material does not fall away and expose fresh unburnt surfaces, hence the material tends to self-extinguish. These are regarded as undesirable due to perceptions of their toxicity. Alumina trihydrate ($Al_2O_3.5H_2O$) releases its water of hydration when subjected to heat, and hence limits the propagation of combustion. However it is added in large amounts to the polymer (50-60%), and may, at the upper limits of addition compromise the biodegradability of the article.

5.4.4 Lubricants

Internal lubricants perform a similar role to that of plasticisers – residing in solution within the polymer, and easing the flow of the long chain molecules relative to each other, hence reducing the apparent viscosity of the melt. When the polymer has solidified into a product, the presence of a lubricant can still make itself felt, as the polymer may exhibit greater flexibility than the unlubricated equivalent. External lubricants include materials that function as mould release agents, the difference in operation being that they migrate to the outer surface of the moulding.

The obvious sustainable origin mould release agents are waxes such as bees wax and Carnauba wax and rapeseed oil. Care must be taken to match the boiling point of these materials to the processing temperature of the polymer melt. These materials can also be effective topical mould release agents, applied directly to the mould surface. The operator must take care to avoid overdosing or underdosing the mould surface. Overdosing leading to a greasy surface of the moulding, and underdosing to difficulties in parting the moulding from the mould surface.

5.4.5 Colorants

Some colourants may be objected to due to inherent toxicity, but environmentally acceptable substitutes of either mineral or vegetable origin are increasingly available. DIN CERTCO promote mineral colorants, certifying as compostable: carbon black, iron oxide, graphite and titanium dioxide up to levels of 49%. Vegetable-based colorants are also available, but there may be issues of resistance to fading associated with the use of these materials.

5.4.6 Blowing (Foaming) Agents

Foaming agents are either liquids, or solids that degrade into gaseous products at temperatures below the processing temperature of the polymer. When boiling occurs, the melt foams, and this structure can be captured as the melt freezes, or cures into a solid. For polyurethane systems, water can be used as a foaming agent exactly as in fossil formulations. Directly injected liquid CO_2 can also be used as a foaming agent, as well a plasticiser. On releasing the constraining pressure, the CO_2 boils, causing the liquid polymer to foam. This typically happens as the melt exits an extrusion die or as it enters a mould cavity.

5.4.7 Crosslinkers

At the moment, crosslinkers or curing agents are usually the same formulations as those used in conventional fossil origin thermoset formulations. Thus articles made from bio-origin thermosets may not be of 100% biological origin, an example of this is the use of plant oil triglyceride origin polyols reacted with non-sustainable origin isocyanates to form polyurethanes. Compare and contrast with the phenolic resin obtained from the shell of the cashew nut (often referred to as cashew nut shell liquid or CNSL). This resin can be reacted with wood alcohol origin formaldehyde to produce a phenolic moulding that is of 100% plant origin. It should however be noted that even though these materials are of sustainable origin this does not mean they are harmless. By definition, thermosetting systems must contain small reactive molecules. This means that the materials are likely to be at best pungent smelling, and possibly harmful if inhaled. Workers must therefore be properly equipped and trained to deal with sustainable thermosets in the same way as with fossil origin equivalents.

5.4.8 Biocides and Antimicrobials

These additives kill or limit the growth of microorganisms, and as such may be useful in extending the life of a biodegradable polymer article. However, the wholesale use of these materials will to lead the evolution of resistant strains of bacteria, fungi and viruses.

Ochs [27] provides a comprehensive list of biocides, and notes the following properties as desirable:

• Low toxicity to higher organisms and the environment

• Easy application (by incorporation into the polymer)

• No negative impact on the properties and appearance of the article

• Storage stability and long lasting efficacy

The method of application of biocides is to mix them into the resin formulation. Over the life of the product, the agent diffuses out on to the surface of the moulding, preventing surface attack by microbes. Once the biocide has leached away, biodegradation starts as normal.

References

1. *Mater-Bi Grade: Y101U Product Datasheet*, Novamont, Novara, Italy, 1998.

2. C.R. Woodings, *Proceedings of the EDANA Technical Conference*, Prague, Czechoslovakia, 2000.

3. A. Turbak in *Concise Encyclopedia of Polymer Science and Engineering*, Ed., J.I. Kroschwitz, John Wiley & Sons, New York, NY, USA, 1998, p.960-962.

4. N.G. McCrum, C.P. Buckley and C.B. Bucknall, *Principles of Polymer Engineering*, 2nd Edition, Oxford University Press, Oxford, UK, 1997.

5. A.K. Bledzki, S. Reihmane and J. Gassan, *Polymer Plastics Technology and Engineering*, 1998, 37, 4, 451.

6. J.H. Schut, *Plastics Technology*, 1999, 45, 3, 46

7. J.A. Brydson, *Plastics Materials*, 6th Edition, Butterworth-Heinemann, Oxford, UK, 1995.

8. A.R. Sanadi, R.M. Rowell and D.F. Caulfield, *Polymer News*, 1996, **21**, 1, 7.

9. D. Maldas and B.V. Kokta, *Journal of Thermoplastic Composite Materials*, 1995, 8, 4, 420.

10. Shimada (2003), manufacturers information from *www.shimada.co.uk*

11. Cincinnati (2003), manufacturers figures from *www.sms-k.com*.

12. D.J. Gardner, *Wood-plastic Composite Extrusion – Processing Systems*, AEWC Center, University of Maine, Orona, ME, USA, 2003, www.aewc.umaine.edu/equipment/woodtruder.htm

13. A. Whelan, *Injection Moulding Machines*, Elsevier Applied Science Publishers, London, UK, 1984.

14. R.J. Crawford, *Plastics Engineering*, 3rd Edition, Butterworth-Heinemann, Oxford, 1998.

15. N.J. Mills, *Plastics – Microstructure and Applications*, 2nd Edition, Edward Arnold, London, UK, 1986.

16. P.C. Powell and A.J.I. Housz, *Engineering with Polymers*, 2nd Edition, Stanley Thornes Publishers Ltd, Cheltenham, UK, 1998.

17. J.M. Charrier, *Polymeric Materials and Processing: Plastics Elastomers, and Composites*, Hanser, New York, NY, USA, 1990.

18. *Concise Encyclopedia of Polymer Science and Engineering*, Ed., J.I. Kroschwitz, Wiley Interscience, New York, NY, USA, 1998.

19. *Pultrusion for Engineers*, Ed., T.F. Starr, Woodhead Publishing Limited, Cambridge, UK, 2000.

20. R.M. Rowell in the *Proceedings of the Natural Fibres Performance Forum*, The Royal Veterinary and Agricultural University, Copenhagen, Denmark, 1999.

21. K. Van de Velde in the *Proceedings of the Fifth World Pultrusion Conference*, Berlin, Germany, 2000.

22. C.D. Rudd, K.N. Kendall, C.G.E. Mangin and A.C. Long, *Liquid Moulding Technologies – Resin Transfer Moulding, Structural Reaction Injection Moulding and Related Processing Techniques*, Woodhead Publishing Ltd., Cambridge, UK, 1997.

23. C.W. Macosko, *RIM: Fundamentals of Reaction Injection Moulding*, Carl Hanser Verlag, Munich, Germany, 1989.

24. F. Cogswell, *Polymer Melt Rheology - a Guide for Industrial Practice*, George Goodwin Ltd, London, UK, for The Plastics and Rubber Institute, 1981.

25. *Modern Plastics Handbook*, Ed., C.A. Harper, McGraw-Hill Publishers, New York, NY, USA, 1999.

26. N. Brooks, B. Willoughby, A.J. Dawson and N. Tucker, *Proceedings of Micro-Moulding 2002 – Advances and Commercial Opportunities in Micro and Miniature Moulding*, Warwick, UK, 2002.

27. D. Ochs in *Plastics Additives Handbook*, 5th Edition, Ed., H. Zweifel, Hanser, Munich, Germany, 2000. Chapter 11, p.647.

6 The Economics and Market Potential for Low Environmental Impact Polymers

Mark Johnson

6.1 Introduction

Within this chapter the economical and market potential for biopolymers will be examined. An overview of the market will be given including a short history. This chapter will also examine issues such as the cost and performance of this new generation of plastics, identification of current markets, and the issues that are driving the uptake and use of them, both currently and in the future. For this Chapter the term 'biopolymer' will be used in place of 'Low Environmental Impact Polymers', because, as discussed later in this chapter, the impact upon the environment of this new generation of materials is potentially greater than that of fossil-origin polymers.

6.2 A Brief History of Biopolymers

The development of biopolymeric materials began in the 1970s. At the time the World had seen two oil crises and Imperial Chemical Industries (ICI) decided to look at biological routes for chemical manufacture in an attempt to insulate themselves from the shock of possible future crises. The outcome of this programme was a thermoplastic called 'Biopol' [1]. The ownership rights to Biopol changed hands from ICI to Zeneca and then to Monsanto, who ceased the development of the material towards production on a commercial scale in 1998 claiming that production costs were uneconomical. Interestingly the US company Metabolix purchased the rights to Biopol in 2001.

The term 'biopolymer' can be used to describe polymers that are biodegradable or based on 'biofeedstocks' [2]. It loosely describes a wide range of materials: from crustacean produced chitosan and plant derived xanthan gum to starch-based packaging for peanuts [3]. Within this chapter the term biopolymer will be used to describe polymers that are both of a biological origin and biodegradable.

There is a wide range of polymers that originate from biological sources. This includes naturally occurring polymers in addition to those materials that undergo chemical and physical modification in order to transform them into biopolymers. This range of materials is shown in **Figure 6.1**.

As can be seen in **Figure 6.1** there exists a wide range of materials that can be used as precursors for the production of biopolymers although the most interesting for the production of biopolymers are polylactate and starch. Starch can be converted into a thermoplastic material by applying a sufficient amount of work and heat to the cereal product to gelatinise all of the constituents completely [4]. Some parties argue that starch is economically competitive with petroleum feedstocks, and has been the precursor for several different biodegradable plastics [5]. Lactic acid, which is the monomer of polylactic acid (PLA), can be produced easily by fermentation of a carbohydrate feedstock [5]. The carbohydrate can be agricultural products such as maize, wheat or waste products from the agricultural industry [7].

6.3 Market Size

The majority of this chapter will, where biopolymers are concerned, focus upon thermoplastic biopolymers, as these are biodegradable, produced in greater quantities and have a larger market potential than thermoset biological origin polymers such as cashew nut shell liquid (CNSL). In 1999, within the UK marketplace, thermoplastic polymers accounted for over 2,950,000 tonnes of the total market for polymers of 4,002,000 tonnes [8]. The approximate production capacity for thermoplastic biopolymers in 2001 is between 30,000 and 40,000 tonnes [9], but it is estimated that the total world manufacturing capacity for biopolymers in the year 2002 is expected to increase to approximately 350,000 tonnes [10]. Even though the production capacity for biopolymers will increase rapidly over the next few, they will still represent a very small proportion of the market in both UK and world terms.

6.4 Classifications and Costs of Biopolymers

Thermoplastic biopolymers can be split into two categories [4]; starch-based polymers and polyesters. Cargill Dow NatureWorks is an example of a polyester thermoplastic [11], and the company announced that it will have a 140,000 tonne per year manufacturing facility dedicated to the production of this PLA based polymer on-stream in 2002 [2]. A more detailed review of Cargill Dow's PLA material is included in this book within Chapter 14. Examples of starch materials are products from National Starch (Eco-FOAM), Avebe (Paragon), Earth Shell (Earth Shell) and Novamont (Mater-Bi) [12]. Many of these materials can be directly substituted for existing plastics and can thus be used in a number of processing routes such as film blowing and injection moulding.

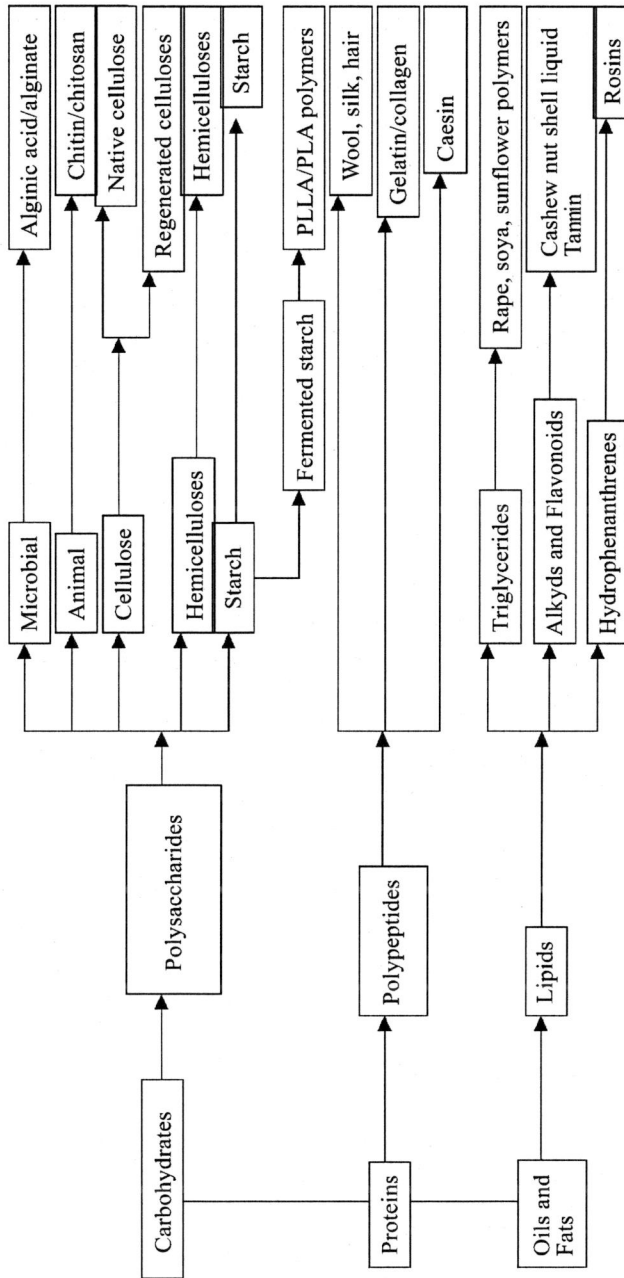

Figure 6.1 Schematic presentation of bio-based polymers based on their origin and method of production [5]

Reproduced with permission from R.M. Johnson, L.Y. Mwaikambo and N. Tucker, Biopolymers, Rapra Review Report No. 159, Rapra Technology, Shrewsbury, UK

A comparison between examples of starch-based (Novamont Mater-Bi) and PLA (Cargill Dow NatureWorks) based materials, in both injection moulding and film grades, with the physical and mechanical properties of conventional polymers is shown in **Table 6.1**. The two polymers that were chosen for the benchmark are polypropylene (PP), which is traditionally used in a wide range of injection moulded products, and low density polyethylene (LDPE) which is used for film blowing. One market for blown film is in the production of packaging bags and sheet.

As can be seen from the data presented in **Table 6.1**, PLA-based biopolymers generally have superior properties to those achieved by starch-based biopolymers. For example, Cargill Dow NatureWorks has superior stiffness (flexural modulus) and tensile strength to that of Novamont Mater-Bi. The stiffness of the PLA biopolymer is something that leads to decreased toughness, and this is borne out by lower tensile elongation compared to Mater-Bi. The stiffness also contributes an element of brittleness to the material, and this is shown by the low values achieved by the PLA materials in the Izod test when compared with the polyethylene and polypropylene. Whilst the performance of some biopolymers approaches, or in some cases exceeds, that of existing polymers such as polypropylene there exist other issues that need to be overcome, the largest of which is cost.

Stephen Cox, the marketing director of PLA polymer producer Chronopol, said that: 'the largest problem with PLA polymers at the moment is the price.' The material cost in 1998 was between £6.20-9.30/kg for PLA, and he estimated that the production volumes attainable at the new Chronopol and Cargill Dow plants could lower costs to £2.30–3.10/kg [3]. This is still in excess of current market prices for plastics such as film-grade LDPE at 1.81 DM/kg and injection moulding grade PP at 1.80 DM/kg which is approximately £0.58/kg for both types of plastic [17]. By comparison, the cost of Novamont Mater-Bi is between DM2.5-8/kg depending upon grade [2], and this equates to a cost of £0.79 to £2.51 per kg. The cost of polyhydroxybutyrate (PHB) biopolymers is €17-20 kg (£10.90-12.80), and the cost of PCL biopolymers is €13-15 kg (£8.30-9.60) [18].

The cost of PLA-based polymers will drop in time with the commissioning of new, larger plants where larger economies of scale can be achieved, but the cost will never approach that of conventional polymers such as polyethylene [3]. The higher cost of PLA-based polymers can be attributed to the processing route being more complex that that of starch-based polymers in addition to the large capital investment that is required to construct a PLA manufacturing facility. The Cargill Dow NatureWorks plant in Blair, Nebraska, USA, is estimated to have cost in the region of $300 million, and has a production capacity of 140,000 tonnes of PLA based biopolymers per year [2].

Table 6.1 Comparison of two types of biopolymers with conventional, petroleum based plastics

Property	Units	LDPE [13]	PP [a] [13, 14]	PLA [b] [11]	PLA [c] [11]	Starch-based [d] [15]	Starch-based [e] [6, 16]
Specific gravity	g/cm³	≤ 0.920	0.910	1.25	1.21	1.33	1.12
Tensile strength at break	MPa	10	30	53	48	26	30
Tensile yield strength	MPa	-	30	60	-	-	12
Tensile modulus	GPa	0.32	1.51	3.5	-	2.1-2.5	0.371
Tensile elongation	%	400	150	6.0	2.5	27	886
Notched izod impact	J/m	No break	4	0.33	0.16	-	-
Flexural strength	MPa	-	-	-	83	-	-
Flexural modulus	GPa	0.2	1.5	-	3.8	1.7	0.18

[a] Data for a PP homopolymer
[b] Data for Cargill Dow NatureWorks 2000D extrusion thermoforming grade
[c] Data for Cargill Dow NatureWorks 3010D injection moulding grade
[d] Data for Novamont Mater-Bi Y101U injection moulding grade
[e] Data for Novamont Mater-Bi ZF03U/A film blowing grade

It is clear that the cost of biopolymers is currently too high for them to compete on price with conventional materials, and thus they must compete on different criteria. One such criteria is that the biodegradability of the material is an essential characteristic of the component to be produced from the biopolymer. Within the next sections we will examine and discuss some of the factors that are currently driving the uptake of these new materials, in addition to market forces that will affect them in the future.

6.5 Current Uses of Biopolymers

There exist a number of end-use applications for biodegradable polymers, but by far and away the most popular is in the manufacture of compost bags. **Figure 6.2** provides an overview of the European markets for biodegradable plastics.

From **Figure 6.2** it can be seen that the majority of current uses for biodegradable plastics are for items that have a short life, and are generally discarded after one use. It is within such sectors that biopolymers currently have the greatest advantage. For example, if biodegradable plastics are used in the manufacture of compost bags then the compost need not be removed from the bag prior to composting, making the composting operation more efficient.

Biopolymers are also finding applications in the medical markets. As detailed in Chapter 10, Metabolix feel that the use of their biopolymer for medical gloves will assist in reducing the number of latex gloves currently consigned to landfill. An additional use are the dissolvable sutures that are currently manufactured from polycaprolactone (PCL). The medical industry sector is an area where the higher unit value and specialised nature of the products allows biopolymers to compete effectively with other materials [20].

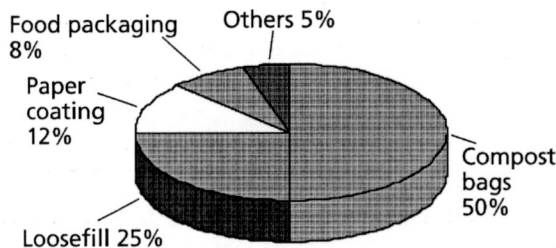

Figure 6.2 The uses of biodegradable plastics within Europe [19]

6.6 Driving Forces

Many parties have different opinions as to what is driving the interest in the use of 'biopolymers'. Some say that increased interest in the production and usage of biopolymeric materials can be attributed to the finite amount of fossil resources and the fact that the consumption of these fossil resources contributes to the damage of the environment [21]. Governmental legislation has lead to end-of-life disposal via the route of land-filling and incineration to become less attractive due to increased taxes, although, as discussed later in this chapter, this is not applicable to all countries. The evolution of new composting technologies in the past ten years has made composting a more favourable end-of-life disposal route for both ecological and economic reasons [22]. All of these issues will now be examined in a 'PEST' framework (shown in **Table 6.2**) to provide a breakdown as to what is, and what will in the future, drive stakeholders towards more rapid uptake of biopolymeric materials.

6.7 Political

6.7.1 Legislation

In the absence of significant market 'pull' other mechanisms must be exploited to encourage the development of this new generation of materials. Currently, one of the largest of these is the application of legislative instruments. Whilst the crux of the majority of these documents is the promotion of recycling, there comes a stage in the supply chain

Table 6.2 Criteria affecting the uptake and acceptance of biopolymers	
Political	Economical
• Legislation • Government initiatives	• Increased disposal costs • Increased competition • Polluter pays • The rising costs of finite resources
Social	Technical
• 'Greening' of consumers • Acceptance of biopolymers	• Economies of scale • 'Organic' *versus* mechanical recycling • Further development • Incorporation of fillers

that recycling becomes unattractive both economically and technically, and these issues will be discussed later in this chapter.

The industry sectors that will be most effected by the ratified legislation are the automotive industry and packaging sectors. The Packaging and Packaging Waste Directive 94/62/EC [23] and the End-of-Life Vehicle Directive 2000/53/EC [24] are two articles of legislation that govern the disposal routes that are available for the automotive and packaging industries. In both cases the disposal strategy leans toward reuse, recovery and recycling although each of these strategies has differing elements.

The packaging directive dictates that member states set up return, collection and recovery systems. In addition, all parties involved in the production, use, importing and distribution of packaging and packaged products must accept responsibility for any waste in accordance with the polluter-pays principle. The establishment of such an infrastructure would be costly as it does not currently exist, and it is highly likely that the costs of creating it would be passed onto users further down the supply chain in the form of increased product prices. The ratification of such legislation means that materials that can be disposed without the need for the creation of such a large and complex infrastructure, (e.g., biopolymers and other biodegradables), become economically more competitive in relation to those materials that require it.

Much of the packaging directive is focussed upon the prevention of the formation of packaging waste, and this means the reduction of the quantity of waste and of its harmfulness to the environment, and particular emphasis is placed upon the development of 'clean' products and technology. The remit of recovery, reuse and recycling has a number of fundamental points. Reuse is the development of packaging that can accomplish a number of trips within its life cycle, and examples of such materials include articles such as milk crates and the 'bag for life' type of carrier bags currently offered by a number of UK supermarket chains.

The recovery of a product is dealt with in Directive 75/442/EEC [25] and states that it is the treatment of waste for reuse. Recovery is also classified as 'energy recovery' where, at the end of life, the packaging is combusted to produce energy. It can be argued that taking into account the total life cycle of polymers *versus* steels in the production of automobiles, plastic materials require three times less energy to produce than steel. If the plastics are combusted at the end of life and the energy recovered then the energy required to produce the steel is five times greater [26].

Recycling purports not only to the reprocessing of waste materials in a production process, but also to 'organic recycling', which is the aerobic and anaerobic treatment of waste to produce stabilised organic residues or methane. The landfilling of waste is not considered to be a form of organic recycling [23]. It can be argued that biodegradable materials can be recycled organically, therefore enabling users of such materials to comply with legislation without the need to create the infrastructure needed for the sorting, treatment and recycling of conventional plastics, although some infrastructure would be needed. This would be in the form of dedicated composting facilities that would provide the microbial activity needed to degrade the biopolymers.

The end-of life vehicle (ELV) directive entered into force on 1st July 2002 and pertains to the disposal of vehicles at the end of their useful life. One of the fundamental aspects of the directive is that preference should be paid to reuse, recycling and recovery instead of disposal. In particular, the directive states that the recycling of all plastics from vehicles should be improved, and the development of markets for recycled materials should be encouraged. The onus of these retrieval and recycling programs and facilities is placed upon the 'economic operators', which includes vehicle producers, distributors and insurers. The directive details targets for the reusability, recovery and recyclability and these are shown in **Table 6.3**.

As is the case with the packaging directive, recycling can be interpreted to include 'organic' recycling, and recovery can include the recovery of energy [24]. The reason why mechanical recycling, (i.e., regrinding and then extrusion into pellets), is favoured for polymeric materials is that 80% of the energy required to produce polypropylene is expended on the polymerisation and associated processes. If the polypropylene can be (economically) recycled to produce other parts at the end-of life then this is deemed to be environmentally better than chemical or energy recycling as the energy does not need to be expended to polymerise the feedstock [26].

The incineration of plastics with energy recovery appears to be an attractive 'stopgap' measure until the relevant infrastructure for more widespread chemical and mechanical recycling can be undertaken. This is not the case. Once again the root of this issue is the dearth of relevant facilities. It can be argued that there are a number of facilities that

Table 6.3 EC directive on end-of vehicle life recovery and recycling weights [24]		
Date	Reuse and recovery (by weight)	Reuse and recycling (by weight)
January 1st 2006	85%	80%
January 1st 2015	95%	85%

could be used to incinerate waste plastics, and among these are the CHP (Combined Heat and Power) plants that handle the large amounts of Municipal Solid Waste (MSW) generated by domestic households.

The truth is that these facilities are designed to burn fuels with a low calorific value, whilst plastic waste tends to have a high calorific value. If this waste is combined and then burnt, leading to a waste-stream with a higher calorific value than MSW, it causes problems in other areas. These drawbacks include a reduction in the rate of the disposal of MSW, and an increase in the gate price for companies wishing to dispose of MSW which leads, in turn, to an increase in the cost of energy generated by the plant. The solution to this issue would be to commission a facility (or number of facilities) dedicated to the disposal of plastic waste. This can be viewed as a long-term solution due to the fact that the lead-time for such a CHP station is estimated to be in the region of eight years. With the end-of-life vehicle directive beginning to affect the auto industry from 2006, this option is unviable.

A further piece of governmental legislation that impacts the disposal routes available for materials is the European Council Directive 1999/31/EC on the landfill of waste [27]. This brought into enforcement the 'polluter pays' principle for waste that is landfilled. This concept is detailed in directive 75/442/EEC [25], in brief the 'polluter pays' principle states that the cost of disposing of waste, less any proceeds derived from treating the waste, shall be borne by the producer or previous handler of the waste. This could have a number of repercussions for industries that generate a large quantity of waste such as the tobacco and food industries. Another facet of the landfill of waste directive is that the amount of biodegradable refuse that is currently disposed of via landfilling is to be reduced over time. **Table 6.4** shows the targets that are required for reductions in the landfilling of biodegradable waste, and this is expressed as a proportion of the landfill waste that was produced in 1995.

The reasoning behind the move away from the landfilling of biodegradable waste is that the degradation of organic matter produces methane gas, which is classified as a greenhouse gas and thus contributes to global warming. As such, it is desirous to reduce the environmental impact of global warming by reducing emissions of the gases that contribute to the phenomenon of global warming.

The strategy for the future disposal of organic waste centres on the creation of dedicated composting facilities. These facilities will allow the waste to be composted in a controlled environment and the methane captured and used as a 'biogas' fuel [27]. Whilst the ratification of the landfill directive will not influence the uptake of biopolymers directly, it will affect it indirectly. The mechanism in which it will aid the uptake of biodegradable plastics is that as time passes and composting facilities become more numerous, the use

Table 6.4 Strategy for the reduction of the amount of biodegradable waste that is land-filled [27]	
Year	Amount (versus 1995 levels)
2006	75%
2009	50%
2016	35%

and disposal of biopolymeric materials becomes more attractive as they can be disposed with the organic wastes that households generate.

6.7.2 Government Initiatives

In addition to the UK and other governments attempting to influence the uptake and development of biopolymers through legislation, the UK government is also attempting to do it by sponsoring a number of initiatives. Programmes such as the Sustainable Technologies Initiative (STI) and bio-wise, both supported by the UK Department of Trade and Industry (DTI), attempt to stimulate activity in the biopolymer arena by offering monies for R&D and marketing, in addition to identifying potential partner organisations. The bio-wise initiative is short-term, and is aimed at companies with products that are close to market, whilst the STI can be classed as medium to long-term. By taking advantage of these initiatives companies can reduce the cost of research and development by gaining grant aid.

6.8 Economic

6.8.1 Increased Disposal Costs

The biopolymers that are currently available in the marketplace are more expensive than fossil origin polymers. The performance of biopolymers such as Mater-Bi and NatureWorks approaches, or even surpasses, that of some commodity polymers such as polypropylene and polyethylene, but they do not warrant the additional cost if purchase price is one of the most important factors in a procurer's decision. Even when total life cycle costs are taken into account, the additional costs that are possibly incurred through the incineration or landfilling of the material are not high enough to warrant the specification of these materials.

A comparison between three different waste management options in three European countries is shown in **Table 6.5**. The table shows how Germany is attempting to use economic instruments to force end-users to recycle and compost waste instead of opting for the more traditional routes of disposal.

By comparison, the cost of landfilling refuse in the UK in 2004 is £13 per tonne, and this rate is going to increase by £1 per year until 2005 [28]. Thus, the UK is still a number of years behind some European countries in attempting to control the build-up of waste via the application of taxes.

Table 6.5 Cost of waste management options in the Netherlands, Germany and Belgium (in US$ per tonne) [22]			
Option	The Netherlands	Germany	Belgium
Composting	60	151	80
Incineration	135	486	110
Landfilling	105	402	75

6.8.2 Increased Competition

As has been mentioned on a number of occasions within this chapter, biopolymers are relatively new materials that are being produced in small quantities. As such, very little competition exists in the market place, and therefore the lack of competition translates into higher prices for the consumer. As the sector matures it will witness increased competition and companies will have to start to compete on price. The same phenomenon can be witnessed at work in the commodity thermoplastics industry, where one of the only differentiators between types of plastic, (e.g., polypropylene), is often price.

6.8.3 Polluter Pays

This principle was mentioned in the section on legislative drivers, and, in essence, is a tax that will be levied against operators within the polymer supply chain. An example of this is the tax that the UK may adopt upon carrier bags. This is similar to that currently in force in Eire, which has levied a tax of €0.15 (roughly £0.10) on each plastic bag used in supermarkets. This has lead to a reduction in the consumption of bags to 2.5% of the level prior to the introduction of the levy [29]. The number of carrier bags used annually in the UK is roughly 8 billion [30], which potentially goes to make up a high proportion of the 7% of landfill that is constituted of plastic film [31], and the introduction of such a tax in the British Isles could affect the uptake of biopolymers.

One of the reasons that Eire has drastically reduced the consumption of carrier bags within its borders is that the tax levied on the bags is passed on to the shops customer. The introduction of biodegradable bags potentially allows supermarket chains in Ireland to negate the tax that is being applied currently.

6.8.4 The Rising Costs of Finite Resources

A further issue that is often quoted is that at some point in the future the cost of fossil fuels will rise, causing polymers that use those feedstocks to increase in price. Of the 3,891 million tonnes of gas and oil produced in 1997 [32] approximately 270 million tonnes were used in the production of polymeric materials [33]. This equates to roughly 7% of the world's fossil fuel reserves being used in the production of plastics, of which approximately 60% is used to fuel the process, although this is dependent upon the type of plastic that is being produced. More current data was unavailable, but these figures represent an indication as to the proportion of oil and gas that is used in the production of polymers.

It must be noted that the majority of oil used is in the transport sector, which in the UK accounts for 82% of the use of petroleum [34]. As such, it appears that finding alternative fuels and power sources for cars is a more pressing need than alternative feedstocks for polymer production, although auto manufacturers have been looking at alternative fuels such as alcohol distilled from sugars (already in heavy use in Brazil) for a number of years.

The estimates of when the supply of oil will run out varies from the pessimistic to the optimistic although many commentators appear to forecast that oil will remain plentiful until between 2025 and 2050 [32], and as we get ever closer to the point at which oil will become no longer economic to use, the price and thus the viability of oil and petroleum products will decrease. Industry will need to have substitutes in place when the time arrives and the timescale for the introduction of polymer materials and the complex infrastructure necessary for their production is long. It took polypropylene, now one of the most popular types of thermoplastic, nearly 40 years to go from discovery through to widespread acceptance in the marketplace [35].

6.9 Social

6.9.1 The 'Greening' of Consumers

For biopolymers to become anything more than a niche product in the current market place requires not only huge investment from the manufacturing companies but also a market pull. This is a scenario where one party will not act without the other; large companies are unwilling to invest huge amounts of capital without a market and consumers will not buy products unless they are available on the market. The market pull could stem from the 'greening' of consumers where people become more conscious of the environment.

This is already occurring in a number of sectors. An example of this is the organic fruit and vegetables that are being sold in a major UK supermarket chain. The organisation reacted to consumer pressure, and has begun to package this food in biodegradable packaging. Further evidence of the growing upswing in consumer 'greenness' is the rise in the number of farmer's markets around the British Isles. In this case, consumers who are concerned with a number of issues such as the application of pesticides and concerns regarding intensive farming are resorting to buying their produce directly from the source.

Convention follows that the same consumers who are concerned with the source of their food will also be concerned with where their waste ends up, and thus will potentially pay a premium for biodegradable packaging. These are the same group of consumers who show concern for the environment by composting putrescible waste at home, and biopolymeric packaging can be handled in the same waste stream. Producers of biopolymers must also consider the concern for the environment that is swelling amongst consumers. Some producers of biopolymeric materials use genetically modified (GM) crops as a feedstock to their process, and whilst this may be acceptable in some countries, in others it is not. The UK is a case in point: consumers have refused to buy produce packaged next to genetically modified organisms (GMO) for fear, perceived or otherwise, that the GMO may leach into the food and cause health problems in the future.

6.9.2 Acceptance of Biopolymers

Many people, especially those involved in manufacturing, display a degree of scepticism whenever the conversation focuses on biopolymers. Much of this doubt is down to a lack of understanding of the mechanisms in which these materials degrade. Some parties assume that because these materials biodegrade then they will be susceptible to attack from elements such as rain, humidity or snow and thus could not be used externally.

This, as most parties with any experience of biopolymers know, is a misnomer. The materials must be biologically attacked by bacteria to initiate the biodegradation, and these organisms are seldom present outside of compost heaps.

In time, and with education from biopolymer producers, this misplaced perception will change and manufacturers may choose to incorporate biopolymers into their products due to environmental benefits. The biopolymer market is beginning to see an upswing in environmental interest from a number of industrial conglomerates in a variety of market sectors. The attraction that these companies have towards biopolymers may not be driven by environmental reasons, but rather the fact that people in higher demographic groups, (i.e., more wealthy), are the consumers that currently are willing to purchase products based purely upon ecological and environmental reasons [36]. This consumer group is more willing to pay a premium for 'green' products, which is obviously attractive to the companies manufacturing and marketing the products, rather than the reasons more traditionally associated with product choice such as cost, quality and brand reputation.

6.10 Technical

6.10.1 Economies of Scale

The production of thermoplastics is extremely capital intensive, and companies involved in their production seek to maximise profit by utilising massive production facilities and taking advantage of the economies of scale that they offer. Michael Porter in his seminal tome on business strategy, 'Competitive Advantage', states that economies of scale arise from the ability to perform operations differently and more efficiently at large volumes, in addition to amortising intangible costs such as research and advertising over a greater sales volume [37].

Biopolymers are relatively new materials on the market, and as such are produced in relatively low volumes in comparison to the commodity thermoplastics that abound in the marketplace. This means that R&D costs are being amortised over a small volume, with the cost being passed on to users, the end result of which is relatively high purchase costs. When, in the future, biopolymers gain acceptance in the market, then costs will surely drop as manufacturers commission larger plants. Already we are seeing some companies, Cargill Dow being one example, construct larger, more efficient manufacturing facilities to leverage these economies of scale. In time, other organisations will follow suit, providing they have a robust process and enough capital, with the end-result that the costs of biopolymers will drop.

6.10.2 'Organic' Recycling versus Mechanical Recycling

There are a number of problems with the current end of life disposal routes for products manufactured from thermoset and thermoplastic polymers. Thermoset polymers, once cured, are notoriously difficult to recycle. This is due to the curing mechanism being chemical and generally irreversible. Thermoset composites such as sheet moulding compound (SMC) and bulk moulding compound (BMC) can be recycled at end-of life, and the process to perform this is to grind the scrap and then add it to virgin material at levels of up to 20% [38].

Other thermoset composite materials such as glass reinforced plastic (GRP) are much more troublesome to recycle. The polymer matrix can be broken down through the use of hydrolysis and the long fibres recovered and utilised in 'cascade' recycling, although hydrolysis technology is not being realised on a large scale [38]. Cascade recycling involves the comminution of the fibres into shorter lengths which can then be utilised as reinforcing fibres for injection moulding compounds.

A large problem with the recycling of thermoplastics is that because of the different types of polymer available, and hence those that are present in the waste stream, separation into the various grades must be performed before effective recycling can be performed. The separation of polymers into different grades is aided by adequate identification of the material and this works with single components such as the drink bottle recycling that is performed in the United States. This is aided by a small financial incentive with each bottle carrying a 5¢ deposit.

Currently the separation of the mixed polymer stream is performed manually, although there exists a large body of academic work in which mineral processing techniques such as water floatation have been applied to polymer recycling [39]. Until there is widespread adoption of this type of technology, the recycling of polymers will be a labour intensive operation, which affects the cost of the recyclate, often making it more expensive than virgin materials. A further option for recycling polymers is to compound the mixed polymer stream, although the resulting material is of a low grade and can only be used in a limited number of applications [40].

Further issues appear when an article that comprises a number of polymeric materials comes to the end of life, and a case in point is the recycling of the plastics from automobiles. There are 16 different types of polymers used in a BMW 5 series and that within these groups there are different grades relating to parameters such as molecular weight that are optimised for performance [38]. The solution to the sorting problem would be to standardise plastic types 'across the board' although this may present other problems such as the increased weight and decreased functionality of the components.

Other problems relating to the recycling of thermoplastics from vehicles, besides separation, are concerns relating to disassembly. A study carried out by the Ford Motor Company [38], on the dismantling of the Escort car confirmed that there were diminishing returns in the recycling of plastics stripped from automobiles. Twenty kg of recyclable plastics were recovered in the first 10 minutes and in the next 10 minutes only 10 kg of plastics were recovered. The dismantling costs were calculated to be 50p/kg after 20 minutes, rising to £12/kg after 50 minutes, the latter figure is far in excess of the prices of many virgin polymers. Until manufacturers standardise on type, and improve 'design for disassembly', the recycling of polymers from automobiles looks to be an uneconomical proposition.

The options that are available for the disposal of both thermoplastic and thermoset components are currently the same options that are available for the disposal of most discarded items: incineration and landfilling. Unfortunately the majority of polymeric waste is currently being land-filled, and approximately 13% of the volume of our landfill sites are occupied by polymeric materials [31]. Of this figure, 6% is comprised of solid plastics, with the remaining proportion of 7% being made up of film plastics such as carrier bags, and it is interesting to note that every man, woman and child within the Organisation for Economic Co-operation and Development (OECD) region is responsible for 640 kg of waste being land-filled annually (in 1999), which means that every person in the developed world is land-filling slightly over 83 kg of polymeric waste annually [31].

Biopolymers solve the problems of the finite of oil reserves and the environmental impact caused by the disposal of plastics at the end-of-life. The precursor to their production is typically an agriculturally-based feedstock such as starch or corn. The feedstocks used in the manufacture of biopolymers are of a sustainable origin. At the end-of-life the products are composted, which eliminates the need to pay any landfill taxes, although as seen in **Table 6.5**, a smaller cost would need to be paid for composting. **Figure 6.3** is a simplified representation of the life-cycle of a biopolymer.

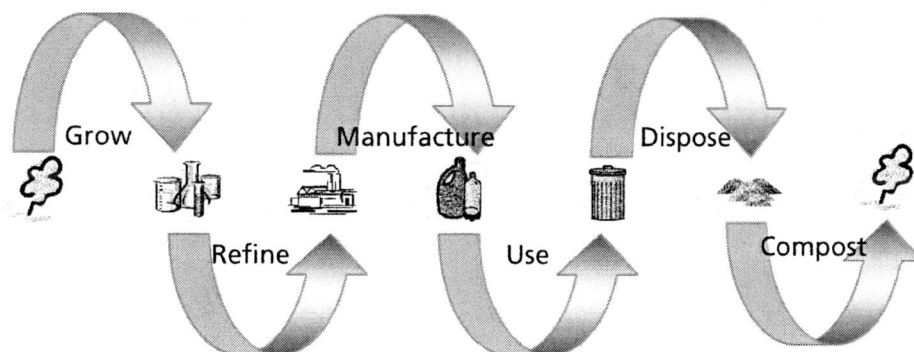

Figure 6.3 The life cycle of a biological origin, biodegradable polymer

As can be seen from **Figure 6.3**, the feedstock to the refinement process is from plant origin in the form of celluloses, starches and triglycerides. It must be noted that the manufacture phase consists of two discrete phases: the refinement of the organic materials into a polymer suitable for processing, (e.g., granules for injection moulding), and the processing of the material into a product. The output of the process is compost, which, whilst not feeding directly back into the growth cycle of the polymer precursors, can be used to aid the propagation of other plants and crops.

6.10.3 Further Development

A further factor is that companies are still on the 'learning curve' in the production of biopolymers, it must be remembered that polypropylene was discovered in 1954 by Giulio Natta, although it was not commercially exploited on a large scale until the 1990s [35]. By comparison, work on biopolymers has only been taking place since the late 1970s, and only in earnest since the early 1990s when the first commercially available thermoplastic biopolymers entered the marketplace.

The fuel costs in the production of biopolymers are higher than those used in the production of some types of standard polymer. Gerngross and Slater state that 1 kg of polyethylene requires 29 MJ of energy to produce, whilst polyethyleneterephthalate (PET) requires 28 MJ. By comparison, 1 kg of Cargill Dow NatureWorks PLA resin utilises 56 MJ in the production process [33]. Therefore, if the total use of energy is taken into account, then biopolymers consume more energy in the production process. With the majority of energy currently derived from fossil resources, the environmental impact of biopolymers through emissions created during manufacture must be greater than conventional polymers.

One of the sole advantages that they have currently is biodegradability, although the answer to the problem of increased energy expenditure during manufacture may be to provide more energy through renewable sources such as biomass or hydro-electric power. This would conserve more of the diminishing fossil fuel reserves although the change to alternative energy supplies could be construed as a short-term measure. In time, as the production process of these polymers evolves, the cost of production in terms of energy will decrease, allowing them to compete both economically and in terms of their environmental impact.

6.10.4 Incorporation of Fillers

Fillers in plastics are, as the name suggests compounded with the polymer to fill out the plastic, thus allowing less (of the more expensive) polymeric material to be used to produce a large volume of mouldable resin [13]. This suggests that fillers are combined with the polymer merely to take up space, but as they can be used to impart improved properties on the resin, this reduces article costs by improving material properties allowing less material to be used. Fillers that are currently available in the market place, such as calcium carbonate, silica and clay, whilst originating from the earth, are not sustainable in origin, nor biodegradable in nature. A filler for a biodegradable polymer must fulfil at least two criteria, the first of which is to reduce article costs and the second of which is to be biodegradable and natural in origin.

The use of a natural fibre filler for plastics can be traced back to 1907 when Baekeland utilised wood flour as a filler in thermosetting phenolic resin [41] to create a material that to this day is known as 'Bakelite' [42]. Research has also been undertaken to investigate as to whether natural fibres would make suitable structural reinforcement instead of simply being used to cheapen a product. A case in point are the series of tests that were performed at Aero Research Limited during the second World War which focussed upon testing the suitability of using Gordon Aerolite (a phenolic resin) and flax fibres to manufacture Spitfire fuselages [43]. The outcome of this series of experiments was that the composite produced could have been substituted for the aluminium that was being used at the time if the need arose.

The market for wood-filled thermoplastic compounds can be classed as relatively mature as there are a number of wood-filled products that are currently available on the market. Such examples are Coexil, a wood flour filled polypropylene (Automotive Industries), Trex and AERT (Advance Environmental Recycling Technologies), which are 50:50 (% by weight) wood flour filled polymers [44]. The matrix polymers used for these materials are polypropylene, high-density polyethylene and low-density polyethylene, respectively. Schut states that industry estimates for the year 1999 pegged the production of wood-plastic compounds at almost 300 million pounds (almost 134,000 tonnes) [45]. The technology to produce natural fibre-filled biopolymers currently exists and could be exploited relatively easily to produce such materials in the very near future.

Work has been published that investigates the use of other natural fibres (excluding wood) for use as a filler and/or reinforcing material in plastics. Examples of such work are: sugar cane bagasse filled thermoset polyester [46], short-fibre jute reinforced polypropylene [47], straw reinforced polypropylene [48], kenaf reinforced polypropylene [49] and flax reinforced polypropylene [50]. From these pieces of work it can be seen that there exists a variety of options open to biopolymer manufacturers and processors who wish to reduce article costs.

In much of the work the reasons cited for the use of natural fibres as a filler/reinforcement was that they are biodegradable, renewable and environmentally friendly [47]. Low in cost [48], abundant [46], and possess good specific properties [50]. The matrix used in much of this previous work was from non-renewable source such as the polypropylene used by Sanadi and co-workers [49] and Karmaker and Youngquist [47]. This may have been due to the availability of biopolymers at the times of their research.

The cost of semi-processed jute fibres that would be suitable for processing into injection moulded composites was estimated to be $500 per tonne in February 2000 [51]. This equates to a figure of approximately £350 per tonne, which is significantly cheaper than the biopolymers currently available in the marketplace. Incorporating this type of material into a composite with a biopolymeric matrix would allow a reduction in article cost to be obtained. Within this handbook there is an entire chapter dedicated to the subject of natural fibre fillers and reinforcements which discusses the subject in greater detail.

6.11 The Future for Biopolymers

It will take time for the drivers detailed in this chapter to have an effect upon the cost and uptake of biopolymeric materials. Within this section the timescales for these drivers will be discussed, they will also be categorised in terms of the timescales related to them. The grouping will be split into three sections: short-term, medium-term, and long-term. Short-term equates to a time scale of between the present and 2 years, medium-term between 2 and 5 years, with long-term equating to 5 years or more. Please note that these timescales are the author's own and are the result of examining the body of knowledge related to the various subjects.

6.11.1 Short-Term

The only technologies or drivers that will be present in the short-term are the incorporation of fillers and some government initiatives. Biodegradable fillers have been present in the market for many years and it would be a relatively easy task to incorporate them into biopolymers. There exists a wide range of options in terms of fibre, ranging from cheap, short fibres such as wood flour to longer reinforcing fibres such as hemp and jute. The only issues that could arise are if there is a difference between the polarity of the fibre and the matrix, although some biopolymers have a natural affinity to wood fibres [52].

The only other issue that may appear in the short-term are certain government initiatives. One scheme currently running in the UK is bio-wise, and as mentioned previously, it aims to encourage the uptake of biopolymers by offering companies grant monies to aid research and development.

6.11.2 Medium-Term

The majority of the driving-forces mentioned in this chapter will begin to influence the biopolymer industry in the medium-term. Of all of these influences, the one that will have the greatest affect upon the industry is legislation. The ratification of legislation that relates to waste management such as the Landfill Directive [27] and the End-of-Life Vehicle Directive [24] will have a multitude of effects upon the uptake and cost of biopolymers.

The first of these is the fact the legislation will increase the cost of the disposal of materials through the conventional routes of landfilling and, to a lesser extent, incineration. Legislation will force economic operators into diverting an increasing proportion of waste to different forms of waste management such as composting and recycling. In turn, significant investment in the relevant infrastructure will be made in an effort for the UK to comply with the legislation. This will lead to the creation of large-scale composting facilities. This leads to 'organic' recycling becoming a viable option for the disposal of compostable waste, whereas the mechanical recycling of polymers has a large number of inherent problems. The largest of these (and the most complex) is the separation of the waste stream. If this is not performed the resulting material has properties similar to those of wood, with a limited number of end-use applications.

A further impact of the ratification of legislation is that the disposal will become more of a consideration for companies, with the cost, under the polluter pays concept, being passed directly to the polluter. This should lead to companies seeking innovative solutions to the issues of end of life disposal for articles. Amongst the options that a company could employ is the use of biopolymers instead of conventional polymers, although they could opt for materials that are easier to recycle. If the auto industry is taken as a case-in-point, then one solution to the impending legislation is to make a higher proportion of cars from steel. Whilst this makes cars more recyclable as steel is easier to reclaim from the shredded remnants of the car at the end of life, it affects environmental impact by making the cars heavier and thus being less economical with fuel. An interesting occurrence recently is that the UK government has allowed auto manufacturers to pass on disposal costs of the car to the party who is the owner at end of life. Under the ELV directive it is the economic operators who have financial responsibility for the disposal of the end-of-life vehicle and with time the liability will pass from the owner to the economic operators.

There is currently a large amount of research being performed into biopolymers, and this is taking place not only in large corporations, but in smaller companies and universities. Government initiatives, such as the STI in the UK provide funds to organisations attempting to develop sustainable technologies and materials such as biopolymers. The STI works in the same manner as the bio-wise initiative mentioned earlier in that companies can reduce the cost of research and development by applying for grant monies.

The final two driving forces that are classified as medium-term are interlinked. With time consumers will become more conscious of the environment and the impact their decisions have upon it. This may be due to a number of reasons such as the impression of legislation upon the costs of products, an increase in the choice of 'green' products or more widespread understanding of the affects that decisions they make have. As consumers become more environmentally conscious then industry must accept these choices and offer products and services that fit the changing profiles of the public. In addition, industry will become more aware of the benefits that biopolymers have over standard polymers.

6.11.3 Long-Term

This is the timescale in which the greatest change should be expected and this is due simply to the fact that many of the drivers that will have an effect upon the cost and uptake of biopolymers will need time to implement. Of these factors three are linked whilst the remaining issue stands alone. The finiteness of oil reserves is an issue that many people are aware of, yet how much oil remains a topic of much debate. It must also be noted that there is a difference between 'oil reserves' and 'recoverable oil reserves'. There comes a point, as with many other things, where the price of production exceeds that which the consumers are willing to pay, and oil reserves are such an entity. As the reserves diminish man will have to look harder and drill deeper for oil, and this may make it uneconomical. Even without cost taken into account, oil may disappear not only as a viable feedstock for polymer production, but also entirely, within 40 years [53], and as we approach this date oil will become more expensive. As such mankind needs to have substitute materials in place not only for polymer materials, but also for the more pressing issues of transport.

The remaining three drivers are all linked. As the market grows and becomes more lucrative more companies will enter the market. This will need further development on products that currently may exist solely in the laboratory to take them to a scale where they become technically and economically feasible. This in turn needs larger facilities to leverage the economies of scale that are possible. The outcome of this process is increased competition in the market, leading to increased choice and reduced cost for the consumer, and this can only be a good thing. **Figure 6.4** is a pictorial representation of the driving-forces that will affect the cost and uptake of biopolymers in the future.

Fillers			
Government initiatives			
Increased disposal costs			
Legislation			
Organic *versus* mechanical recycling			
Polluter pays			
The greening of consumers			
Acceptance			
Economies of scale			
Further development			
Increased competition			
Rising costs of fossil fuels			
◄──── Short-term ────►◄─ Medium-term ─►◄─── Long-term ───►			

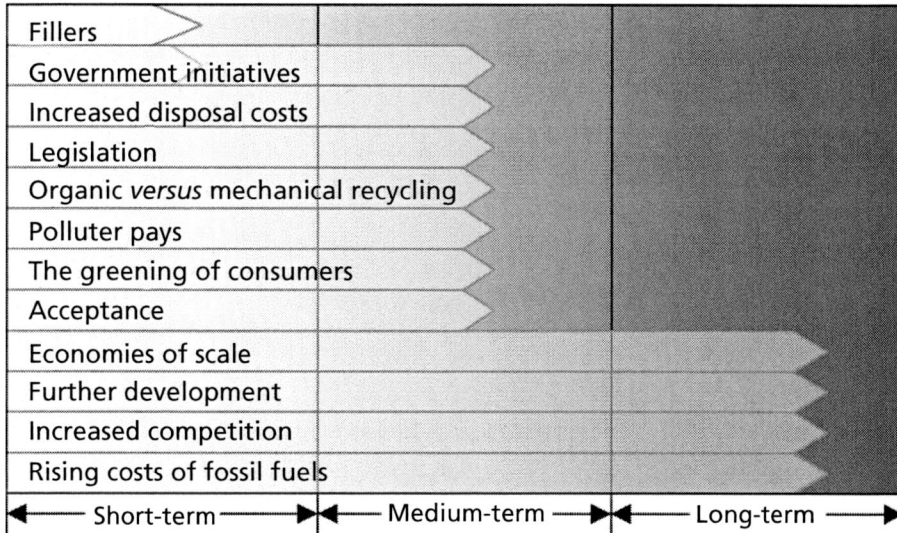

Figure 6.4 Driving forces influencing the cost and uptake of biopolymers

6.12 Conclusions

The current biopolymers available on the market do not have any advantages over conventional plastics bar biodegradability. They are more expensive and whilst the mechanical properties of some biopolymers are good (NatureWorks), the cost for such performance is currently too high. The opening of new manufacturing facilities will go some way to redressing the cost equation as some economies of scale will occur, although biopolymers in the near future will only form a fraction of the total thermoplastics market. In the long-term and as the market expands we should also expect to see increased competition, a result of further development work being undertaken. The incorporation of fillers goes some way to equalising the cost premium that currently apply to biopolymers and government initiatives reduce the financial strain of developing new materials.

By the application of legislation, the whole life cycle costs of polymers become more influential and biopolymers will become more competitive. **Figure 6.2** shows the uses of biopolymers in the current market and it is worth noting that the majority of uses are for items that are used once and then disposed of such as packaging, and it is in these markets where biopolymers will find their greatest opportunities. In the long-term customer acceptance should prove to be a non-factor as fossil origin materials disappear from the marketplace, although in the short to medium term it is important as customer demand will drive producers to formulate biopolymers. As the oil reserves diminish producers of polymeric materials will turn to alternative feedstock sources making biopolymers a reality.

References

1.   N. Uttley, *Manufacturing Chemist*, 1985, **56**, 10, 63.

2.   A. Scott, *Chemical Week*, 2000, **162**, 13, 73.

3.   I. Kim, G. Ondrey and T. Kamiya, *Chemical Engineering*, 1998, **105**, 7, 43.

4.   C. Bastioli in *Proceedings of the 7th Annual Meeting of the Bio/Environmentally Degradable Polymer Society*, 1998, Cambridge, MA, USA.

5.   C. Bastioli, *Polymer Degradation and Stability*, 1998, **59**, 1-3, 263.

6.   R. van Tuil, P. Fowler, M. Lawther and C.J. Weber in *Biobased Packaging Materials for the Food Industry*, Ed., C.J. Weber, The Royal Veterinary and Agricultural University, Frederiksberg, Denmark, 2000, Chapter 2.

7.   A. Garde, A.S. Schmidt, G. Jonsson, M. Andersen, A.D. Thomsen, B.K. Ahring and P. Kiel in *Proceedings of Food Biopack Conference*, Copenhagen, Denmark, 2000, p.50.

8.   *The UK Plastics Industry*, Mintel International Group, London, UK, 2000.

9.   C.J. Weber, *Personal Communication*, 26th June 2001.

10. R. van Erven, *Personal Communication*, 27th June 2001.

11. R.E. Drumright, P.R. Gruber and D.E. Henton, *Advanced Materials*, 2000, 12, No. 23, 1841.

12. V.J.J. Marron in *Biobased Packaging Materials for the Food Industry*, Ed., C.J. Weber, The Royal Veterinary and Agricultural University, Frederiksberg, Denmark, 2000, Chapter 7.

13. R.J. Crawford, *Plastics Engineering*, 3rd Edition, Butterworth-Heinemann, Oxford, UK, 1998.

14. H. Domininghaus, *Plastics for Engineers: Materials, Properties, Applications (translated and revised version)*, Hanser, Munich, Germany, 1993.

15. *Mater-Bi Grade: Y101U Product Datasheet*, Novamont, Novara, Italy, 1998.

16. C. Bastioli in the *Proceedings of Recycle '94, Global Forum and Exposition 1994*, Davos, Switzerland, Paper No.4.

17. *European Plastics News*, 2001, **28**, 7, 18.

18. D.V. Plackett and T.L. Andersen in *Proceedings of the 23rd Risø International Symposium on Materials Science: Sustainable Natural and Polymeric Composites – Science and Technology*, Risø National Laboratory, Roskilde, Denmark, 2002, p.299.

19. A. Warmington, *European Plastics News*, 2000, **27**, 8, 56.

20. R. Chandra and R. Rustgi, *Progress in Polymer Science*, 1998, **23**, 7, 1273.

21. M. Heyde, *Polymer Degradation and Stability*, 1998, **59**, 1, 3.

22. B. De Wilde and J. Boelens, *Polymer Degradation and Stability*, 1998, **59**, 1, 7.

23. *The Official Journal of the European Communities*, 1994, L365, 10.

24. *The Official Journal of the European Communities*, 2000, L269, 34.

25. *The Official Journal of the European Communities*, 1975, L194, 39.

26. N.G. McCrum, C.P. Buckley and C.B. Bucknall, *Principles of Polymer Engineering, 2nd Edition*, Oxford University Press, Oxford, UK, 1997.

27. *The Official Journal of the European Communities*, 1999, L182, 1.

28. *HM Treasury UK Budget Report 2002*, The Stationery Office, London, UK, 2002.

29. S. Bagshaw, *Plastics & Rubber Weekly*, 10th May 2002, 7.

30. E. Brockes, *The Guardian*, 22nd May 2002, 6.

31. *OECD Environmental Outlook for the Chemicals Industry*, OECD, Paris, France, 2001.

32. V. Vaitheeswaran, *The Economist*, 2001, 358, 8208, 5.

33. S.C. Slater and T.U. Gerngross, *Scientific American*, 2000, August, 24.

34. *DTI Digest of United Kingdom Energy Statistics*, The Stationery Office, London, UK, 2001.

35. *Polypropylene Handbook: Polymerisation, Characterisation, Properties, Processing, Applications*, Ed., E.P. Moore, Carl Hanser Verlag, Munich, Germany, 1996.

36. *The Green and Ethical Consumer*, Mintel International Group, London, UK, 1999.

37. M.E. Porter, *Competitive Advantage: Creating and Sustaining Superior Performance*, Free Press, New York, NY, USA, 1985.

38. J. Maxwell, *Plastics in the Automotive Industry*, Woodhead Publishing Ltd., Cambridge, UK and the Society of Automotive Engineers, Inc, Warrendale, PA, USA, 1994.

39. H. Shent, R.J. Pugh and E. Forssberg, *Resources, Conservation and Recycling*, 1999, 25, 2, 85.

40. A. Smith, *Plastics, Additives & Compounding*, 2001, 3, 3, 18.

41. D. Maldas and B.V. Kokta, *Journal of Adhesion Science*, 1991, 5, 9, 727.

42. B. English in *Proceedings of ARC 1998, SPE Annual Recycling Conference*, Chicago, IL, USA, 1998, p.193.

43. J.E. Gordon, *Aero Research Technical Notes*, Bulletin No. 34, Duxford, Cambridge, UK, 1945.

44. J.A. Youngquist, *Forest Products Journal*, 1995, 45, 10, 25.

45. J.H. Schut, *Plastics Technology*, 1999, 45, 3, 46.

46. S.N. Monteiro, R.J.S. Rodriguez, M.V. De Souza and J.R.M. D'Almeida, *Advanced Performance Materials*, 1998, 5, 3, 183.

47. A.C. Karmaker and J.A. Youngquist, *Journal of Applied Polymer Science*, 1996, 62, 8, 1147.

48. M. Avella, R. dell'Erba, E. Martuscelli, B. Pascucci, M. Raimo, B. Focher and A. Marzetti in *Proceedings of ICCM/9, Volume 2: Ceramic Matrix Composites and Other Systems*, 1993, Madrid, Spain, p.864.

49. A.R. Sanadi, R.M. Rowell and D.F. Caulfield, *Polymer News*, 1996, **20**, 1, 7.

50. K.P. Mieck and T. Reubmann, *Plastics Southern Africa*, 1995, **25**, 5, 30.

51. G.C. Ellison and R. McNaught, *Research and Development Report NF0309: The Use of Natural Fibres in Nonwoven Structures for Applications as Automotive Component Substrates*. Ministry of Agriculture, Fisheries and Food, London, UK, 2000, p.88

52. C. Bastioli, *Personal Communication*, 23rd July 2001.

53. B. Lomborg, *The Skeptical Environmentalist: Measuring the Real State of the World*, Cambridge University Press, Cambridge, UK, 2001.

7 Ecodesign

Tracy Bhamra and Vicky Lofthouse

7.1 Introduction

It is now widely recognised that industry should reduce the environmental impact of its activities. The most advanced companies are moving beyond the compliance mentality and being proactive in shaping future markets, consumer needs and influencing legislative developments. They see environment as an opportunity rather than a threat, recognise that 'prevention is better than cure' and are attempting to 'design out' rather than simply manage the problems.

Sustainable development was first publicised by the Brundtland report in 1987 [1] as being:

> '... development that meets the needs of the present without jeopardising the ability of future generations to meet their needs.' [1]

Ecodesign is one strategy being used to move towards a more sustainable future, understood to be, the systematic integration of environmental considerations into the design process across the product life cycle, from cradle to grave (manufacture to disposal) [2, 3, 4]. There are various internal and external influences driving organisations towards implementing ecodesign, these are:

- *Cost savings*: integrating environmental issues into product development can result in cost savings such as less raw materials used, less waste produced, energy efficiency and water efficiency.

- *Legislative regulations*: these are becoming more and more important to companies as they are increasing in both the country in which they operate and to which they export.

- *Competition*: pioneering companies have realised that they may gain some competitive advantage by considering ecodesign.

- *Market pressure*: Ecodesign can be an effective way to improve an organisation's environmental performance and therefore help to meet the increasing market pressure associated with the environment.

- *Industrial customer requirements*: many suppliers are now being asked to meet their customer's environmental requirements.

- *Innovation*: new market opportunities can be opened up by integrating environmental issues into product development as this stimulates product and process innovation.

- *Employee motivation*: introducing environmental considerations in an organisation is an effective way to engage and motivate employees so that they can actively contribute and improve their working environment.

- *Company responsibility*: Many more companies are becoming more aware of their responsibility towards the environment and the role they need to play in a more sustainable society.

- *Communications*: Many organisations are using the environment as an effective communication tool with all stakeholders. It is providing a new mechanism to promote both the organisation and their products or services.

This mix of internal and external pressure has resulted in a real drive in many industrial sectors to consider ecodesign as an integrated part of all product development.

7.2 Development of Ecodesign

Traditionally, ecodesign was considered to be solely in the realm of the scientists who provided environmental support and guidance for production. By the late 1990s it was recognised that to be most effective ecodesign should also be considered earlier in the product development process during design. It had been recognised that waste generation occurs as a direct consequence of design decisions [5]. It is clear that these early design stages are of critical importance in determining the environmental impacts of products, as it is estimated that between 80-90% of a product environmental and economic costs are determined here [6, 7]. Also, the cost of any environmental intervention is at its lowest and most flexible at these early stages - also known as 'front-loading' [8]. The 'early stages' were seen as important because it was here that the design brief is most flexible and the most critical decisions with respect to: cost, appearance, materials selection, innovation, performance, environmental impact, and perceptions of quality (longevity, durability, reparability), are made [7, 9]. Thus the earlier consideration of ecodesign issues reduces the costs involved and improves the effectiveness of the changes made. In addition to this it was also recognised that to be most effective the practice of ecodesign needs to be an integral part of the product development process (PDP) (in the same way as quality and safety) rather than being something which is added on at the end [2, 4, 10]. The DEEDS Project (DEsign for Environment Decision Support) – a 3 year collaborative research project between Cranfield University and Manchester Metropolitan University sponsored by the EPSRC (Engineering and Physical Sciences Research Council) - aimed to assess and describe the ways in which industry was integrating ecodesign [11]. A key conclusion from the research was that ecodesign needs to be integrated 'early' and included within or before the product specification.

> *'A number of companies discussed this early stage of design and highlighted how important it is to ensure that the environment is considered as early as possible. There was recognition that beyond a certain point in the design process, it is extremely difficult to alter certain product features that are key to the environmental performance.'* [4].

Bakker [9] draws similar conclusions about design involvement at the early stages, but describes this as 'product planning', rather than pre-specification. This requires involvement in, not only the operational stages of detailing designs, but also in more strategic decision making, such as what product to design at all. Van Hemel [12] has correctly pointed out that this form of innovative product development is more closely associated with 'new business' than to traditional product design.

It was the influence of this research finding that encouraged the focus of ecodesign to move up the product development process from: end of pipe considerations – where waste and pollution are 'managed' after they have been created; towards the earlier stages of product development – where waste and pollution (and increased costs created with this) are 'designed out' in the first place [13]. **Figure 7.1** illustrates this shift.

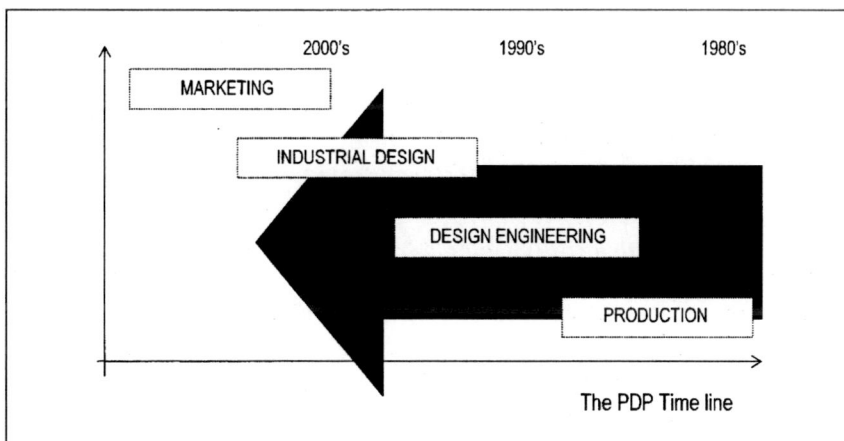

Figure 7.1 Illustration of how consideration and concern for environmental issues has moved towards the earlier stages of product development [14]

7.2.1 Ecodesign Theory

There are a variety of different ways to understand and conduct ecodesign. These may require different approaches and may achieve differing levels of environmental improvement.

7.2.1.1 Definitions and Descriptions of Ecodesign

The last two decades have seen a proliferation of terminology relating to the incorporation of environmental considerations into design. This has included ecodesign [15]; environmentally conscious design [16]; Design for the Environment [12]; Life Cycle Design [17]; EcoRedesign [18]; and green, ecodesign and sustainable design [10]. Keoleian and Menerey [17] point out that these differing terminologies may be explained by the traditions from which they developed. Design for the Environment developed out of the Design for X (DfX – where X can be assembly, manufacture, etc.) concept which is prevalent in engineering. Life Cycle Design in contrast, having developed from environmental sciences and environmental management terminology has a more scientific orientation [17]. Each of these, though similar, may have specific nuances and practices, or represent varying degrees of environmental integration or innovation. In the literature, the term ecodesign has both a specific and generic usage.

The earliest attempts to describe and define ecodesign tended to focus on its integration into existing design practice, in balancing it with other design considerations such as cost, quality, functionality, etc. It was also extensively promoted as a means of enhancing design and company profitability, public perception, and reducing costs. Keoleian and Menerey [17] reflect this, in describing Life Cycle Design as a:

> '... *systems-orientated approach for designing more ecologically and economically sustainable product systems which integrate environmental requirements into the earliest stages of design. In Life Cycle Design, environmental, performance, cost, cultural, and legal requirements are balanced.*' (p.650)

Other environmentally related design terms come from Dewberry and Goggin [10]. They propose three ecodesign approaches as: green design, ecodesign, and sustainable design:

- *Green design:* has a single-issue focus, perhaps incorporating the use of some new material, such as recycled or recyclable plastic, or consideration of energy consumption.

- *Ecodesign:* adopts the lifecycle approach, exploring and tackling all or the greatest impacts across the product's lifecycle.

- *Sustainable design:* would take a more broad and holistic approach, including: questioning/addressing needs, concern for ethics and equity, services and leasing, dematerialisation, empowerment, caring and sharing, as well as incorporating ecodesign best practice.

There are clearly a variety of definitions and ways to understand ecodesign. Definitions do not merely describe differing types of ecodesign but different approaches, which allude to different knowledge and operating domains.

7.2.1.2 Characteristics of Each Approach

Literature from a variety of authors indicates distinct differences in approaches to ecodesign. For the purposes of this chapter these will be characterised as 'innovative' and 'incremental' approaches to ecodesign innovation:

- *Incremental (improvement, evolutionary) approach:* where environmental issues are incorporated into design in an evolutionary approach. It uses existing products, business models or forms of development as its starting point. These are viewed as technical or technological problems, which will lead to sustainable development in the more long-term. This tends to deal with the following factors: optimisation, efficiency, technology, new materials, and existing product redesign. It has a single product or environmental issue focus.

- *Innovative (radical, revolutionary) approach:* where environmental considerations are used as the driver for new and more radical concept development. This uses a more revolutionary approach arguing that existing products and patterns of production and consumption can and would never lead to sustainability. It is viewed as a marriage of technology, culture and nature (though many authors select two of the three). It tends to deal with the following factors: effectiveness, innovation and creativity, mimicking natural principles and ecological models, and engaging cultural and lifestyle factors. It is multi-disciplinary, extending beyond single or traditional product and company boundaries.

7.2.2 Ecodesign Models

7.2.2.1 Life Cycle Thinking

The life-cycle concept is a 'cradle to grave' approach to thinking about products, processes and services. It recognises that all product life-cycle stages (extracting and processing raw materials, manufacturing, transportation and distribution, use/reuse, and recycling and waste management) have environmental and economic impacts.

Government, business and non-governmental organisations can apply the life-cycle concept to their decision-making processes related to environment and product policy, design, and improvement. The life-cycle approach can also be used as a scientific tool for gathering quantitative data to inventory, weigh and rank the environmental burdens of products, processes and services.

A product's life cycle (illustrated in **Figure 7.2**) is made up of the activities that go into making, using, transporting and disposing of that product. The life cycle is commonly shown as a series of stages, from 'cradle' (raw material extraction), through manufacturing, packaging, transportation, consumption, and recycling, to the grave (disposal).

Organisations are usually directly involved in one - or perhaps more - of these stages. Why is the life cycle concept relevant to the environment? The environmental problems

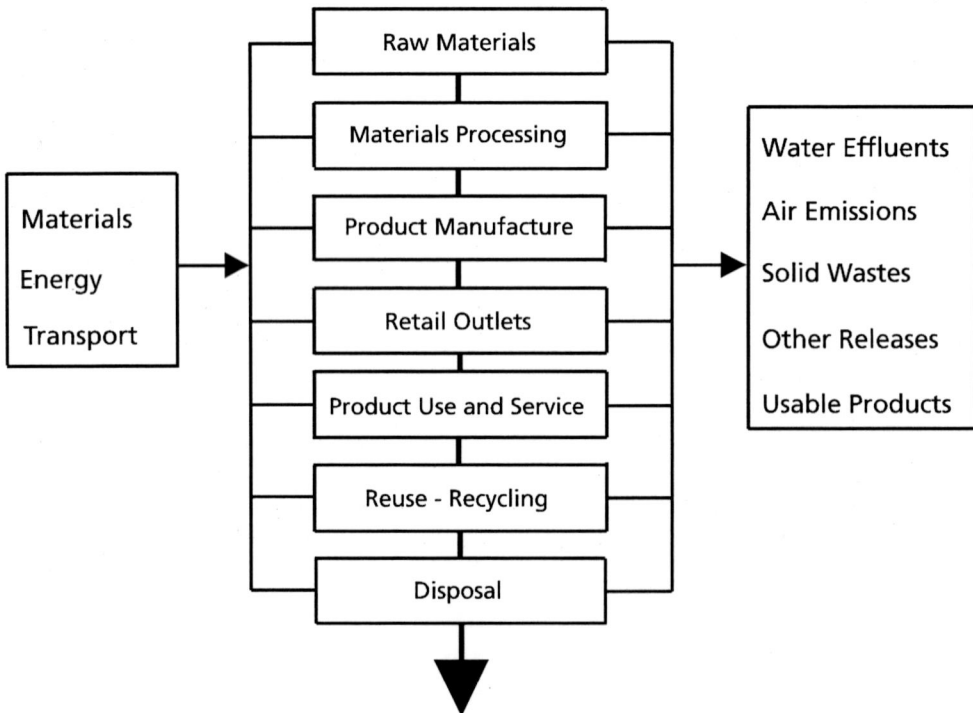

Figure 7.2 The product life cycle

associated with a given product can be traced back to the inputs that go into the product (land, materials, water, energy), and the outputs generated, (e.g., air emissions, liquid effluents, solid wastes), at each stage in the life cycle.

Life cycle thinking is a useful way of starting ecodesign as it can help to identify and prioritise the actions needed for environmental improvements.

7.2.2.2 Ecodesign Innovation

This section considers models of ecodesign. Brezet [19] proposes a four-step model of ecodesign innovation consisting of differing design criteria and considerations. The four steps are described as:

- *Product improvement:* The improvement of existing products with regard to pollution prevention and environmental care. Products are made compliant.

- *Product redesign:* The product concept stays the same, but parts of the product are developed further or replaced by others. Typical aims are increased reuse of parts and raw materials, or minimising the energy use at several stages in the product life cycle.

- *Function innovation:* Involves changing the way the function is fulfilled. Examples include a move from paper-based information exchange to e-mail, or private cars to 'call-a-car' systems.

- *System innovation:* New products and services arise requiring changes in the related infrastructure and organisations. A changeover in agriculture to industry-based food production, or changes in organisation, transportation and labour based on information technology.

Figure 7.3 illustrates this model and shows that to move from level 1 to level 4, increasing amounts of time and complexity are required, which has the potential to lead to higher eco-efficiency improvements. This model suggests that these more complex ecodesign innovations will only be achieved over a significant time period, say 10-20 years.

7.2.3 Ecodesign Practice

Current and 'best' practice in ecodesign lags some way behind the maturing theoretical and research framework. This chapter aims to describe this and in doing so it relates various studies on 'best' ecodesign practice back to the theoretical framework.

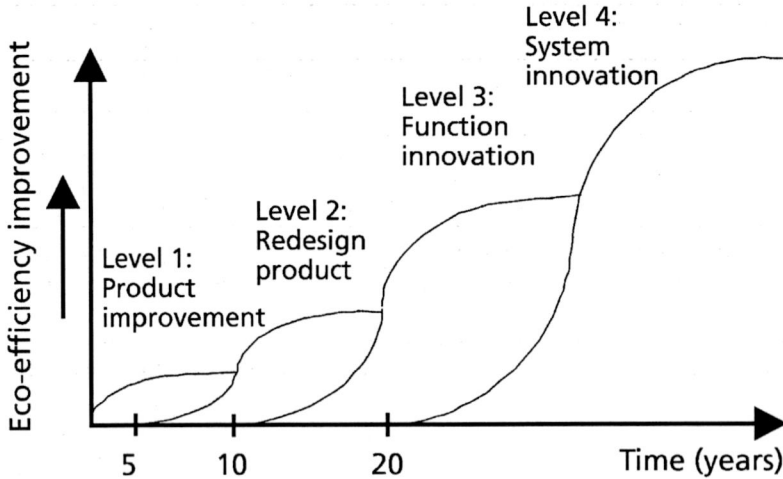

Figure 7.3 Four-stage model of ecodesign innovation [19]

7.2.3.1 Design Practice

Surveys of UK ecodesign practice by Dewberry [3] and Sherwin and Chick [20], highlight its immaturity. Both in-house and design consultancies demonstrated poor levels of ecodesign awareness and little practice. Indeed, Dewberry's [3] original survey aimed to assess UK 'best' practice, but had to be revised to 'better' practice as ecodesign was so poorly developed. She concluded that the most advanced companies were nowhere near 'sustainable design' practice, with most demonstrating only random examples of 'green design or ecodesign'. Her key conclusions are shown in **Table 7.1**.

Van Hemel [12] and Brezet [19], both describe ecodesign 'best' practice as at level 2 (product redesign) of Brezet's earlier ecodesign innovation model (**Figure 7.2**). Here the product concept remains the same, but parts of the product are developed further or replaced by others. This ties in strongly with both the surveys mentioned previously, as ecodesign practice is currently at very best incremental. Van Hemel [12] states that ecodesign integration and practice tend to be evolutionary in their nature or more 'incremental' in the approaches defined here.

An Open University study on Green Product Development [21], found that leading ecodesign products were generally developed without any specific environmental targets

Table 7.1 Conclusions from UK ecodesign survey [3]	
Issues	Conclusions
Current state of the art for UK designers' attitudes to ecodesign	Generally reactive; confused; sympathetic but have little understanding of environment; frustrated and cynical; cope and comply with clients/management
Opportunities and constraints for ecodesign	Time is critical factor; cost savings are seen as opportunity; ecodesign increases design criteria and creativity
Information for ecodesign	New information is needed, which needs to be: specific and focused; understandable in presentation; accessible; 'hands-on'
Positioning of ecodesign	Ecodesign is not a priority; designers are not involved in strategic decisions within product development; there is a lack of communication

or goals. It was often after the development process had commenced, that environmental benefits were recognised and exploited:

> '... it is not surprising that most of the companies adopted an incremental or ad hoc 'green design' approach to the incorporation of environmental factors in product development...

None of the firms in this study routinely adopted a systematic 'ecodesign' approach to product development in an attempt to reduce/balance environmental impacts over the whole lifecycle of the product from 'cradle to grave', including both product and process impacts' [21]

7.2.3.2 Ecodesign Integration

A comprehensive overview of how industry integrates ecodesign comes from McAloone [16]. He proposes an integration model for the electrical and electronics sector, consisting of five factors critically important to success. These were initial and sustained motivation, communication/information flow, whole-life thinking, hands-on environmentally conscious design, and positioning in the world. Each in turn has a number of criteria and characteristics as shown in **Figure 7.4**.

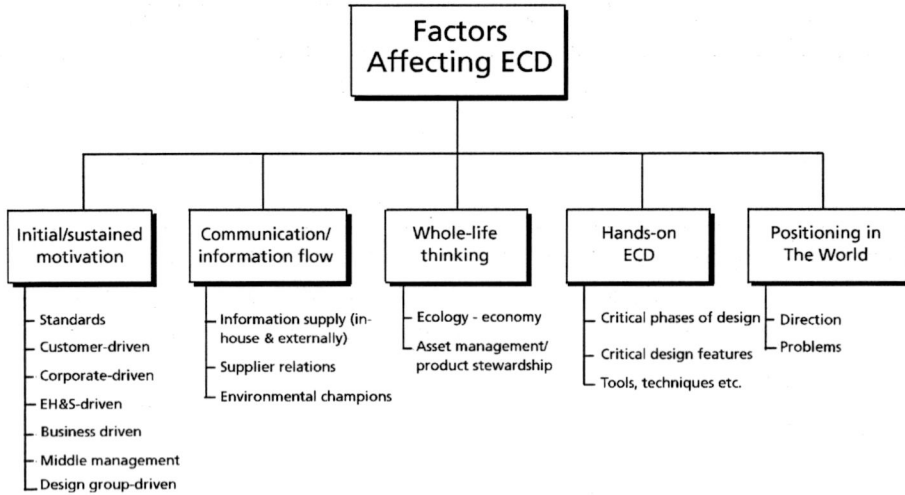

Figure 7.4 Model of ecodesign integration [16]. ECD: Ecodesign

McAloone [16] explores several of these integration themes in greater detail. Three factors describing the way companies integrate ecodesign are illustrated next:

- *The timing of environmental decisions is the key to environmentally conscious design:* Environmental decisions made in the pre-specification stage of design have greater impact on the product.

- *Enthusiasm is the key to environmentally conscious design:* Involvement of decision-makers that have the enthusiasm to find solutions to problems is more important than detailed environmental subject knowledge (companies tend to appoint an environmental champion).

- *Senior management commitment is the key to environmentally conscious design:* Without top management commitment (manifest through the provision of resources, company environmental visions statement, the commitment to achieve recognised environmental standards, the support of environmental training schemes, and corporate membership of external environmental forums) environmentally conscious design does not become an integral part of the design process.

Many obstacles to more successful and innovative ecodesign are often stated as external to design [3]. These can include the structure of the organisation, the profile and role of the design team, or the nature of design task specified. Dewberry [3], Bakker [9] and Sherwin and Chick [20] all concluded that designers work mostly in an operational role. Here design briefs, product specifications and important decisions were made by clients or senior management, and designers have little influence over these more strategic, 'early' design decisions that are so critical to successful ecodesign innovation:

> *'Surprisingly, a large proportion of the designers interviewed indicated that for many of the design projects they undertake, the design brief is basically decided before it is even handed to them. One might presume that the design decision process is taking place at a management level within the company or by the client.'* [3]

7.3 Implementing Ecodesign

A wide range of processes and tools have been developed over the last decade, to support the integration of ecodesign [11, 16, 22-27]. Within this section the authors intend to provide an introduction to the types of processes and tools that are available.

In the process of integrating ecodesign into the PDP, companies usually go through a number of distinct stages [27]:

1. Identifying the product.

2. Identify goal and ecodesign targets.

3. Develop first ideas.

4. Research and compile environmental data sheets.

5. Evaluation.

6. Realise and launch.

It is the second of these stages, 'identifying the goals and ecodesign targets' that we specifically wish to focus on.

There are a number of different ways of identifying the most appropriate ecodesign goals or targets for a given product. As ecodesign targets are product specific, it is not possible to state, for example, that it is always better to consider the reduction of energy consumption as the primary target. Instead, decisions with respect to the functionality and expected life of the product have to be taken. A rough analysis of an existing or similar product is often enough to allow a design team to identify the most important issues to tackle for the product they are working on. For example, for a cooker, the use phase is the most environmentally intensive, so reduction of the energy consumption is considered to be an important issue.

7.3.1 LiDS Wheel

Van Hemel [12] with Brezet [22] propose a series of ecodesign strategies and principles described within the LiDS wheel (LIfe Cycle Design Strategies). As an integrated tool, the LIDS wheel aims to represent strategies for both the incremental and the innovative approaches to ecodesign.

The principles and strategies are presented as a directional wheel numbered from 1 to 8. Importantly, van Hemel points out that the ecodesign strategy wheel is hierarchical and relates to the various stages of the product development process. Thus as one moves through strategies 1 to 8, one moves from the late to the very early stages of the product development. Strategy 8 (New concept development) represents the very earliest (product planning or strategic) stages, where the 'need' is defined and where there are highest degrees of freedom. Strategy 1 (selection of low impact materials) deals more with later (detail design) product development stages:

> 'This development, from product system level, via establishment of the functional structure at product structure level, through to the search for design options for product details at product component level, is reflected in the Design for the Environment (DFE) strategy wheel.' [12]

As such, the LiDS Wheel (see **Figure 7.5**) is a comprehensive summary of ecodesign strategies at every stage of product design and development and represents all (or many) types of ecodesign.

Once goals and targets have been identified, the team will generate potential solutions using a range creativity techniques, including brainstorming, stimulus (case studies, competitors products). For example, in the case of the cooker, designers might consider solutions such as – alternative energy supplies to existing fossil fuel, such as solar or kinetic power; reduced energy requirements, through improved product efficiency or through increased insulation; approaches to encourage improved user education, to capitalise on the fact that ovens can be switched off 15 minutes before the end of the cooking time without reducing the heat of the oven. **Table 7.2** presents a number of different ecodesign issues that might be considered.

Once a number of ideas have been generated, they can be evaluated in terms of their economic and environmental potential. It is suggested that all potential options are assessed

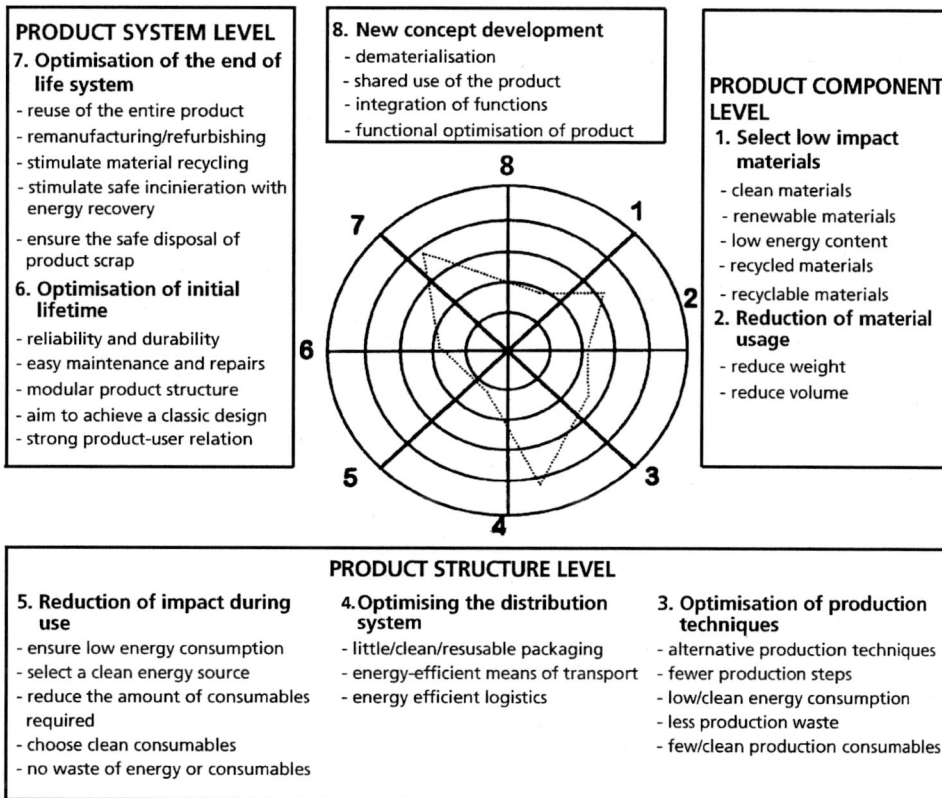

PRODUCT SYSTEM LEVEL
7. Optimisation of the end of life system
- reuse of the entire product
- remanufacturing/refurbishing
- stimulate material recycling
- stimulate safe incinieration with energy recovery
- ensure the safe disposal of product scrap

6. Optimisation of initial lifetime
- reliability and durability
- easy maintenance and repairs
- modular product structure
- aim to achieve a classic design
- strong product-user relation

8. New concept development
- dematerialisation
- shared use of the product
- integration of functions
- functional optimisation of product

PRODUCT COMPONENT LEVEL
1. Select low impact materials
- clean materials
- renewable materials
- low energy content
- recycled materials
- recyclable materials

2. Reduction of material usage
- reduce weight
- reduce volume

PRODUCT STRUCTURE LEVEL

5. Reduction of impact during use
- ensure low energy consumption
- select a clean energy source
- reduce the amount of consumables required
- choose clean consumables
- no waste of energy or consumables

4. Optimising the distribution system
- little/clean/resusable packaging
- energy-efficient means of transport
- energy efficient logistics

3. Optimisation of production techniques
- alternative production techniques
- fewer production steps
- low/clean energy consumption
- less production waste
- few/clean production consumables

Figure 7.5 The LiDS wheel [22]

as to their suitability by considering their environmental benefits, market opportunity, technical feasibility and financial feasibility. Trade-offs then have to be made between these four areas when selecting the option to take forward [28].

Providing designers with qualitative guidelines is a common support mechanism for ecodesign. The literature shows that guidelines are presented in a variety of ways, such as 'rules of thumb', 'checklists', 'guidelines', 'questions' or 'statements' which relate to either a specific topic (such as packaging) within the life cycle, or the whole product life cycle. The aim of these 'guidelines' is to remind designers of the important ecodesign issues that need to be considered during the development process (see **Table 7.3** for an example).

In addition to checklists, recent research has shown that designers also need to be provided with examples of case studies to stimulate creative thinking and supportive, appropriate information [14].

Table 7.2 Potential ecodesign issues for consideration	
Raw Materials Acquisition and Processing	• Conservation of natural resources, land, water, air • Protection of natural habitats and endangered species • Waste minimisation and pollution prevention • Transportation • Use of renewable resources; sustainable use of resources • Use of degradable resources • Use of recycled materials • Energy consumption
Manufacturing and Distribution Issues	• Minimal use of materials • Toxic use/release • By-product/waste generation and handling • Energy consumption • Water use • Emissions to air, land and water
Product Use and Packaging Issues	• Energy efficiency • Conservation of natural resources such as water required for the use of the product • Consumer health and environmental safety
After-Use/Disposal Issues	• Recyclability, ease of reuse, remanufacture and repair • Durability • Biodegradability/compostability • Safety when incinerated or landfilled

7.4 Examples of Ecodesign Projects

A number of the more proactive companies are starting to make good advances in the implementation of ecodesign principles, this includes companies such as Sony, Philips, Electrolux, Xerox, IBM and Kodak. This section will provide a number of short case studies of industry-based ecodesigned products and outline the practical ways in which improvements have been made in these proactive companies.

7.4.1 Case Study 1: Philips NV

Since the setting up of the Eco Vision programme in 1998, Philips NV have become world leaders in the commitment that they show to environmental considerations of their product development. They do this by aiming to embed environmental issues into

Table 7.3 Abridged Summary check: Design for Recycling [31]				
Design for recycling materials and components	Y	N	N/A	Comments
Has the number of different materials been limited where practical?				
Are all plastic parts greater than 50 g marked to ISO 1043 [29] and 11469 [30]?				
Has surface coating been avoided?				
Has the product part count been reduced and by how much?				
Have pigmented plastics been avoided where aesthetics are not a concern?				
Have recycled plastics been used wherever practical?				
Has the product been designed so that it can be upgraded?				
Can any value be obtained from the return of parts to the supplier?				
Is there a disposal manual or information indicating which parts are recyclable and how to process them?				
Does the manual/information give special instructions for disposal of hazardous material?				

product policy and strategy from the very beginning [32]. They focus at four different levels:

- product development – creating an environmentally sound product that satisfies customer needs and can be efficiently produced;

- supplier involvement – contracting and cooperating with suppliers on environmental aspects;

- production – developing and establishing environmentally sound processes; and

- marketing – addressing eco-performance parameters in market-driven activities.' [32]

Rather than selecting single products as flagship, or hero products, they try to apply environmental considerations to the whole product range. The latest Environmental

Report 2001, states that 51% of their products are now considered to be ecodesigned. To become a 'Green Flagship' three focal areas need to have been considered and must offer better environmental performance than predecessors/competitors in two or more focal areas. Philips' 'Green Focal Areas' are: weight, hazardous substances, recycling and disposal, energy consumption, and packaging [32].

Examples of Philips Green Flagships, include their 28 inch widescreen TV which are on average 10% lighter and use 49% less energy than competitors and the recently launched Philishave Quadra action 6000 Series Shaver (a waterproof, electric razor) which weighs 38% less and uses 16% less energy than the nearest competitor [33]. These products benefit the environment by using less resources and less energy over their lifetime.

7.4.2 Case Study 2: Dishlex

Dishlex (a subsidiary of Email Major Appliances, Australia's largest manufacturer and distributor of white goods) took part in the EcoReDesign Programme which was run at Royal Melbourne Institute of Technology (RMIT) in 1997 to develop a better understanding of how to design dishwashers which would be more environmentally responsible [34]. Dishwashers have a number of environmental problems including high energy use due to heating of water for the washing cycles and hot air to dry, pollution caused by the detergents that are used and high water use.

The project had a high level of management buy-in, involved a substantial team from Dishlex and the Centre for Design and RMIT, with a high level of commitment and interaction being maintained throughout the project [34]. Those involved in the project, recognised that the people were the key to the project's success as the:

> '... haphazard pursuit of ecodesign principles and specification of benign or recyclable materials in isolation, without the will to break unproductive and environmentally indifferent attitudes to product development, rarely generates sustainable outcomes.' (p.312) [34]

The final dishwasher design had some fairly revolutionary environmental improvements. These included:

- A six-star energy rating: actual energy consumption of 256 kWh compared to competitors, which ranged from 273-330 kWh.

- 'Wash programmes compatible with enzyme based detergents, making low-temperature cycles highly efficient at cleaning while using minimal energy.'

- A water conservation rating of A, consuming 17.9 litres on the most efficient wash programme

- A four stage micro-filtration system which contributed to its ability to use less water more efficiently.

- 'Major components designed for easier disassembly and recycling'.

- Plastic components coded to ease disassembly.

- Fewer material types.

- Reduced weight, 7 kg less than previous models [34].

7.4.3 Case Study 3: Kodak's Recyclable Camera

The Kodak Single Use Camera programme provides a good example of an ecodesign project which focused on the redesign of the product, to reduce its environmental impact. The programme set up in 1990-1991 set out to redesign single-use cameras to facilitate recycling and reuse of parts. The new design 'featured parts that were easier to inspect and to reuse, and had an easier means for reloading film' [35].

In conjunction with the redesign, the recycling concept took into consideration that consumers must take the single-use cameras to be developed, which provided an ideal opportunity for the company to take back their product for recycling. Kodak set out to capitalise on this and set up a system to get the camera shells back to the company once the customer's film had been removed. The following system was implemented. Once the cameras are collected all the packaging, front and back covers, and the batteries are removed. The cameras are cleaned and visually inspected. Old viewfinders and lenses are replaced with new ones for quality purposes. For flash models, fresh batteries are inserted. Many small parts are reused, including the thumbwheels for advancing the film and the counter wheels for keeping track of exposures taken. The entire camera frame, metering system and flash circuit board are re-used following rigorous quality testing. Sub-assemblies are then shipped to manufacturing plants for final assembly into new product. A new roll of film, battery (where necessary) and outer packaging made from recycled materials (with 35% post-consumer content) are added [35].

The plastic front and back covers are checked for traces of metal and then shipped to be re-ground into pellets to be remoulded into cameras, etc., and cardboard packaging is recycled. The batteries which are removed from the recycled cameras are tested and, if good, are:

- used internally at Kodak, i.e., in employee pagers

- donated by the company to various organisations as part of a gifts-in-kind programme

- sold through third parties on the wholesale or retail markets as recycled batteries

Recently, Kodak have also developed labels which are 100% recyclable (made from a squeezable opticite material with high-quality printing and adhesion characteristics, and also can be reground, and recycled), saving the time previously needed to remove the labels [35].

7.4.4 Case Study 4: Eco Kitchen

One final example of an ecodesign project, which is slightly different in nature to those outlined previously, was the collaborative Ecodesign kitchen of the future project, carried out by the Electrolux Industrial Design Centre and Cranfield University. This project was entirely conceptual in nature and hence did not result in 'actual' products, instead it led to the development of seven visionary concepts that aimed to raise customer awareness and to improve understanding of the benefits of ecodesign, within the company [36]. **Figure 7.6** provides an illustration and description of one of the concepts generated, 'The Smart sink' [36].

Figure 7.6 The Smart sink [13]

The Smart sink is the centre of household water management. A membrane sink of expanding material grows when filled, to help minimise water use and the smart tap switches from jet, to spray, to mist, to suit your needs. A consumption meter and a water-level indicator in the main basin give feedback on rates and level of water usage. Household grey water is managed visibly by an osmosis purifier and a cyclone filter located in the pedestal, both linked to the household grey water storage.

The approach used within this project differed quite considerably from those used within the other case studies. Instead of focusing on improving elements of the product, it took a much more holistic view of the 'problem' and considered the function of the kitchen as a whole. It focused on using background information and scenarios, such as:

- 'Information on environmental impacts in the kitchen (waste, water and energy use, household purchasing, etc).

- Specific life cycle analyses of products, or information on specific consumer lifestyles issues in the kitchen.

- Food, health and dietary tips and suggestions.

- Environmental targets and goals facing designers (resource reduction targets, etc).

- Examples and case studies of ecodesign products for:

  - The kitchen
  - Industrial Design based ecodesign

- Future trend suggestions and requirements (such as re-introduction to biological processes, sustainable values, etc).' (p.100) [37]

This example of ecodesign helps to indicate that it is possible to go a stage further than has been suggested so far in this chapter to begin to use environmental issues for product innovation rather than the 'corrective' activity. It suggests that there is added value in moving beyond the view that the environment should be integrated into the project brief in the same way as other design criteria, to placing it at the top of the list of design priorities as a driver for innovative design [13].

7.5 Conclusions

This chapter has shown how ecodesign has the potential to play an important role in the move towards sustainability. It has outlined both the current theory and practice of ecodesign highlighting some successful case studies of how ecodesign has been applied. Ecodesign itself can be implemented using very simple techniques which enable organisations to use their creativity to look at reducing the environmental impact of products or services often leading to more innovative and cost effective solutions.

In the future, if organisations are to move beyond ecodesign towards more sustainable products there will need to be a more revolutionary approach to the way in which products and services are developed. New strategies will need to be used in design that create outcomes that are environmentally acceptable, socially desirable and economically appropriate.

However, many organisations are at the start of this journey, and are yet to begin considering the environment during product development. For these organisations there is one simple message: start now and start simply. By considering a few of the areas of ecodesign it is possible to start making real improvements to products, services, processes and organisations.

References

1. G. Brundtland, *Our Common Future: The World Commission on Environment and Development*, Oxford University Press, Oxford, UK, 1987.

2. T.E. Graedel and B.R. Allenby, *Industrial Ecology*, Prentice Hall, Upper Saddle River, NJ, USA, 1995.

3. E.L. Dewberry, *EcoDesign - Present Attitudes and Future Directions, The Design Discipline Technology Faculty*, Open University, Milton Keynes, UK, 1996. [PhD Thesis]

4. T. Bhamra, S. Evans, M. Simon, T. McAloone, S. Poole and A. Sweatman, *Proceedings of EcoDesign '99: First IEEE Symposium on Environmentally Conscious Design and Inverse Manufacturing*, Tokyo, Japan, 1999.

5. S. van Weenan and J. Eeckles, *The Environmental Pressure*, 1989, **11**, 231.

6. P. Burall, *Product Development and the Environment*, Gower, Aldershot, UK, 1996.

7. T. McAloone, *Co-Design*, 1996, 05-06, 76.

8. *Electrolux The Global Appliance Company - Environmental Report 1997*, Electrolux, Stockholm, 1997.

9. C. Bakker, *Environmental Information for Industrial Designers*, Delft University of Technology, Delft, The Netherlands, 1995. [PhD Thesis]

10. E. Dewberry and P. Goggin, *Co-Design*, 1996, 05 06, 12.

11. M. Simon, S. Evans, T. McAloone, A. Sweatman, T. Bhamra and S. Poole, *Ecodesign Navigator - A Key Resource in the Drive Towards Environmentally Efficient Product Design*, Cranfield University, UK, 1998.

12. C. van Hemel, *EcoDesign Empirically Explored - Design for Environment in Dutch Small and Medium Sized Enterprises*, Delft University of Technology, 1998. [PhD Thesis]

13. C. Sherwin, and T. Bhamra, *Proceedings of Ecodesign '99, First International Symposium on Environmentally Conscious Design and Inverse Manufacturing*, Tokyo, Japan, 1999.

14. V.A. Lofthouse, *Facilitating Ecodesign in an Industrial Design Context: An Exploratory Study*, Cranfield University, 2001. [PhD Thesis]

15. *Proceedings of an Eco2-IRN Workshop on Defining 'Ecodesign'*, Manchester, UK, 1994.

16. T. McAloone, *Industry Experiences of Environmentally Conscious Design Integration: An Exploratory Study*, Cranfield University; 180, 1998. [PhD Thesis]

17. K.A. Keoleian, and M. Menarey, *Journal of the Air & Waste Management Association*, 1994, **44**, 644.

18. C. Ryan, *Eco Design*, 1996, **4**, 1, 5.

19. H. Brezet and C. van Hemel, *Industry and Environment*, 1997, **20**, 1-2, 52.

20. C. Sherwin and A. Chick, *Ecodesign Survey: A Study of the Chartered Society of Designers (CSD) membership*, The Centre for Sustainable Design, Farnham, UK, 1997.

21. M.T. Smith, R. Roy, and S. Potter, *The Commercial Impacts of Green Product Development*, Milton Keynes, Open University, Design Innovation Group, Milton Keynes, UK, 1996, 1-57.

22. H. Brezet and C. van Hemel, *Industry and Environment*, 1997, 19, 1-2, 59.

23. Philips Electronics NV, *From Necessity to Opportunity - Corporate Environmental Review*, Philips Electronics NV (CEEO), Eindhoven, The Netherlands, 1997.

24. RMIT, *Introduction to EcoReDesign - Improving the Environmental Performance of Manufactured Products*, Melbourne, Victoria, Australia, 1997.

25. *1999 Environment, Health and Safety Progress Report - Towards Sustainable Growth*, Xerox Corporation, Stamford, CT, USA, 1999.

26. M. Charter in *Sustainable Solutions - Developing Products and Services for the Future*, Eds., M. Charter and U. Tischner, Greenleaf Publishing Ltd., Sheffield, UK, 2001, 220-242.

27. U. Tischner in *Sustainable Solutions - Developing Products and Services for the Future*, Eds., M. Charter and U. Tischner, Greenleaf Publishing, Sheffield, UK, 2001, 263-281.

28. J. Cramer and A. Stevels in *Sustainable Solutions - Developing Products and Services for the Future*, Eds., M. Charter and U. Tischner, Greenleaf Publishing, Sheffield, UK, 2001, 326-339.

29. ISO 1043, *Plastics – Symbols and their Abbreviated Terms*, 2001.

30. ISO 11469, *Plastics – Generic Identification and Marking of Plastic Products*, 2000.

31. T. Clark and G. Adams, *Eco-Design Checklists For Electronic Manufacturers, 'Systems Integrators', and Suppliers of Components and Sub-assemblies*, Version 2, The Centre for Sustainable Design, Farnham, UK, 2002, 1-38.

32. *Philips Environmental Improvement Programme*, Koninklijke Philips Electronics NV, 2002.

33. *Philips Environmental Policy*, Koninklijke Philips Electronics N.V. 2002.

34. J. Gerstakis in *Sustainable Solutions - Developing Products and Services for the Future*, Eds., M. Charter and U. Tischner, Greenleaf Publishing Ltd., Sheffield, UK, 2001, 118-138.

35. Eastman Kodak, *A Tale of Environmental Stewardship: the Single-Use Camera*, 2002.

36. P. Thompson and C. Sherwin in *Sustainable Solutions - Developing Products and Services for the Future*, Eds., M. Charter and U. Tischner, Greenleaf Publishing Ltd., Sheffield, UK, 2001, 349-363.

37. C. Sherwin, *Innovative Ecodesign - An Exploratory and Descriptive Study of Industrial Design Practice*, School of Industrial and Manufacturing Science, Cranfield University, Cranfield, UK, 2000. [PhD Thesis]

8 Casein Adhesives

Dennis Price

8.1 History

Casein is the main protein in cow's milk, and depending on the treatment route, can be used to make either glues and plastics, or cheese.

The calcium caseinates form an insoluble white curd when acidified. This acid casein is used widely in wood-working adhesives, water paints, for coating paper, and in printing textiles and wallpaper. In neutral solutions the enzyme rennin (EC 3.4.23.4) can convert caseins to an insoluble curd; this being the dominant protein in cheese. When treated with formaldehyde the curd can be formed into casein plastic. These casein plastics were amongst the first commercially developed plastics, with a German patent being issued in 1899 to Kirsch and Spitteler for a 'plastic composition'. The plastics were used to make washable 'white boards' for use in schools to replace slates which kept getting broken and because paper was just too expensive for children to practice their writing on. Applications in the field of composites manufacture are limited as the compound is usually made up into stock shapes for machining, as it is not well suited to moulding processes.

This section will therefore concentrate on the application of casein adhesives as a composite matrix material. Like much naturally occurring glue, casein has been used for thousands of years, for example, as the pigment binder in ancient Egyptian wall paintings, and more recently in Europe in tempera paintings. In these early times it was often used in the crude form of precipitated curd. The commercial marketing of casein for glue seems to have had its beginnings in the early nineteenth century in Germany and Switzerland, and by the middle of the century patents for formulated glues began to appear in the USA. The beginning of the twentieth century saw the marketing of ready formulated dry powder mixes for industrial use, mostly for wood to wood bonding. The need for quality water-resistant wood glues during World War II gave casein-based glues a high level of commercial importance. At its peak the industry was producing 50,000 tonnes per year of casein powder for conversion to various adhesives. This lasted until after World War II by which time the development of lower cost synthetic adhesives such as phenol-formaldehyde, resorcinol-formaldehyde and urea-formaldehyde began to significantly reduce the casein wood glue market which was dealt a final blow by the introduction of the poly vinyl acetate emulsions. While caseins were replaced by the synthetics for cost and ease of use it must not be forgotten that caseins have a long history of sustained bond strength with water resistance up to BS EN 204/205 D3 [1, 2] (or the old BS 1204 MR [3]) [4]

Casein still continues to be used but in the form of a liquid or paste for the labelling of glass bottles. Here it is still successful due to its combination of high wet tack, good water resistance of the dry film but coupled with the useful property of being easily washed off in the mild alkali conditions of most bottle washing systems: something not yet matched by synthetic adhesives. In combination with synthetic latex, casein is a very successful adhesive for the bonding of wood or paper to metal.

8.2 Manufacture

The casein is isolated from skimmed cows milk by acidifying it to pH 4.5: under these conditions a curd separates and after washing, drying and grinding, yields raw casein at a rate of about 3 wt% from the original milk. This low yield goes some way to explaining the high cost of casein powder. A superior process, producing a higher yield and a better product, uses the addition of bacterial cultures producing lactic acid to adjust the pH to the desired level. A further alternative is to use hydrochloric acid as an acidifying agent.

Many factors affect the quality of the final casein product. These range throughout the manufacturing process, starting with the quality of the dairy cows, their grazing and hence the quality of the milk. Once the milk has been transported to the processing plant, the quality and thoroughness of the washing and the choice of method of acid precipitation, are especially important. The matching of agricultural and industrial processes make the manufacture of high quality casein a particular challenge for the industrial chemist.

8.3 Types of Casein Glues and Their Uses

8.3.1.

A typical formulation for casein powder glue for wood-working is given in **Table 8.1**.

This is mixed 2 parts water to 1 part powder (again by weight) and thoroughly mixed for a minute It is left to stand for 10-15 minutes and mixed again. This allows time for the following reactions to take place.

Table 8.1 Formulation of a typical wood glue	
Constituent	Weight, wt%
Casein	55
Hydrated lime ($CaCO_3.6H_2O$)	20
Sodium carbonate (Na_2CO_3)	15
Wood flour	6
Preservative	1
Paraffin	3

The lime, which is initially in excess, reacts with the sodium carbonate to produce sodium hydroxide; this alkaline solution then dissolves the casein. The blend shown in **Table 8.1** would give a working pot life of about 5 hours at 20-25 °C, a closed assembly time of up to 45 minutes and a pressing time of about 1 hour. These times will however vary depending on environmental conditions such as substrate moisture content, pressing pressure, and process temperature.

As a rule of thumb, an increasing excess of lime decreases the working life of the mix, but the increases water resistance of the adhesive. The water resistance is controlled by the level of conversion of soluble sodium caseinate into insoluble calcium caseinate by reaction with the excess lime as the adhesive bond forms. Other sodium salts such as phosphate, sulfite or fluoride can replace the sodium carbonate. Each will bring a slightly different property to the final mix. For example, to stick paper backing to aluminium foil, some surface etching of the substrate is required and for this purpose the fluoride salt is preferred.

The other components each have a part to play in the bonding process. The wood flour lowers cost, and prevents over penetration into many substrates by controlling the flow properties of the mixture. Thus chemical and processing properties of the mix can be tuned to suit particular applications.

Addition of preservative is extremely important otherwise the mix could be attacked by bacteria or fungi (particularly if stored in damp conditions), similarly biological attack on the finished joint will weaken it.

The paraffin has a duel purpose, initially it suppresses the dust from the dry mix during handling, and later as a component of the aqueous solution aids wetting out and moderates the interaction between the reactive components.

While the previous blend does not have or require it, plasticisers in the form of glycerol or sulfonated oils may be added to reduce the hard brittle nature of the dry glue line.

Due to the high cost of casein it is possible to substitute part of it with the lower cost soya flour. This will then require some reformulating of the blend to achieve a higher pH, increasing the polar nature of the soya flour, otherwise its adhesive properties are poor. While it is possible to do either part or even whole substitution of the casein with soya, practice has shown that the best compromise is about a 50:50 ratio of the two [5, 6].

There are some drawbacks to the use of casein glues based on this type of formulation. They have limited life once the powder is mixed with the water. Usually if any is left over at the end of the day it can be 'blended off'(any mix made but not used at the end of the working day can be saved and added to the first mix of the next day, provided that it has not thickened or become contaminated with bacteria of fungi) in the first mix of the next day, but care must be taken that it has not gone too far and is beginning to gel or has picked up a bacterial infection. The later can be detected by an associated terrible smell, like sour milk initially but if not dealt with promptly then going on to become even more powerful and rancid. For this reason the wash down water from the spreaders and mixers needs to be well diluted and even treated with additional biocide. Partially due to the high pH of the mix and the nature of the casein this type of glue will easily stain any wood surface that it contaminates and is not easily removed, so requires care to ensure clean working on susceptible surfaces.

8.3.2 Label Pastes

Casein-based adhesives are still favoured for this application as they combine a set of properties that are not readily found in synthetic adhesives. These are high wet tack, a level of water resistance in the dry film that allows bottles to be stored in chill cabinets or immersed in iced water and the ability to resolubilise in warm mild alkali which means the labels can easily be washed off, so that the bottle can be reused. Unlike the powder adhesives these are made as a liquid with a good storage life by the manufacturer.

Simply, this is achieved by a combination of adjusting the pH to between 8-9 with ammonia and heating to around 80 °C. The actual practice is a little more complex and is done in a number of stages. This process gives the basic casein solution, which then requires further additions and modifications. The water resistance is achieved by the addition of a metallic salt or oxide, the most favoured being zinc oxide. Other materials added can be viscosity modifiers such as urea and rheology modifiers/extenders such as wheat flour and china clay [7, 8].

To fully appreciate how critical the properties of the adhesive are, the process of labelling a bottle on a high speed machine will be described next. The casein adhesive is pumped from a container by the side of the machine to the top of a roller applicator. It then flows down by gravity between the rotating roller and a fixed doctor blade, which adjusts to set the level or coating weight of the adhesive. At the bottom of the roller it then runs into a tray and back down into its original container. First, the adhesive has to flow well under both high shear conditions in a pump and freely under low shear gravity feed. However, it is essential that sufficient remains behind the doctor blade to provide a continuous coherent film but not so much as to flood the machine and starve the original container.

In the next process a ribbed rubber pad passes over the application roller and is coated with a film of adhesive. This pad then passes over the inside of the front label of a cassette of labels and picks it off. Then via a small cut out at one end of the pad the label is picked off by a small pair of mechanical pincers and swung round into the path of the bottle to be labelled. At the point where the label comes in contact with the bottle the pincers release the label and it transfers to the bottle. The label is now on the bottle but standing out from the surface, so a set of brushes then wipes the label down onto the surface of the bottle. With modern machines this all happens in 1 or 2 seconds so the adhesive has to adapt very quickly. In addition immediately after this operation the labelled bottle exits the machine and can undergo various other stresses such as rubbing against metal guide rails that could move or dislodge the label. The adhesive must not let this happen.

As can be seen the process and end-use demand a great deal from the adhesive in terms of control of rheology and tack and all the required properties are provided by casein-based systems.

8.3.3 Casein Latex

Initially these systems were developed for bonding paper to aluminium foil for packaging. They were used either to give a bright background to the print on the outside of the pack or to provide an hermetically sealed pack to retain the freshness and taste of some food products like powdered soups, etc. When initially described the system sounds quite simple. But the process of combining an ammoniacal solution of casein, (similar to the base solution for the label adhesive) with styrene-butadiene or neoprene has to be carefully regulated because it is possible to destabilise one or other of the components and end up with a coagulated solid. The blends have to be formulated to meet fairly stringent rheological properties as these materials are run on high speed roller application machines so they must not 'throw', 'string', or foam all of which will damage the coating or dirty the machine and prevent it running. In addition they must not destabilise and coagulate under the shear of the rollers. As these adhesives are use in direct contact with food they must be free from odour, colour and toxic ingredients [9, 10].

With the advent of surface aluminised paper and polyester, the need for such laminating adhesives started to decline. They have however found a new use for bonding aluminium to wood for panels for caravans and fire engines or as a primer for steel and aluminium for bonding with other types of adhesive such as resorcinol-formaldehydes and urea formaldehydes where different properties are needed from the finished item.

8.4 Current and Future Markets

Casein based adhesive systems have declined in traditional areas over the years more on the basis of cost than performance. They remain only in certain areas where as yet the synthetics cannot match the performance needs. However work is in progress in these areas and even these could be lost eventually. The cost of casein has risen as milk production has been diverted to other more lucrative areas such as milk powders for pre-processed ready-to-use-foods.

However, the particular properties of casein adhesive, namely biological origin and controlled biodegradability, with concurrent benefits in terms of end-of-life disposal may well mean that this material is due a well deserved renaissance along with other little used adhesives such as dextrin and other animal or protein-based systems. This is because the relatively high material cost can be offset against the low cost of disposal by composting, as compared to the increasing cost of landfill, and the environmental unattractiveness of incineration.

For more information on the history of casein-based plastics the reader is referred to the web site of the historical plastics society *http://www.plastics-museum.com.*

References

1.   BS EN 204, *Classification of Thermoplastic Wood Adhesives for Non-Structural Applications*, 2001.

2.   BS EN 205, *Adhesives - Wood Adhesives for Non-Structural Applications - Determination of Tensile Shear Strength of Lap Joints*, 2003.

3. BS 1204, *Specification for Type Mr Phenolic and Aminoplastic Synthetic Resin Adhesives for Wood*, 1993.

4. D.F.G. Rodwell, *Ageing of Wood Adhesives - Loss in Strength with Time, Building Research Establishment*, Glasgow, UK, 1984.

5. C.N. Bye in *Handbook of Adhesives*, 3rd Edition, Ed., I. Skeist, Van Nostrand Reinhold, New York, NY, USA, 1990, p.135.

6. W.D Detlefsen in *Adhesives from Renewable Resources*, Eds., R.W. Hemingway and A.H. Connor, ACS Symposium Series No.385, ACS, Washington, DC, USA, 1987, p.445.

7. M.P. McKenna, *Paper, Film & Foil Converter*, 1988, **62**, 1, 38.

8. L. Broich, B. Herlfterkamp and H. Onusseit, inventors; Henkel KGAA, assignee; US 5455066, 1995.

9. H. Klein, *Plastverarbeiter*, 1977, **28**, 9, 480. [In German]

10. *Converter*, 1981, **18**, 1, 18.

9 PHA-Based Polymers: Materials for the 21st Century

Marcia Miller, James Barber and Robert Whitehouse

9.1 Introduction

The current practice of extracting energy and petrochemicals from fossil carbon is not sustainable. Fossil carbon use generates high emissions of CO_2 that are already significantly increasing atmospheric CO_2 levels, and will contribute to global warming with consequent real impact on climate change. Most of the industrially developed countries are not self-sufficient in the 'cleaner' forms of fossil carbon such as gas and oil. The mining of coal is environmentally offensive, and its use contributes to the buildup of atmospheric CO_2 more than any other fossil carbon source. The increasing gap between demand for fossil carbon for both strategic raw materials and energy, and domestic supply has led to greater reliance on foreign sources of fossil carbon, in particular the politically unstable Middle East, and has profound implications for global foreign policy. The trend to explore other global regions as additional sources of oil and gas focuses on areas that are climatically less hospitable (e.g., North Sea, Alaska, Siberia, off-shore South America), are, in some cases, more environmentally fragile, are less well understood, and may require new and untried recovery technologies.

A sustainable and environmentally benign approach to supplying society's needs for energy and materials is to harness alternative natural resources such as solar, wind and geothermal. Solar energy combined with atmospheric carbon dioxide is the basis of plant life and, ultimately, all life. The current rapid development of biotechnology will enable transformation of atmospheric carbon dioxide into useful energy and materials other than food products.

Polyhydroxyalkanoates (PHA) offer a viable alternative to traditional, petrochemically-derived plastics. These polymers are produced in low levels by many living organisms and comprise a large family of polymeric materials whose properties range from stiff, highly crystalline materials to rubbery elastomers to amorphous, tacky substances.

Premature efforts to commercialise PHA foundered because of the twin deficiencies of natural microbial producers: low efficiency (hence high cost) and a very limited range of polymer architecture. However, the proprietary, groundbreaking industrial genome engineering of Metabolix, Inc., has enabled both the highly efficient production of PHA in microbial strains and a broad range of polymer architectures and properties. Metabolix, located in Cambridge, Massachusetts, USA has developed new microbial strains that produce high levels of PHA through fermentation. No other family of polymers based on renewable resources spans even a fraction of the property space of PHA, which can, in some cases, extend even beyond the performance characteristics achievable with conventional polymers. This makes them potential candidates to replace over 50% of the polymer materials now synthesised from oil and natural gas. In the future, direct production in non-food plant crops will bring production costs down even further.

9.2 History of PHA

The first PHA identified in nature for possible industrial use was polyhydroxybutyrate (PHB). PHB is a highly crystalline thermoplastic sharing many properties with polypropylene (PP). However, the processing window between melting of the polymer and thermal degradation is relatively small. Over the past five decades, over a hundred other PHA polymers have been described in the scientific literature. The properties of these polymers range from stiff, highly crystalline materials like PHB to soft thermoplastics like polyhydroxyoctanoate to other PHA that are highly elastic with good tensile set properties and more like vulcanizeised rubber.

9.3 Production

There are two general approaches to the manufacture of PHA: production in microbial biofactories and directly in plants. Current production uses fermentation to produce PHA. Metabolically engineered, highly efficient microbial systems can incorporate a range of co-monomers into the fermentation process by co-feeding sugar and other low cost carbon sources to produce a wide range of well-defined polymer compositions. **Figure 9.1** shows the current fermentation process (top) and direct production in plants (bottom).

Microbial production is typically carried out in aerobic fermenters, feeding sugars or oils and, optionally, other co-feeds (**Figure 9.1**). Fermentation combines virtually unlimited flexibility in PHA structure with economics suitable for many specialty applications with manufacturing costs approaching $1.3 per kilogram. A PHA fermentation typically lasts less than 40 hours. In the early part of the fermentation process, bacterial cells are allowed to multiply without producing PHA. Then, carbon utilisation in the cell is shifted from growth to polymer production. At the end of the fermentation, the broth consists of cells

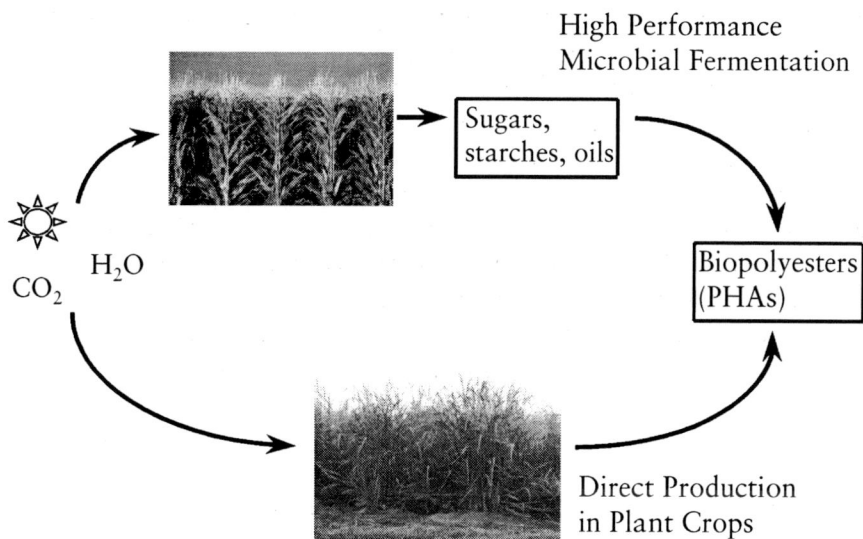

Figure 9.1 Current and future PHA production (via fermentation directly in plants)
© *Metabolix, reprinted by permission*

containing between 80% and 95% of polymer by dry weight. Subsequent processing can yield the PHA product in the form of resin or latex.

Metabolix uses transgenic *Escherichia coli* as the biological platform for the majority of its current fermentations for several reasons:

- It grows quickly without special nutrient requirements.

- Its entire genome sequence is known.

- Its metabolism is well-characterised.

- It is relatively easy to break open.

- It does not possess enzymes which could degrade PHA, simplifying downstream processing.

- It lends itself to many compositions, simplifying fermentation strategies.

Very high volume industrial markets are quite price sensitive, and will eventually be served by PHA produced directly in non-food crop plants. Metabolix foresees that achieving high yield production of PHA in plants will enable the production of bio-polyesters and derived chemicals at costs competitive with those of the existing petrochemical industry and has the potential to provide a renewable foundation for plastics and chemicals. The company has developed proprietary transgenic technology enabling PHA to be produced directly in plants, using sunlight, carbon dioxide, and water, and is overcoming the deficiencies of earlier efforts, which have fallen far short of demonstrating an economic system.

In October 2001, Metabolix received a US Department of Energy 'Industry of the Future' award to develop PHA production directly in switchgrass (*Panicum virgatum*), a leading candidate for bio-based energy production in North America because of its high yield and the fact that it can be grown on land of marginal use for other crops. Switchgrass also fixes about 2 kg of carbon dioxide in its root system for every kg of biomass above the ground. Once the PHA are extracted, the resultant biomass can be used as a fuel source for energy production, thus it is a net energy producer. Yields of switchgrass average about 5 tons per acre in the US today, and experimental plots have demonstrated yields of about 25 tons per hectare over a six year period, with yields as high as 37 tons per hectare observed. It can be expected that the application of the tools of modern biotechnology will enhance these yields further. Other plant targets, such as sugarcane, which yields about 75 tons of biomass per hectare, will also be attractive sources of PHA and energy.

The transfer of proprietary PHA genes into plants results in an accumulation of PHA polymers in specific sites, such as seeds or green tissue. Copolymers in plants may vary in composition because of changing climatic conditions and soil nutrient compositions in the growing regions. Through post-chemical modification of the plant-based PHA, optionally with fermentation-derived polymers, a range of consistent product grades can be produced (**Figure 9.1**).

A key question in any biomass-based energy scheme is whether meaningful amounts of energy can actually be generated agriculturally. The answer to this question is a very plausible yes. For example, switchgrass has an energy content of about 1.6×10^{10} J/dry ton. Assuming that yields of 25 to 37 tons per hectare can be reached (and there is no reason to believe that they will not be), then roughly 40.5 million hectares of switchgrass will provide the energy content equivalent to much of the oil currently imported into the US, i.e., 9 million barrels per day, or 3.2 billion barrels per year. To put this land use into perspective, the US has about 174 million hectares designated as cropland today, and the Federal government pays farmers **not** to cultivate about 16 million hectares (calculated from US Department of Energy statistics) [1].

A major problem for agriculture in the developed world today is that farmers produce far more than is needed of traditional crops, and are subsidised in all countries through one mechanism or another. Advanced, industrialised nations subsidise exports of farm produce to developing countries where agriculture is a value adding activity appropriate to their early industrial stage of development. New, value-generating crops are needed, and PHA producing plants can fill that need.

9.4 Applications

Applications for PHA cover a wide range and the market potential is profound. Through the design and introduction of new metabolic pathways, Metabolix has produced a wide range of PHA polymers and copolymers based on the hydroxyacids of generic structure [-OCH(R)(CH$_2$)$_n$CO-]. By varying fermentation conditions and carbon feeds, Metabolix has the ability to change the alkyl side group R from hydrogen to dodecyl and the number of methylene units in the main chain from 1 to 3, and can make homopolymers, copolymers and terpolymers. The pathways for co-monomer incorporation have been engineered and demonstrated for several copolymer families, and others are under development. Although the company is still in the process of investigating the broad design space associated with this range of structures, it has produced polymers with glass transition temperatures (T$_g$) varying from +6 °C to –55 °C and melting points (T$_m$) from around 190 °C to 40 °C. This design space is comparable to that of the acrylic polymers being marketed by companies such as BASF and Rohm & Haas. PHA, however, have additional benefits over acrylics of ultra high molecular weight and crystallinity. Metabolix can provide polymers with molecular weights ranging from less than 1,000 to 1,000,000 or more. By varying their composition, the crystallinity of these polymers can be changed from around 10% to as high as 60% or 70%. Adhesion properties can be modified by varying the polymer structure to change hydrophobicity: the surface energy of PHA can be varied between that of polyethylene terephthalate (PET) (considered to be polar) and PP (considered to be non-polar).

Table 9.1 summarises the physical properties available from Metabolix's PHA polymers. The physical property design space available with PHA creates many and varied market opportunities. Figure 9.2 illustrates this breadth of design space.

The market segments for PHA fall into two broad categories:

- Specialty product markets based on fermentation

- Very large volume markets based on direct plant-based production.

Table 9.1 Summary of physical properties of PHA ©Metabolix, reprinted by permission		
Property	Units	PHA
Glass transition, T_g	°C	+5 to −55
	°F	+9 to −67
Melting point, T_m	°C	60 to 186
	°F	140 to 367
Density	g/cm³	1.10 to 1.25
Tensile strength	MPa	7 to 70
Elongation @ break	%	5 to >1200
Tensile modulus	GPa	0.8 to 2.5
Flexural modulus	GPa	0.2 to 3.5
Moisture vapor transmission rate	g/50 m/m²/day @ 23 °C/90% RH	20 to 150 (medium barrier)
RH: *relative humidity*		

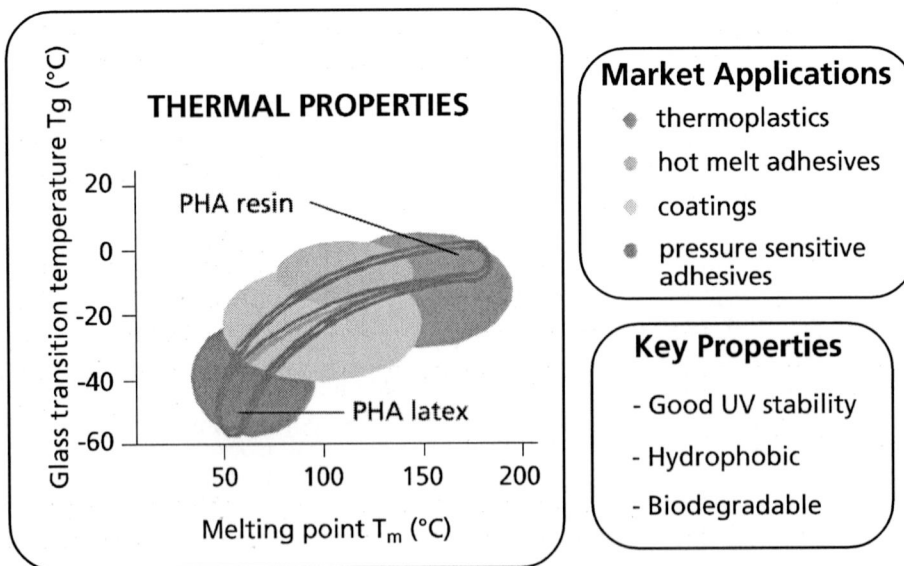

Figure 9.2 Thermal characterisation of PHA relative to market opportunities

© *Metabolix, reprinted by permission*

PU are a good illustration of the diverse markets and applications served by an existing family of specialty polymers with a broad range of physical properties. In 2000, over 2.2 billion kilograms were used in North America alone, in markets as diverse as construction, transportation, furniture, carpet, appliances, packaging, bedding, footwear, and many others. In meeting the needs of these markets, PU are used as flexible and rigid foam, moldings, coatings, binders, adhesives, sealants and elastomers. The acrylate family of polymers provides another good example of a common technology platform that has been exploited in a broad range of product forms.

The design spaceshown in **Figure 9.2** illustrates that PHA are extremely versatile products that may be used in a wide range of applications. PHA are unique candidates for complete packaging solutions because not only can they provide film and moulded products but also can provide coatings, adhesives and inks that are biodegradable.

Moisture vapour permeability is critical in food packaging where product longevity is required. PHA have moisture vapour barrier resistance that is an order of magnitude higher when compared to other natural and synthetic biodegradable polymers, and are comparable to PET (as shown in **Table 9.2**). Biopol grades (poly-3-hydroxybutyrate-*co*-valerate) have food contact approval in Europe.

There are no current hydrophobic degradable hot melt adhesives that can meet the high speed laminating and technical performance required in current packaging applications. Starch-based adhesives are extremely sensitive to water, can fail under high humidity environments, and are difficult to recycle. Metabolix PHA can potentially be formulated to compete with the existing high speed hot melt laminating adhesives based on ethylene vinyl acetate, styrene isoprene styrene and styrene-butadiene rubber resins, and also offer the potential to be recycled through paper pulping processes or be 'organically recycled' via composting (domestic or managed sites).

PHA are also stable for long periods of storage, but will degrade in marine, fresh water, soil, and composting environments and are compatible with repulping processes. PHA can be produced in both resin and latex forms, and are ideal for applications such as film, high gloss paper and board coatings, degradable inks and adhesives. They can outperform other sustainable products because of their combination of hydrolytic stability, water vapour barrier, and mechanical properties.

Non-woven fabrics made of cost-effective, biodegradable, but hydrolytically stable PHA will finally enable 'flushability' to be offered in sanitary, (e.g., feminine hygiene products and diapers), and cleaning, (e.g., wipes), markets. There are no acceptable technical solutions yet for products that can provide the level of hydrophobicity needed during use and the ability to break down and biodegrade under defined waste treatment processes. In 1998, the total market for fibres used in non-woven applications was around 1 billion

Table 9.2 Comparison of moisture vapour barrier resistance. ©Metabolix, reprinted with permission	
Polymer	g/m²/day @ 23 °C/90% relative humidity 50 µm film
Biaxially orientated PP	3-5
Biaxially orientated PET	10-15
Nylon 6	15
Metabolix PHA (depending on morphology)	20-150
Annealed polylactic acid (crystalline)	3400
Polycaprolactone	3600
Ecoflex (BASF)	3400-3600
Ecostar (Eastman)	3400-3600
Bionelle (Showa Denka)	6600
BAK 1095 (Bayer)	14000
Cellulose acetate	58400
Cellulose proprionate	34000
Data compiled by Metabolix	

pounds broken down by segment into 41% hygiene, 12% medical, 14% wipes and 33% industrial. Hygiene and wipes markets are projected to grow at a rate 5-6% and 10-12%, respectively. Major suppliers of feminine and other personal hygiene products and consumer wipes are evaluating PHA-based non-woven materials for these applications.

Green solvents can be made from PHA derivitives such as methyl and ethyl 3-hydroxybutyrate that are highly polar solvents with high boiling points and low volatility. Water miscible, they are strong candidates for paint strippers, and cleaners for microelectronics. They can be converted into a wide range of chemical intermediates such as alkanoic acids, hydroxyacid esters and dihydroxy oligomers (used for polyester, polyurethanes and radiation curing coatings). They can also be used as a source for beta hydroxy acids such as glycolic acid, lactic acid and other compounds used in skin care applications.

An important family of applications for PHA-use water-based dispersions of polymer particles, called latex. Latex products are designed to provide good film-forming characteristics. A particularly attractive segment for latex type products is as water-resistant coatings for paper and board, where improved water resistance coupled with the ability to recycle the waste product is important. Currently, water resistance is obtained by using non-biodegradable hydrophobic coatings based on polyolefin waxes or styrenic and acrylic latexes. The need to dispose of these items either by incineration or landfill routes is limiting the growth of this market. 41 billion kilograms of paperboard products are produced annually in the US alone, requiring 0.4-1.1 billion kilograms of binder. To date, no acceptable hydrophobic polymers have been identified that allow easy recycling or composting of coated board products.

Medical gloves and other protective articles make up another attractive market segment. The spread of AIDS and infectious hepatitis has led to a massive increase in the use of short-service-life protective gloves. However, this increased usage has resulted in a corresponding increase in the number of people who exhibit allergic reactions to certain proteins present in traditional natural rubber latex gloves or to a number of the chemicals used to cure and modify the rubber. The market in short-service-life gloves is large and annual growth rates are increasing rapidly: the US consumed over 35 billion gloves using around 600 million pounds of rubber latex in 1999; over 5 billion were surgical gloves, valued at around $250 million and using over 80 million pounds of rubber latex. Worldwide, surgical gloves used over 120 million pounds of natural rubber latex. Current alternatives are deficient in physical properties such as tear and puncture resistance, uncomfortable to wear, or do not provide the surgeon with the right 'feel'. Bringing PHA latex products to these markets will have real benefits.

A specific, large volume segment for latex products is in the paints and coatings industry. In 2005, the North American coatings industry alone is projected to consume approximately 8 billion gallons of product with a value of $73 billion. Approximately 40% of that volume would be used in aqueous interior and exterior coatings. PHA latex can be used directly or readily blended with conventional acrylic or vinyl acetate latex products to improve abrasion resistance for architectural coatings.

PHA-can be also be used to enhance the performance of other polymers, including such environmentally-friendly materials as polylactic acid and starches. For example, flexible and elastomeric PHA can be compounded with fairly stiff PLA to enhance its ductility. Water resistant PHA can be blended with water-sensitive starches for more water-resistant adhesives and coatings, or with thermoplastic starches for packaging film that is less likely to leak.

Another benefit of PHA is its reduced disposal and environmental impact. Despite great efforts to recycle plastics and chemicals in industrial countries, most plastics eventually wind up in landfills. Benign PHA biodegradation will be an advantage in packaging and disposable goods (e.g., in 2000, over 17 billion pounds of polyethylene, PP, and polystyrene were used in film, blow-moulded items, and other forms of packaging) and in durable goods such as automobiles. Usage of 0.75 billion and 3.75 billion kilograms of PHA per year that could be recycled through composting sites would conserve in excess of 0.9 million and 4.5 million cubic metres of landfill, respectively, now occupied by non-degradable plastics. The carbon dioxide generated through composting of PHA is originally derived from atmospheric CO_2, and hence does not negatively impact the CO_2 biocycle. Alternatively, the burdens of plastic waste disposal to both industry and to communities struggling with available landfill volume can be reduced by PHA which can be cleanly incinerated, and their energy content recovered.

One industry that will benefit from the biodegradability of PHA is the automotive industry, which must increasingly take responsibility for the end-of-life fate of the automobile. PHA fibres, combining stability to water, high-performance, and value-adding properties with compostability, offer solutions other materials cannot. In addition, blending PHA with certain natural fibres, like cotton, jute, flax, etc., will allow them to become very competitive and full life-cycle costs will be attractive. Headliners and trunk insulation panels are just two specific examples of applications.

Green solvents can be made from PHA. For example, methyl and ethyl 3-hydroxybutyrate are highly polar solvents with high boiling points and low volatility. They are water miscible and are strong candidates for use as paint strippers and cleaners for microelectronics. They can be converted into a wide range of chemical intermediates such as alkenoic acids, hydroxyacid esters and dihydroxy oligomers (used for polyesters, PU and radiation curing coatings).

In the near future PHA will be found in a diverse range of performance products, including feminine hygiene, adhesives, paints and coatings, medical products, food packaging, paper treatments and automobiles. Metabolix also sees a potentially huge agricultural market in regions like North America, Europe and Asia where PHA can be used for mulch film and as geomembranes in soil stabilisation projects.

Although PHA can be cost-effective solutions to customers' problems, experience shows that unless PHA are comparable to or better than current products in technical performance and overall cost in use, no additional value is associated with either 'renewable-ness' or biodegradability. Therefore, the size of prospective markets for PHA, and indeed for any candidates that might play a significant role in a "post-petroleum" marketplace, is directly related to the cost of production. While biodegradability will be a strong selling point for some consumers and manufacturers, performance and value will bring in customers. Although PHA and other natural polymers such as polylactic acid (PLA) may not entirely replace petroleum-based polymers, there is a huge market awaiting sustainable plastics that prove to be cost-effective with equal or superior qualities.

Metabolix estimates that sales of PHA could reach several hundred million dollars within the next five years as the biopolymers find applications in more and more products.

PHA are a transforming technology for the polymer and chemical industries, putting them on a sustainable basis, decreasing their environmental footprint, and answering the need for an effective alternative to the current use of materials whose manufacture and use cause significant damage to an increasingly burdened global environment. Sustainable development was eloquently defined by the World Commission on Environment and Development as 'development that meets the needs of the present world without compromising the ability of future generations to meet their own needs.' Sustainable solutions are those that are viable over the long-term and set a standard for protecting environmental quality, improving economic well-being, and promoting higher quality of life. They use a 'best practice' approach to increase energy efficiency and manufacturing productivity, and the use of innovative technologies. Metabolix is actively working to develop real, sustainable alternatives to the use of conventional materials derived from fossil carbon. Replacing traditional plastic products with biopolymers can cut down on the energy needed for manufacturing and help reduce waste, improve air quality, conserve petroleum supplies and reduce cleanup costs, thus increasing the options for plastics management in our world. Is the promise of biopolymers real? The answer is yes, and Metabolix has developed the technology base to make this promise an exciting reality.

Reference

1. *The Technology Roadmap for Plant/Crop-Based Renewable Resources 2000*, US Government Department of Energy, Washington, DC, USA, 1998.

10 Renewable Resource-Based Polymers

Mike O'Brien

A new polymer platform derived entirely from natural-based resources is creating new product possibilities on a global scale. The polymer, NatureWorks (Trademark of Cargill Dow LLC) PLA, was developed from a revolutionary process to produce lactide and polylactide (PLA). Made by Cargill Dow LLC, located in Minnetonka, Minnesota, USA, NatureWorks is changing the industry and the future of bio-based products. The company now offers a family of polymers that compete directly with hydrocarbon-based polymer materials on a cost, production and performance basis.

NatureWorks PLA is a commercially viable alternative to traditional materials. For consumers, this means new alternatives for apparel, household and industrial fabrics using a more sustainable material that bridges the gap between natural fibres, such as silk and wool, and synthetic fibres. Consumers also have the opportunity to benefit from packaging that is natural-based and fits all current disposal options, including being able to degrade in municipal compost facilities.

The project to develop PLA was started by Cargill Inc., in 1988 to add value to starch processed by the company. Dr. Pat Gruber, now Vice President and Chief Technology Officer for Cargill Dow, was the initiator and project champion, and along with a small group of scientists and engineers at Cargill, invented the key lactic acid to polylactide step.

Competing technologies for PLA did exist but were regarded as higher-cost options. Cargill also had key technology patents for stabilising the polymer, leaving only very low concentrations of monomer in the final product. This technology is critical for a polymer to have the high-temperature melt processability required for conversion.

Cargill worked with various polymer partners in the period of 1989-1994, but proceeded for the most part on their own. In 1994, the company built an eight million-pound/year PLA facility in Savage, Minnesota, to prove lactic acid to PLA technology on a larger scale.

This plant was then used to perfect the manufacturing technology and allow development of a commercial market for PLA. While construction of the new 140,000 metric tons facility in Blair, Nebraska, USA, was underway, the Savage facility operated at full capacity to keep up with market demand for NatureWorks PLA.

In early 1995, Cargill realised it needed a partner with a presence in the polymer market. Decision makers felt Cargill alone did not have the necessary credibility, especially when several other PLA producers were perceived as leading the way. Cargill therefore assembled a list of potential partners and The Dow Chemical Company emerged as the best candidate. A joint evaluation agreement was entered into and the Dow team quickly realised that the cost-effective manufacturing of PLA, with a range of properties and applications was achievable.

Customer interest was also very strong, so in 1997 the 50/50 Cargill Dow joint venture was launched. The expertise of both Dow and Cargill were critical, as the project moved forward. One partner's competences in one area would be backed up by the other's competences in another.

10.1 NatureWorks PLA – The Technology

NatureWorks PLA was developed on the principles of green chemistry and is made through a process of simple fermentation, distillation and polymerisation of natural plant sugars. The process uses fermentation to make two chiral isomers of lactic acid from dextrose, a conventional fermentation route impossible with chemical synthesis, and these are then chemically cracked to form three lactide isomers. Various combinations of the lactides are combined to generate a range of polymers.

The polymer chains come in a number of shapes and lengths, and making useful materials from them requires that the amount of each in the mixture be controlled, as the properties of the resulting material depend on the proportions of its components. The conventional and more expensive way of doing this was to exploit the fact that different PLA have slightly different solubilities. However, it has been recently discovered that the soluble polymers also have different boiling points. That means they can be separated by distillation and Cargill Dow has perfected an inexpensive way to do this which is shown diagrammatically in **Figure 10.1**.

Cargill Dow is currently using the sugars derived from corn (maize) as the raw material for PLA. Eventually, NatureWorks PLA will be made using other sources of biomass (plant stalks and leaves) as feedstock. Using a less expensive feedstock than dextrose would cut the cost of making PLA, as well as allow the manufacture of more novel products.

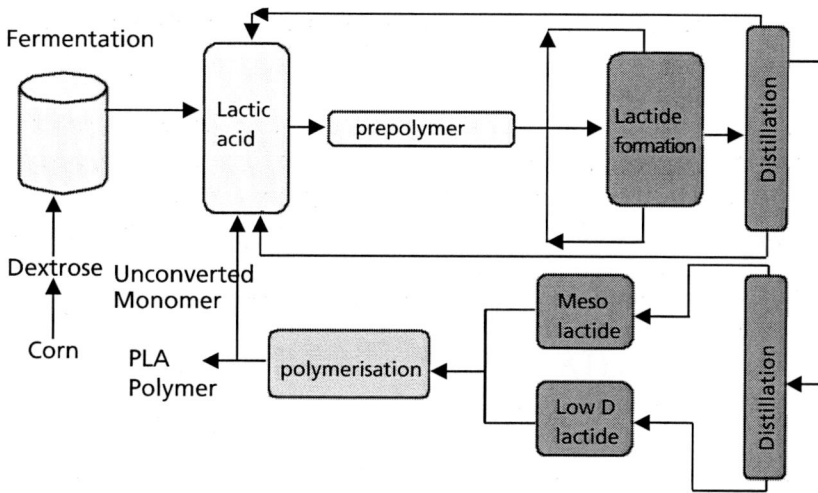

Figure 10.1 Non-solvent process to prepare polylactic acid

10.2 Performance Without Sacrifice

What truly makes NatureWorks PLA a reality for converters, mills, manufacturers, brand owners, retailers and consumers is its exceptional performance in the fibre and packaging markets. In fact, NatureWorks PLA performs equal to or better than traditional materials in many key performance areas, including consumer goods ranging from clothing and furnishings to food containers and single-serve bottles.

When NatureWorks PLA resin is converted into fibre, it provides a bridge between natural fibres, such as wool, cotton and silk, and conventional synthetics. It is melt processable, complements natural products and combines the performance advantages of both natural and synthetic materials. The result is a unique property spectrum allowing the creation of products with superior feel, drape, comfort, moisture management, UV resistance and resilience. The performance benefits provided by NatureWorks fibres allow for a wide range of applications in the apparel, industrial and institutional fabrics, industrial carpet tiles and non-wovens industries.

The excellent performance capabilities of NatureWorks PLA – combined with a unique blend of physical properties – makes it equally well suited for a wide range of films, rigids and single-serve bottle applications in the packaging industry. For example, most technical experts often mistake NatureWorks for polystyrene or cellophane based on its gloss, clarity and stiffness. NatureWorks also has a low temperature heat seal layer and/

or flavour and aroma barrier in co-extruded structures where its combination of properties allows layer simplification or replacement of polyamides.

In addition to outstanding gloss and clarity, the relative ease of processing that NatureWorks PLA exhibits in thermoforming enables it to be used for both conventional and form-fill-seal applications. Its stiffness enables more efficient down-gauging *versus* incumbent materials.

With a 2.1% haze, the material is ideal for applications like window envelope films. The United States Postal Service conducted optical reader tests on envelopes with windows made from NatureWorks PLA films – the results demonstrated equal performance to petroleum-based films and far superior performance to envelopes made with glassine windows. The high clarity also makes the film an excellent choice for protection of printed surfaces.

10.3 Environmental Benefits and Disposal Options

Made from annually renewable resources, NatureWorks uses 20 to 50% less fossil fuels than is required by conventional plastic resin. And, because carbon dioxide is removed from the atmosphere in growing corn, the overall carbon dioxide emissions and greenhouse gases are lower than comparable plastics. Also, the solvent-free process for making PLA ensures that there are no hormone disrupters – which are commonly found in traditional thermoplastics.

There are also a number of waste management options because the products are fully compatible with all standard waste and recycling management practices and are fully compostable in municipal and industrial facilities.

NatureWorks PLA incinerates cleanly and with a reduced energy yield (18480 btu/kg) compared to traditional polymers. PLA polymers also contain no aromatic groups or chlorine and burn much like paper, cellulose and carbohydrates. Combustion of PLA produces few by-products and 0.01% ash. In areas where capacity is limited, this is an advantage in that the lower heat output permits a higher incinerator facility throughput.

Municipal composting is a method of waste disposal that allows organic materials to be recycled into a product that can be used as a valuable soil amendment. And, because PLA is made from an annually renewable resource, the compost can be used to grow the crops to produce more PLA. Extensive testing, at laboratory and pilot scales according to international standards, and in actual composting facilities, demonstrates PLA polymers are fully compostable according to ISO, CEN, ASTM and DIN draft regulations. DIN-

Certco Compost Certification has been awarded for PLA polymer use in Germany. The compostability of these products is similar to that of products made from paper or cotton.

PLA polymers compost in a two-step process. First, chemical hydrolysis reduces the molecular weight of the polymer, then microorganisms degrade the fragments and lactic acid into carbon dioxide and water. Heat and water stimulate degradation of PLA polymers.

For post-consumer recycling, the following conditions need to be met in order to recycle PLA:

- There is a sufficient quantity of material in a waste stream; a disciplined collection system is put into place for pick up;

- The product is clearly marked and physically easy to separate;

- And, there are outlets desiring to purchase the recycled feedstock.

This infrastructure does not currently exist for PLA recycling. The impact of PLA on existing recycle streams also depends on the above factors and needs to be studied on a case-by-case basis. Because PLA polymers hydrolyse with water to generate lactic acid, it would be straightforward to completely degrade PLA into lactic acid and recover the monomer.

For pre-consumer recycling or regrinding options, thermoforming and injection moulding studies using PLA regrind have successfully produced material at the same rate as existing materials. Seven passes through the extruder at 100% regrind showed minimal loss of physical properties, as long as the PLA was dried before processing.

Under agitated water conditions of repulping, PLA easily peels away and releases from paper fibers, permitting fibre recovery. Studies show PLA film remains in much larger pieces than conventional films when repulped under normal conditions of temperature, pH and consistency. This allows PLA to be separated from repulped paper slurry more easily.

10.4 'Committed to Sustainability Options'

Sustainability is key to Cargill Dow. Sustainability is viewed not as an endpoint, but a path, and a significant step has been made with NatureWorks towards creating truly sustainable products. In fact, the goal is to create a manufacturing and product-use system that will change the world by, in effect, not changing it at all. NatureWorks is proof that a product can be environmentally sound and cost effective.

Low Environmental Impact Polymers

By reducing fossil energy and carbon emissions, our goal is a stride closer. Through this ideal came the technology to develop PLA on a large-scale, which resulted in an efficient manufacturing process that provides a polymer which is quality and cost-comparable to mass-market plastics. The cost and performance of NatureWorks PLA enables it to meet the desires of consumers and the needs of converters and manufacturers that are currently limited by polymers made from fossil fuels. Cargill Dow continues these efforts by actively working with developers and industry leaders to bring a wider variety of applications made from NatureWorks packaging and fibres to market.

11 Polyhydroxyalkanoates: the Next Generation of Bioplastics

Jean-Charles Gayet and Laurent Masaro

11.1 Introduction

11.1.1 Scientific Achievements

The presence of what were initially assumed to be lipid inclusions in bacteria was first mentioned in the literature one century ago [1]. Granule inclusions inside *Bacillus megaterium* were characterised in greater detail by Lemoigne only twenty years later [2-5]. More specifically, Lemoigne demonstrated that this intracellular material was a polyester having the formula $(C_4H_6O_2)_n$, and named it '*acide β-oxybutyrique*'. Today, this polyester is commonly called poly(3-hydroxybutyric acid) (PHB) and it belongs to the polyhydroxyalkanoate family (PHA).

Over the next few decades, academic research brought a greater understanding of this particular phenomenon. Thus, by the end of the 1950s it was clearly evident that intracellular bacterial inclusions were a form of carbon energy storage [6-8]. More importantly, it was then well understood that their accumulation can be enhanced by limiting the supply of nitrogen in the growth medium while providing an excessive carbon source [8]. In addition, the storage of PHB as energy material was reported to be a widespread phenomenon occurring mainly in Gram-negative bacteria [9]. In 1974, the existence of hydroxyalkanoate monomer units other than 3-hydroxybutyric acid (3HB) was reported by Wallen and Rohwedder [10]. The new hydroxyalkanoate monomer unit found was a 3-hydroxyvalerate (3HV), while a minor constituent, 3-hydroxyhexanoate (3HHx), was also reported. Since then, other authors have reported additional monomer units such as 3-hydroxyheptanoate (3HHp) [11] and 3-hydroxyoctanoate (3HO) [12]. To date, more than 90 distinct monomer units have been identified. These are produced by many micro-organisms that are grown using various feeding strategies [13]. The discovery of the many monomers has been a major development as it brings the opportunity to produce heteropolymers (formed of distinct monomer units). As a consequence, the previously limited range of physico-chemical and mechanical properties associated with the initial brittle PHB homopolymer is suddenly unlimited leading to the multiple practical product applications for these heteropolymers.

While research trends in the 1980s focused more on the identification of existing hydroxyalkanoates monomers [13-15] as well as on bacteria (Gram-negative and Gram-positive) capable of producing them [16, 17], research in the 1990s focused instead on genetic engineering of bacterial strains [18, 19] and plants [20] to produce PHA. These methods of PHA production will be discussed in further detail in the following sections.

11.1.2 Commercial Developments

The first commercial development of PHA was achieved by WR Grace and Co., in the USA in the early 1960s. They developed sutures and prosthetic devices [21] as well as plastic laminates [22]. However, low fermentation yield combined with numerous difficulties in producing highly purified products led to the abandonment of the project. Nevertheless, this pioneering work showed the great potential of PHB for biomedical and thermoplastic applications.

The driving force in developing biopolymers from renewable resources was the energy crisis of the 1970s. ICI had developed a biopolymer by fermentation using *Ralstonia eutropha* [23]. They were able to increase the yield with an enhanced patented purification process where PHB represented 70% of the dry biomass [24]. In order to improve mechanical properties of brittle pure PHB, ICI had also developed and patented production of copolymers of 3HB and 3HV (PHBV), which they named Biopol [25]. Commercial exploitation of Biopol was launched in the beginning of the 1990s with a biodegradable shampoo bottle made by Wella AG, in Germany. Coated board drinking cups, water denitrification filter balls and biodegradable commercial credit cards are other examples of Biopol products. Approximately 50% of the production was sold in Japan. Although ICI had made major advances in the production yield and purification process, production costs for Biopol remained much higher than those of synthetic polymers.

In the last decade, three major chemical companies pursued development of PHA biopolymer technologies. First, Zeneca followed up the project after splitting from ICI. Their development focused mainly on extraction and purification. Second, Monsanto started a project to produce PHA biopolymer by genetically modifying plants [26]. Finally, Procter & Gamble also had a project to use PHA biopolymers for plastic applications [27]. However, due to high production costs and extraction/purification problems, none of these three companies were able to sufficiently develop their projects to complete commercial development.

Recently, Procter & Gamble finalised a licensing agreement with Keneka Corporation to produce Nodax [poly(3-hydroxybutyrate-*co*-3-hydroxyhexanoate)] (PHBH). Under this agreement, Keneka will have to complete the application and processing of PHBH within the next three years, to end in 2004.

11.1.3 Environmental Concerns

Environmental concerns have become a major issue and attention to this issue has been rising as the population of developed nations face a growing realisation of these factors. Problems pertaining to air, water and soil pollution, overpopulation, limited food resources, genetically modified organisms, greenhouse gas effects and deforestation are a few examples of threats that will face future generations and must be addressed immediately.

For example, Japan has only sufficient capacity to compost solid waste for the next ten years. In addition, they are restrained in the release of carbon dioxide due to the 1997 Kyoto Protocol which limits incineration. Development of biodegradable plastics from renewable resources would be a relief in this particular case, as well as for many other countries.

Therefore, fully biodegradable polymers represent the ideal solution to address the growing environmental issues of solid waste. PHA is a potent candidate for this purpose because it responds to all requirements. Trends for the biodegradable polymer market show volume growth of 29% per year [28], which is very attractive. For this reason, chemical companies that are industry leaders are developing and commercialising new polymers which are a mix of aliphatic polyesters that can be regarded as biodegradable and aromatic polyesters that are not. These new materials are marketed as fully biodegradable, which is not accurate from a scientific point of view. Therefore, a clear definition of what should be considered as biodegradable is necessary as well as government incentives aimed at encouraging industries to adopt fully biodegradable polymers.

The object of this chapter is to introduce PHA, their methods of production and their properties. This chapter focuses more on practical PHA industrial applications identified through filed patents. For a more detailed basic scientific review of the literature about PHA, readers are invited to read the following review articles [29-32].

11.2 Production of PHA

11.2.1 Fermentations

PHA and the most common of these polymers, PHB, occur in a large variety of microorganisms including Gram-negative and Gram-positive bacteria, as well as in blue-green algae. PHB is stored inside the bacteria in the form of granules which can account for up to 80% of the cell's weight. These granules act as carbon energy storage for the bacteria and are biosynthesised in adverse conditions where an essential nutrient such as nitrogen, oxygen or phosphorus is limited. Under such conditions, the bacteria can no longer grow or proliferate and they switch their metabolism to the production of PHB in order to have a usable carbon source when conditions revert back to normal.

The biochemical pathway of PHB biosynthesis is quite well understood [31]. It first implies the transformation of the carbon source to acetate, an important metabolite in the bacterial metabolism, followed by the activation of acetate into acetyl-CoA. Then, three key enzymes lead to the synthesis of PHB: two molecules of acetyl-CoA are condensed by beta-ketothiolase (EC 2.3.1.16) to give acetoacetyl-CoA which is further reduced to (R)-beta-hydroxybutyryl-CoA by NADP dependent acetoacetyl-CoA reductase (EC 1.1.1.36), and the latter metabolite is polymerised by PHA polymerase (EC 3.1.1.75).

As one can see, the main advantage of the biosynthesis of PHB is its stereospecificity. Bacterial PHB is 100% stereoregular with the R(-) configuration and the resulting polymer is isotactic. In contrast, chemical synthesis of PHB seldom leads to stereoregular polymers even if enantiomerically pure butyrolactone is used.

Currently much research is being done with recombinant microorganisms like *Escherichia coli*. A better understanding of the metabolism and genetics of PHA accumulating bacteria has allowed scientists to develop recombinant bacteria that are capable of using a broader range of carbon sources, accumulate a greater amount of PHA or produce new polymers from different monomers [33]. Such recombinant bacteria can be obtained by incorporating into the *E. coli* genome, the genes that hold the sequence for the enzymes responsible for PHA synthesis.

11.2.2 Production in Plants

As mentioned in the introduction, production of PHA in plants has attracted much attention in recent years. Basically, a recombinant genome containing one or more genes specifying enzymes critical for the polyhydroxyalkanoates biosynthetic pathway occurring in specific microorganisms, is introduced into plants. An example of production of PHA in plants is given in a patent by Bright and co-workers [26].

Recovery of the biopolymer from plants is neither simpler nor cheaper than recovery from microorganisms. Liddell provided a protocol in five steps:

(a) grinding and cooking the seed;

(b) enzymically solubilising carbohydrate;

(c) allowing the resulting product to separate;

242

(d) decanting each phase of the global fraction, and

(e) recovering the biopolymer from its particular phase [34].

11.2.3 Chemical Synthesis

PHB and it's copolymers can be chemically synthesised using γ-butyrolactones [35]. The ring opening polymerisation of such molecules could lead to more or less optically active PHA depending on the butyrolactone and the catalyst used. In fact, chemical synthesis of PHA requires the use of rather exotic metal-based catalysts and rather expensive and toxic lactone monomers. Moreover, the yields are usually low.

11.2.4 Extraction and Purification

Extraction and purification steps have always been, and are still, a concern in the production of PHA because polyester granules are accumulated intracellularly. In addition, these steps negatively affect the production costs which have a direct impact on the competitiveness of the biopolymer.

Whether natural or recombinant bacteria are used for the biosynthesis of PHA, the bacterial cells must be broken apart and the PHA collected from the cellular debris. This can be accomplished by extracting the PHA from the biomass with chloroform followed by filtration and precipitation with cold ethanol [36]. However, large amounts of solvents are required with this method, therefore several other methods have been developed. One of these approaches is to disrupt the bacterial cell and solubilise the whole biomass by chemical (hypochlorite) or enzymic hydrolysis [24], thereby isolating the native PHA granules. Some mechanical or thermal treatment can be added to help these separation processes. The chloroform method could be used for the polishing step if highly pure PHA is required. All these methods of extraction and purification of the biopolymer are both costly and time consuming. There is still much room to enhance extraction and purification steps in order to improve the biopolymer quality while reducing related costs.

11.3 General Properties

11.3.1 Physico-Chemical Properties

The molecular weight of PHA depends to a great extent on the microorganism and the conditions used for its synthesis. Depending on the bacteria, the molecular weight could range from 50,000 to 2,000,000 g/mol. For a given bacteria, the carbon source used or the composition of the culture medium could affect the molecular weight of the final polymer. The extraction process, i.e., use of hypochlorite, could further degrade the polymer, leading to a shorter chain length than that of the native PHA. Chemically synthesised PHA has a smaller molecular weight, in the range of 20,000 to 400,000 g/mol.

PHA are partially crystalline polymers showing a glass transition temperature (T_g) for the amorphous phase and a melting temperature (T_m) for the crystalline phase. These temperatures greatly depend on the monomer composition and the crystallinity of the polymer. Pure PHB shows a melting temperature of 177 °C which decreases to 140 °C when the ratio of valerate monomer is 25%. This is due to the destructive effect of the valerate content on the crystallisation capacity of the PHB. The same could be observed with medium-chain-length PHA (mcl-PHA) where various monomers with a longer aliphatic side chain are present. These mcl-PHA are almost amorphous and show a melting temperature in the range of 30-60 °C.

PHA are thermoplastic polymers and can be extruded, blow-moulded, injected, threaded and processed as any other regular thermoplastic material. However, some care should be taken with PHB as it is not very thermostable with degradation starting at 200 °C leading to crotonic acid formation and it is totally degraded at 300 °C. As the melting and degradation temperatures of PHB are close, the use of copolymer with valerate or other monomers decreases the melting point thereby making it easier to process the thermoplastic.

As with the thermal properties, the mechanical properties are greatly affected by the monomer composition. Pure PHB has tensile strength and elastic modulus similar to that of polypropylene (PP) but its elongation at break is markedly lower and it is also a more brittle polymer. This brittleness is also due to its crystallinity and could be alleviated by the incorporation of a comonomer. PHB becomes much more elastic as the valerate content increases and mcl-PHA could show elongation at break of two orders of magnitude greater than that of pure PHB.

Some classical properties of PHA are shown in **Table 11.1**. PHA also has interesting properties such as resistance to UV light and common chemical products (acids, bases, solvents except halogenated ones), as well as having good barrier properties to oxygen and carbon dioxide.

11.3.2 Degradation

The most important property of the PHA family is undoubtedly the fact that it is 100% biodegradable. Since bacteria are able to biosynthesise PHA as an energy reserve, they

Table 11.1 Physical properties of some PHA				
Polymer	T_m (°C)	T_g (°C)	Tensile strength (MPa)	Elongation at break (%)
PHB	177	10	40	6
PHB/3%HV	170	8	-	-
PHB/9%HV	162	6	25	20
PHB/15%HV	150	4	-	-
PHB/20%HV	145	-1	20	50
PP	176	-10	38	400

also have the necessary enzymes to degrade it. These enzymes, known as PHA depolymerases (EC 3.1.1.76), catalyse the reverse reaction to PHA polymerases. Many bacteria and fungi secrete PHA depolymerases and use the hydrolysed PHA fragments as well as the resulting hydroxyalkanoates monomers as a carbon source. The rate of degradation is affected by a number of factors such as the monomer composition, the crystallinity, the shape and physical texture of the polymer object. Another important fact to consider is that no harmful molecules are generated by PHA biodegradation since 3HB is a natural metabolite found in most microorganisms, higher animals and humans. This explains the strong interest in PHA as biomaterials for medical applications [29]. A good review on biodegradation of PHA is given by Jendrossek and co-workers [37].

11.4 Industrial Applications

11.4.1 Compounding

Due to its brittle properties, pure PHB has limited industrial applications. Therefore, numerous efforts have been devoted to define compositions containing biodegradable polyester and plasticiser. Hammond and co-workers [38] report the use of a series of 69 commercial plasticisers with PHB and PHBV (several HV ratios, 11% maximum). All of them showed improved mechanical properties and elongation at break (+ 163%). However, plasticisers that provide best results for mechanical properties were different than the ones improving elongation at break.

Asrar and Pierre [39] used the same series of 69 plasticisers and incorporated nucleant and thermal stabilisers for the production of shaped PHA articles. Authors report that pellets should contain PHA of molecular weight greater than 470,000 g/mol in order to achieve good results in processing a desired shape.

Hammond and Bal [40] used oligomers of PHA as plasticisers. The oligomers can be synthesised by reducing the molecular weight of a similar polymer or by chemical synthesis from the cyclic monomer, a lactone. While the main advantage of this innovative plasticiser is its compatibility with the structural polymer, the major disadvantage is its price. In fact, starting materials are quite expensive.

In summary, all these patents claim that compounding is useful in the production of PHA end products by extrusion, moulding, coating, spinning and calendering operations.

11.4.2 Coating and Packaging

Coating of paper/cardboard as well as film for food packaging are the most important developments associated with PHA. One plausible explanation could be that products associated with such developments are of single-use with a short life and could be disposed of in a compost site after their use. In addition, PHA have very good oxygen barrier properties, which is the main property needed in these applications. It is interesting to note that most of the PHA patents concern coating and packaging applications with the pure PHB, even though its elongation to break is very low.

Coating processes are generally achieved starting from latex solutions. Examples of PHA latex solution production and application are given in [41, 42].

Production of films is generally achieved with conventional processing equipment. Waddington proposed a process to support or cast PHA films by applying a layer of molten biopolymer [43]. Later, the same author reported a method for preparing films and coating starting from high melt flow of PHA [44]. Asrar and co-workers [45] provided a method during which degradation of the biopolymer is observed. In fact, they reported that the molecular weight of the biopolymer in pellets was in the range of 200,000 to 400,000 g/mol while for extruded coated objects the molecular weight was in the range of 125,000 to 150,000 g/mol. No explanation is provided. Waldock defined a process to produce oriented PHA that requires less stringent temperature control than the conventional process [46].

Kemmish and Montador described a biodegradable film comprising of two layers, cellophane and PHA [47].

11.4.3 Plastic Food Services Items

Another interesting application for PHA is their use in disposable items for the food service industry since these biopolymers possess many of the required properties for such applications: the PHA biopolyesters are biodegradable, present good mechanical resistance, are hydrophobic, and are resistant to liquid and grease. In addition, this is an interesting market because of its enormous potential. However, these applications require biopolymers with enhanced properties that pure PHB does not have. Thus, Noda and Satkowski [27] have developed copolymers made of PHB and polyhydroxyoctanoate or PHHx to manufacture utensils, plates, cups, cup lids, cup holders, trays, toothpicks, straws and sticks. Despite the fact that the proof of concept was a success and that plasticisers permit similar development with pure PHB, no further developments have followed.

11.4.4 Toner

A unique application of PHA as a biodegradable polymer is its use as a toner. Fuller and co-workers [48] in a patent application for Xerox, demonstrated that semicrystalline polyester resin particles, especially PHB, are suitable to replace non-degradable synthetic resins such as styrene or methacrylate.

The main advantage of this application lies in the de-inking process where non-degradable resins, that usually accumulate in sludge and need further treatment, are replaced by biodegradable ones. Although highly beneficial for the environment, no other application regarding toner was found. To our knowledge Xerox did not develop this application to a commercial level.

11.4.5 Paint

Paint based on mcl-PHA from unsaturated vegetable fatty acids was developed by Buisman and co-workers [49]. According to this invention, the biopolymer acts as the paint binder, eliminating the need for non-degradable polymer resins. In addition, the composition is solvent free containing only water thereby improving both the drying process and time when compared to existing non-degradable commercial compositions.

11.4.6 Food Applications

Yalpani [50] reported that PHA can be used as a flavour delivery system for low-fat and no-fat food. In this particular situation, PHA is used in the form of microspheres that carry food ingredients or substitutes. Later, Yalpani reported that PHA can be directly used as a fat substitute [51]. Interest in PHA lies in its particular ability to stimulate fat-like taste in the mouth. Therefore, it can be used to reduce or eliminate fat in food composition.

11.4.7 Other Applications

One of the first patented applications or intentions for the use of PHA as a commercial product was as cigarette filters [52] by British American Tobacco (BAT), which was never pursued on a commercial basis.

According to Kemmish, PHA was reported to have adhesion properties [53]. In fact, PHA do not have strong adhesion properties, however, the author claims that some adhesives do not need to provide an extremely strong bond. In the case of weaker or temporary bond requirement, PHA can be an interesting candidate because of its biodegradability. Kauffman and co-workers [54] described using PHB and PHBV as a base for hot melt adhesive.

Several patents describe methods to modify PHA to change its physico-chemical and mechanical properties thereby allowing for additional new applications. For example, Hammond enhanced PHA properties by using a transesterification catalyst [55], Wnuk and co-workers reported copolymerisation of PHA with polylactic acid, polyurethane and polycaprolactone [56].

Biomedical and pharmaceutical fields are domains where many applications and products based on PHA have been developed because they are biocompatible, bioresorbable and biodegradable polymers. These characteristics made them potential candidates for numerous applications. However, the brittleness of the PHB is still a limiting factor in its applications while limitations due to the price are less of a concern because of the high quality of polymer required.

11.5 Conclusion

PHA are a great biopolymer family providing biopolyesters with unimaginable properties. No other biopolymer or synthetic polymer group or family can provide such a variety in their physico-chemical and mechanical properties.

Today, the limiting factor for industrial applications of PHA is its price which remains higher than synthetic resins. Although PHA are currently still more expensive than non-biodegradable polymers (petroleum-based or synthetic) they are well within the price range of other biodegradable biopolymers. In addition, progresses in research and development will lead to improvements in the manufacturing process that will allow PHA to be produced at a competitive prices in the near future. BioMatera is optimising its production process to achieve this goal.

References

1. M.W. Beijerinck, *Bakteriologie*, 1901, **11**, 650.

2. M. Lemoigne, *Comptes Rendus de l'Académie des Sciences*, 1925, **180**, 1539.

3. M. Lemoigne, *Annales de l'Institut Pasteur*, 1925, 38, 144.

4. M. Lemoigne, *Bulletin de la Societe Chimie Biologique*, 1926, 8, 770.

5. M. Lemoigne, *Annales de l'Institut Pasteur*, 1927, **41**, 148.

6. D.H. Williamson and J.F. Wilkinson, *Journal of General Microbiology*, 1958, **19**, 198.

7. R.M. Macrae and J.F. Wilkinson, *Journal of General Microbiology*, 1958, **19**, 210.

8. M. Doudoroff and R.Y. Stainer, *Nature*, 1959, **183**, 1440.

9. W.G.C. Forsyth, A.C. Hayward and J.B. Roberts, *Nature*, 1958, **182**, 800.

10. L.L. Wallen and W.K. Rohwedder, *Environmental Science and Technology*, 1974, 8, 576.

11. R.H. Findlay and D.C. Right, *Applied Environmental Microbiology*, 1983, **45**, 71.

12. M.J. DeSmet, G. Eggink, B. Withold, J. Kinhma and H. Wynberg, *Journal of Bacteriology*, 1983, **154**, 870.

13. A. Steinbüchel and H.E. Valentin, *FEMS Microbiology Letters*, 1995, **128**, 219.

14. M. Kunioka, Y. Nakamura and Y. Doi, *Polymer Communications*, 1988, **29**, 174.

15. Y. Doi, A. Tamaki, M. Kunioka and K. Soga, *Makromolecular Chemie Rapid Communications*, 1987, 8, 631.

16. A.J. Anderson and E.A. Dawes, *Microbiological Reviews*, 1990, **54**, 450.

17. A. Steinbüchel in *Biomaterials-Novel Materials from Biological Sources*, Ed., D. Byrom, Macmillan, Basingstoke, UK, 1991, 125.

18. S.C. Slater, W.H. Voige and D.E. Dennis, *Journal of Bacteriology*, 1988, **170**, 4431.

19. P. Schubert, A. Steinbüchel and H.G. Schlegel, *Journal of Bacteriology*, 1988, **170**, 5837.

20. Y. Poirier, D.E. Dennis, K. Klomparens and C. Sommerville, *Science*, 1992, **256**, 520.

21. J.N. Baptist and J.B. Ziegler, inventors; WR Grace & Co., assignee; US 3,225,766, 1965.

22. J.N. Baptist and F.X. Werber, inventors; WR Grace & Co., assignee; US 3,107,172, 1963.

23. P.A. Holmes, *Physics and Technology*, 1985, **16**, 32.

24. P.A. Holmes and G.B. Lim, inventors; Imperial Chemical Industries PLC, assignee; US 4,910,145, 1990.

25. P.J. Senior, S.H. Collins and K.R. Richardson, inventors; Imperial Chemical Industries PLC, assignee; EP 0,204,442A2, 1986.

26. S.W.J. Bright, D. Byrom and P.A. Fentem, inventors; Monsanto Company, assignee; US 6,175,061 B1, 2001.

27. I. Noda and M.M. Satkowski, inventors; The Procter & Gamble Company, assignee; WO 0149770A1, 2001.

28. J. Dwivedi, *Spectrum Performance Chemicals and Materials*, Decision Resources Inc., Waltham, MA, USA, 1999, Study 5, Page 1.

29. C.R. Hankermeyer and R.S. Tjeerdema, *Reviews of Environmental Contamination and Toxicology*, 1999, **159**, 1.

30. P.J. Hocking and R.H. Marchessault in *Biopolymers from Renewable Resources*, Ed., D.L. Kaplan, Springer-Verlag, Berlin, Germany, 1998, 220.

31. K. Sudesh, H. Abe and Y. Doi, *Progress in Polymer Science*, 2000, **25**, 1503.

32. *Biopolyesters, Advances in Biochemical Engineering/Biotechnology*, Eds., W. Babel and A. Steinbüchel, Volume 71, Springer-Verlag, Berlin, 2001.

33. L.L. Madison and G.W. Huisman, *Microbiology and Molecular Biology Reviews*, 1999, **63**, 21.

34. J.M. Liddell, inventor; Monsanto Company, assignee; WO 9717459, 1997.

35. H.M. Muller and D. Seebach, *Angewandte Chemie – International Edition in English*, 1993, **32**, 477.

36. J.F. Stageman, inventor; Imperial Chemical Industries PLC, assignee; US 4,562,245, 1985.

37. D. Jendrossek, A. Schirmer and H.G. Schlegel, *Applied Microbiology and Biotechnology*, 1996, **46**, 451.

38. T. Hammond, J.L. Liggat, J.H. Montador and A. Webb, inventors; Zeneca Limited, assignee; WO 9428061A1, 1994.

39. J. Asrar and J.R. Pierre, inventors; Monsanto Company, assignee; US 6,127,512, 2000.

40. T. Hammond and J.S. Bal, inventors; Zeneca Limited, assignee; WO 9411445 A1.

41. J.M. Liddell, N. George and P.D. Turner, inventors; Monsanto Company, assignee; US 5,891,936, 1999.

42. N. George, T. Hammond, J.M. Liddell, R. Satgurunathan and P.D. Turner, inventors; Monsanto Company, assignee; US 5,977,250, 1999.

43. S.D. Waddington, inventor; Zeneca Limited, assignee; US 5,578,382, 1996.

44. S.D. Waddington, inventor; Monsanto Company, assignee; US 6,111,006, 2000.

45. J. Asrar, J.R. Pierre and P. D'Haene, inventors; Monsanto Company, assignee; WO 9904948A1, 1999.

46. P. Waldock, inventor; Monsanto Company, assignee; WO 9722459A1, 1997.

47. J.D. Kemmish and J.H. Montador, inventors; Zeneca Limited, assignee; WO 9515260A1, 1995.

48. T.J. Fuller, R.H. Marchessault and T.L. Bluhm, inventors; Xerox Corporation, assignee; US 5,004,664, 1991.

49. G.J.H. Buisman, F.P. Cuperus, R.A. Weusthuis and G. Eggink, inventors; Instituut Voor Agrotechnologisch Onderzoek, assignee; US 6,024,784, 2000.

50. M. Yalpani, inventor; The Nutrasweet Company, assignee; WO 9209210A1, 1992.

51. M. Yalpani, inventor and assignee; US 5,229,158, 1993.

52. W. Wiethaup, K. Möller, B. Hauser, E. Rittershaus and T. Hammond, inventors; BAT Cigaretten-Fabriken GmbH and Imperial Chemical Industries plc, assignees; EP 0,454,075A2, 1991.

53. D.J. Kemmish, inventor; Zeneca Limited, assignee; WO 9502649, 1995.

54. T. Kauffman, F.X. Brady, P.P Puletti and G. Raykovitz, inventors; National Starch and Chemical Investment Holding Corporation, assignee; US 5,169,889, 1992.

55. T. Hammond, inventor; Zeneca Limited, assignee; WO 9411440, 1994.

56. A.J. Wnuk, D.H. Melik and T.A. Young, inventors; The Procter & Gamble Company, assignee; WO 9734953, 1997.

12 Thermoset Phenolic Biopolymers

Leonard Y. Mwaikambo and Martin P. Ansell

12.1 Introduction

Thermosetting resins are the most predominant matrices for high performance composite materials; this is due to their combination of low viscosity for processing and intrinsic high strength and high temperature performance. It is reported that the formation of thermoset resin materials by the reaction of phenols and aldehydes had been known a long time prior to their commercialisation in the early 1900s. Their first reported formulation was in 1872 by Baeyer [1] using phenols and aldehydes. Such work lead to thirty years of trial formulations leading to the polymerisation of phenols with formaldehyde; these were the first commercial formulations. Their synthesis was first patented in 1907 by L.H. Baekeland and became a commercial product in 1910 [1].

The polymerisation of phenol-formaldehyde is a two stage process, which leads to a soluble and fusible intermediate resin by attack of a phenol-derived species on the carbonyl-derived species. The first phenol-formaldehyde formulation was base-catalysed and is called the resole process and requires molar ratio of less than one, respectively. The second and/or newer formulation involves an acid catalyst and is called the novolak process and requires a molar ratio of greater than one phenol-formaldehyde [2, 3, 4]. The products obtained from these two processes are called resole and novolak resin. The intermediate resins are converted to the final crosslinked solid by acid-catalysed second stages.

Whilst these early polymers used fossil-based raw materials, natural monomers are also in existence. Since the most available of these naturally occurring resinous materials are phenolic-based, they form an obvious basis for the manufacture of biological origin thermoset resins.

The most abundant naturally occurring resinous materials are the phenolic based tannins and cashew nut shell liquid (CNSL). Further, controlled synthesis of natural oils has resulted in the formulation of precursors for the production of other natural polymers (also called biopolymers) as both thermoset and thermoplastic materials [5, 6].

253

Thermosetting biopolymers are increasingly receiving worldwide interest both at the research and industrial levels. The advantages of thermosetting bioresins are potential lower cost and reduction in the release of carbon dioxide (CO_2), and the resultant rise in global temperature.

Thermosetting phenolic resins undergo an irreversible polycondensation reaction (curing) and once set become infusible. This results in a stiff material but with little resistance to crack propagation. To combat this brittle behaviour the resin can be toughened by the addition of fibrous or particulate reinforcement. The resin then becomes the matrix in a composite material. The role of the matrix can be summarised as:

(a) Bind the fibres into stable solid and rigid structures,

(b) Conferring resistive compressive and shear forces on fibre bundles that would otherwise withstand tensile strains only,

(c) Transfer stresses into the fibres,

(d) Separating the fibres from each other, thereby preventing crack propagation at fibre interfaces,

(e) Protect the fibres from abrasion damage and environmental attack.

The obvious choice for reinforcement in biopolymer-based composites are natural fibres and currently, research in composite materials is being directed at developing natural matrix materials that provide optimum performance and at the same time cost less than the traditional synthetic matrices. Natural plant-based chemicals have been found to provide such qualities and their monomeric compounds can be crosslinked with or without using other chemicals to produce environmentally friendly resin matrices which are also cost effective and renewable.

The use of plants as sources of resin and the development of phenol-based CNSL from plants has strong environmental advantages. The application of CNSL as a full replacement for synthetic resins may be of immense interest in these days of diminishing petroleum reserves. Although the reaction mechanism of CNSL is quite similar to that of phenols, an understanding of its curing characteristics is essential in order to determine the process conditions for composites. Crude and distilled CNSL-formaldehyde, and phenol-CNSL-substituted formaldehyde have been polymerised by using sulfuric acid as well sodium hydroxide as catalysts to which about 10 to 15% hexamethylenetetramine was added as a curing agent [7]. Thermal properties of the resins were influenced by the amount of the curing agent. Similar observations were found when concentrations of the polymerising agents were varied. However, *in situ* preparations of the resins for cellulose-based composites tend to favour alkaline media because of the adverse hydrolytic effect of the acids on the cellulose. The following section discusses some of the thermoset phenolic resin with more detailed discussion about the CNSL-based resins.

12.2 Natural Plant-Based Resins

Natural resin precursors are obtained from plants with the exception of a few such as lac or shellac secreted by an insect known as *Laccifer lacca* and will not be discussed. The most common natural resins are phenol-based namely tannins and CNSL. Others are non-phenolic, such as rosins. Plant starch and soybean, can also be used to make adhesives.

Tannins are defined as any group of polyphenolic plant products that can be used to tan animal skins to produce leather [8]. There are two major classes of tannin that have been recognised namely hydrolysable and condensed [9]. Hydrolysable tannins consist of a carbohydrate core, often glucose esterified with gallic acid or ellagic acid. Condensed tannins are polymers of flavan-3-ols. However, other tannins have been identified which do not fit into these two classes [10]. Tannins can be condensed into adhesives and used as lubricants for drilling in oil wells, thus providing industry with a source of renewable natural products in place of chemicals that are derived from fossil fuel [11]. It is the condensed tannins, which will be described in this section.

Condensed tannins are crosslinked using formaldehyde forming thermosetting polyflavonoid-formaldehyde resins and they have been used as phenolic exterior grade adhesives for wood. Industrial polyflavonoid tannin extracts are mostly composed of flavan-3-ols repeating units, and are smaller fractions of polysaccharides and simple sugars. Polyflavonoids in such tannin extracts present phloroglucinol or resorcinol A-rings and catechol or pyrogallol B-rings as shown in **Figure 12.1**.

The types of commercial tannin extracts are the black wattle or mimosa tannin (*Acacia mearnsii*), which is a bark extract, quebracho tannin (*Schinopsis balansae*, variety *chequeno*), which is a wood extract, pine tannin (*Pinus radiata*), which is a bark extract, pecan tannin (*Carya illinoensis*), which is nut pith extract and gambier tannin (*Acacia catechu* and *Uncaria gambir*), which is a leaf and shoot extract [12]. All the five tannin extracts are suitable for the preparation of tannin formaldehyde polycondensates, although some differences among them exist. The five tannins can be divided into two classes (a) mimosa and quebracho and (b) pecan nut, pine, and gambier tannins. The latter class has a much higher reactivity towards formaldehyde due to the phloroglucinol nature of the A-ring of the predominant type of the flavonoid repeat unit [12].

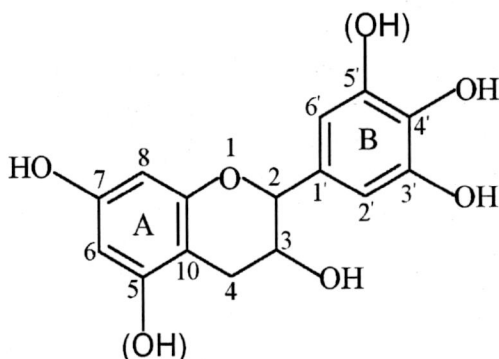

Figure 12.1 The basic structure of condensed tannins consisting of linked flavanol links having rings A and B

The repeating units are linked to each other in the configurations C_4-C_6 or C_4-C_8, the former predominating in tannins mostly composed of fisetinidin (resorcinol A-ring; catechol B-ring) repeating units. The C_4-C_8 interflavonoid linkage instead predominates in tannins composed of catechin (phloroglucinol A-ring; catechinol B-ring) and gallocatechinin (phloroglucinol A-ring, pyrogallol B-ring) repeating units. When the polymeric tannins are composed of fisetinidin/rebinetidin units the polymers are called profisetinidin/prorobinetidin, respectively; when they are composed of catechin/gallocatechin the polymers are called procyanidin/prodelphinidin, respectively. The free C_6 and C_8 sites on the A-ring are the sites which react with formaldehyde to form adhesives, under the usual conditions under which these materials are used. Some participation in the reaction with formaldehyde by the B-ring can occur when such a ring is pyrogallolic in nature. Although the reaction of the formaldehyde with the tannin extract occurs at reactive sites on the phenolic nuclei of the tannins, polysaccharides and simple sugars can also create conditions that influence the mechanism and route taken by the extract/formaldehyde reaction.

12.2.1 General Reactions of Phenols

Phenolic resins are produced by a condensation reaction between phenol and formaldehyde and they are divided into two broad classes, resoles and novolak resins. As a general rule each formaldehyde bridge is shared by two phenols, the ratio of phenol to formaldehyde that must react to form a three-dimensional network is 1:1.5 [13]. The formulation of resole and novolak resins is a two-step process starting with a soluble and fusible intermediate, which in the second stage it is acid-catalysed to convert to a crosslinked solid resin-matrix.

12.2.1.1 Resole Resin Matrices

Equations 1 to 5 shows the base-catalysed phenol-formaldehyde reaction mechanism leading to an intermediate resole resin (**Figure 12.2**). The first step shows the formation of the phenolate anion followed by a reaction that involves attack on the carbonyl group (Equation 12.2). The next step involves nucleophilic displacement of an hydroxyl from the hydroxymethyl group (Equation 12.2). The reaction then proceeds with a ring-substitution (Equation 12.3) to form methylene bridges (Equation 12.4) or ether linkages between rings (Equation 12.5) [2, 14]

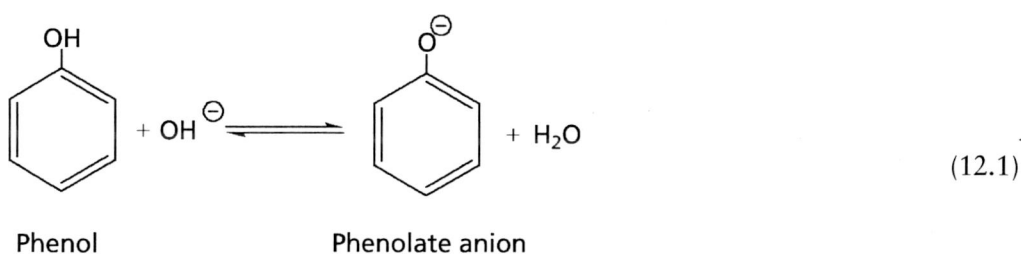

(12.1)

Phenol   Phenolate anion

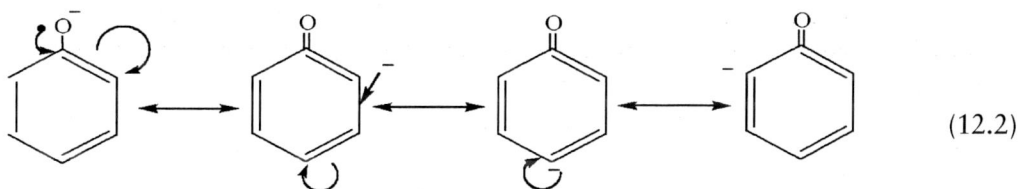

(12.2)

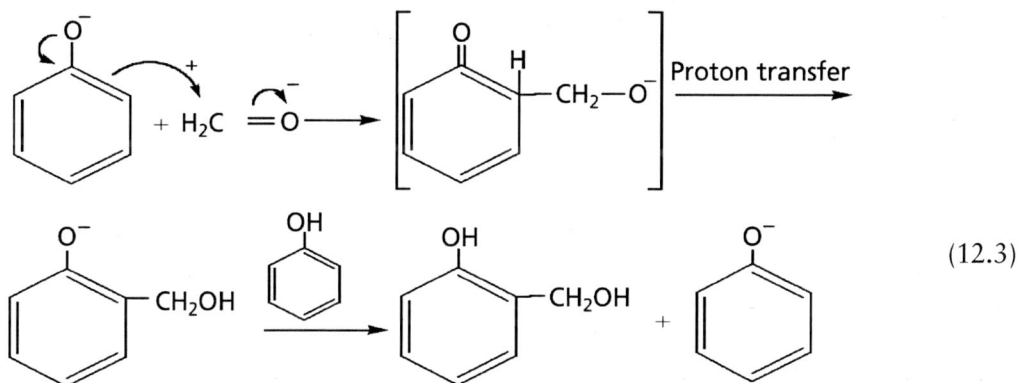

(12.3)

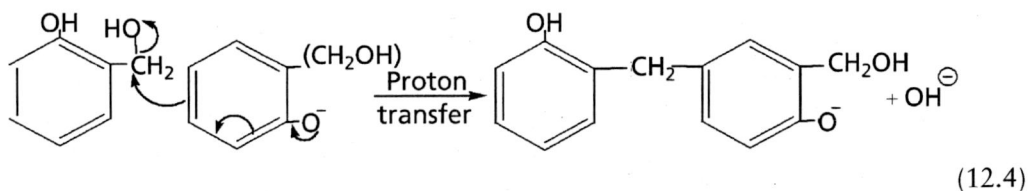

(12.4)

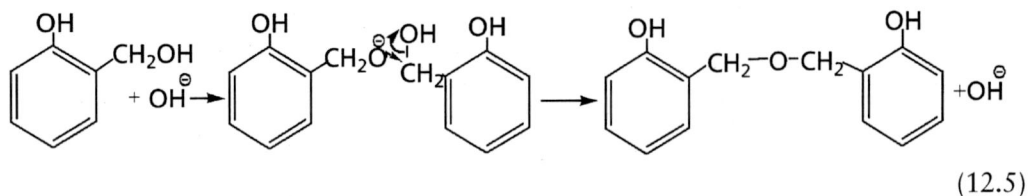

(12.5)

In order to avoid the formation of an insoluble gel, the formulation of the resole resin proceeds in two stages, and due to the presence of excess methanal, the ends of groups in the intermediate resin are CH_2OH shown in **Figure 12.2**. A three-dimensional crosslinked network (**Figure 12.3**) is obtained by processing the intermediate resin in acid and heating to 150 °C [15, 16].

In practice the condensation of two methylolphenols proceeds more quickly than the condensation of a methylolphenol and phenol [17]. The formation of ether linkages is very unlikely under alkaline conditions, but resols are generally neutralised before crosslinking takes place. Under these conditions ether linkages tend to predominate up to around 130-150 °C [18], after which methylene bridges become predominant. The crosslinking density depends on a variety of factors, including the formaldehyde-phenol ratio, whereby high formaldehyde-phenol ratios and high pH favour crosslinking, and the type of base used as a catalyst [19]. The increase in temperature and the small size of the oligomeric prepolymers implies that there will be few steric barriers, which could account for the tight crosslinked structure of the resol. Likewise, the prevalence of the stable methylene linkages accounts for the thermal stability and fire resistance.

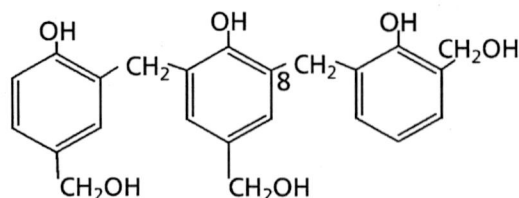

Figure 12.2 Possible structure of a base-catalysed intermediate resol resin [2, 14]

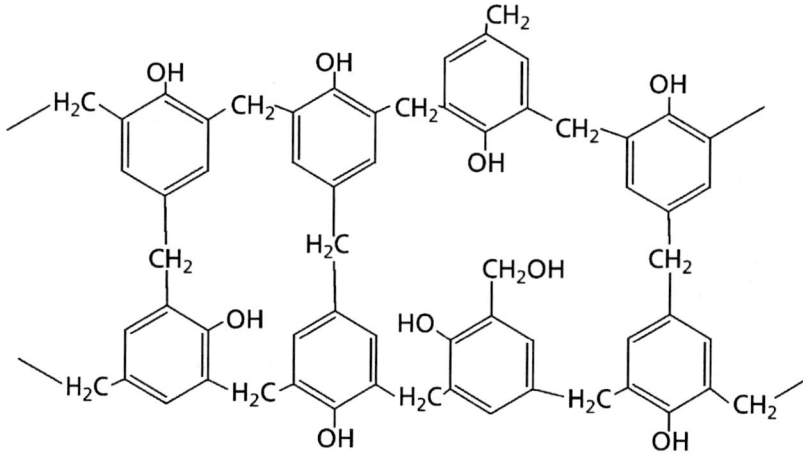

Figure 12.3 Possible structure of second stage cured resol resin [14]

12.2.1.1 Novolak Resin Matrices

The formation of novolak resin in the phenol-formaldehyde formulation uses excess phenol under acid catalyst [2, 14]. The process under which the reactions proceed is shown in Equations 12.6 to 12.10. The reaction starts with the protonation of the methanal (Equation 12.6) followed by primary ring substitution (Equation 12.7), dehydration of methylol groups (Equation 12.8), formation of methylen bridges (Equation 12.9) and formation of the ether bridges (Equation 12.10), respectively. It is believed that due to excess phenol, the methanal species are all utilised to give an intermediate resin shown in **Figure 12.4** thus limiting further reactions by the end-groups. The second step involves the addition of a crosslinking agent to the intermediate resin. Hexamethylenetetramine (HTMA), the most common crosslinking agent is added at levels of between 8-15% and under controlled temperature at around 150 °C, the reaction proceeds to a fully crosslinked and insoluble resin (**Figure 12.5**) in about 5 minutes [4].

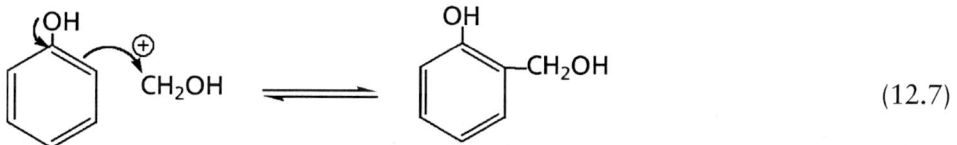

$$H_2C = O + H^{\oplus} \rightleftharpoons H_2{}^{\oplus}COH \qquad (12.6)$$

$$(12.7)$$

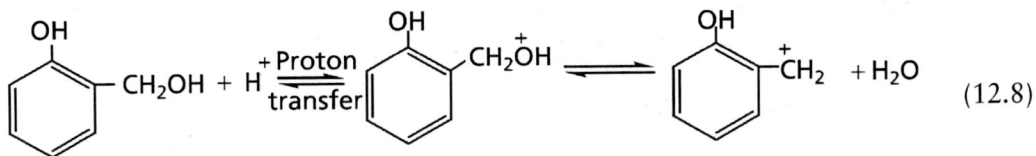

$$(12.8)$$

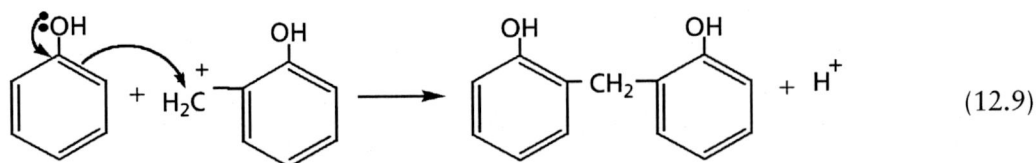

$$(12.9)$$

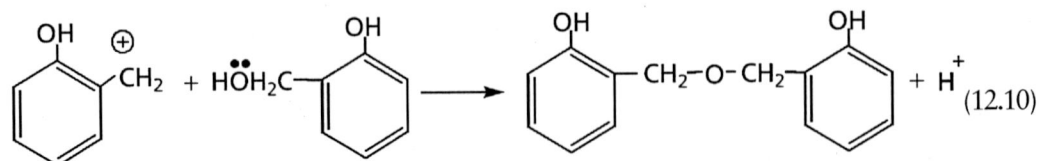

$$(12.10)$$

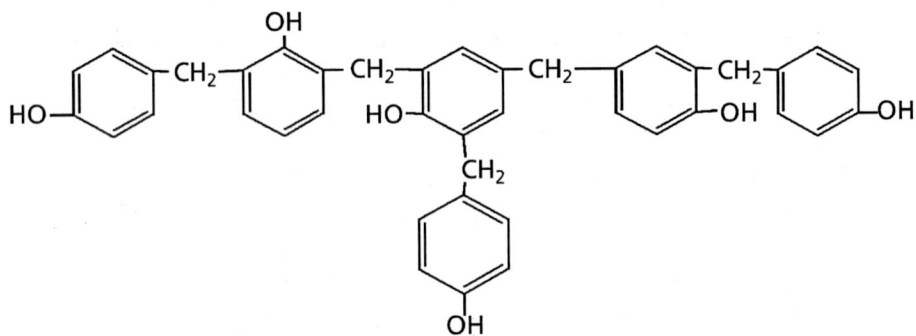

Figure 12.4 Representation of novolak phenol-methanol resin formed by acid catalysis with excess of phenol [2]

The monomeric methylol derivatives are present only transiently in very small concentrations [14]. Under acidic conditions, methylol groups are converted to benzylic carbonium ions, which react quickly with other phenolic nuclei by electron substitution reactions to form dihydroxydiphenyl methanes (Equation 12.9).

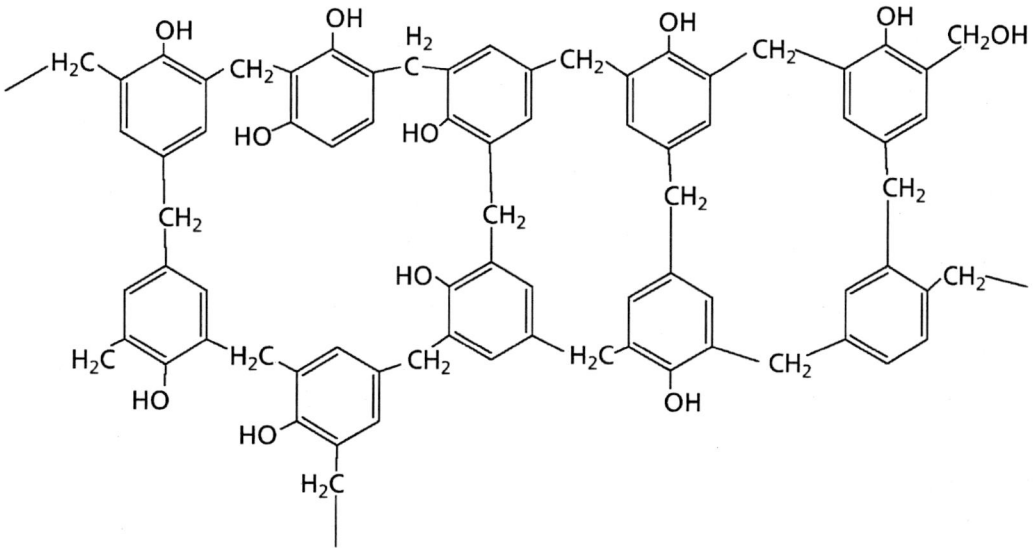

Figure 12.5 Possible structure of a second step novolak resin cured using HMTA [14]

Methylene bridge formation takes place at both the *ortho* and *para* positions, but the *para* linkages are favoured. The slight excess of phenol prevents uncontrolled crosslinking, resulting in relatively linear pre-polymers that are much larger than those of the resole [14]. Cure of the soluble intermediate (**Figure 12.4**) is achieved via the methylene groups provided by a crosslinking agent. The use of HMTA results in the formation of both methylene linkages and benzylamine linkages, and the release of ammonia. Mwaikambo and Ansell [20] have demonstrated that HMTA eliminates the water resulting from the polycondensation reaction. A fully crosslinked cured novolak resin is shown in **Figure 12.5**. At temperatures above 140 °C, the benzylamine groups largely decompose, but even at 190 °C, the nitrogen content of the products may still be above 6%.

Having discussed the general reaction of phenols it is now important to concentrate on a particular phenol type, namely cashew nut shell liquid, which occurs naturally in plant seeds. CNSL is the most common of the natural plant based monomers used for resin manufacture and is readily available. The next section therefore will discuss the sources, chemical reactions and applications of CNSL.

12.2.2 Cashew Nut Shell Liquid

The shell of the cashew nut (*Anacardium occidentale* L.) contains an alkylphenolic oil named CNSL amounting to nearly 25% of the total weight of the nut [21]. This oil is composed of anacardic acid (3-*n*-pentadecylsalicylic acid), and lesser amounts of cardanol (3-*n*-pentadecylphenol), cardol (5-*n*-pentadecylresorcinol), and methylcardol (2-methyl-5-*n*-pentadecylresorcinol), the long aliphatic side-chains are saturated, mono-olefinic (8), di-olefinic (8, 11), and tri-olefinic (8, 11,14) with an average value of two double bonds per molecule [21]. The world production of cashew nuts is nearly 500,000 tons per year, with Brazil as the largest producer. Other countries, which produce CNSL include India and Tanzania with the potential of producing 8,000 tonnes per year. CNSL is a dark-brown and partially polymerised by-product, especially when derived from the most diffused roasted mechanical processing of the cashew nuts. CNSL may represent both a dangerous source of pollutant and a low-cost, widely available and sustainable method of obtaining pure cardanol, which is used in fine chemical processes. Thermal treatment of cashew nuts and CNSL induces the partial decarboxylation of anacardic acid, which is completed by subsequent purifying distillation. The result is a 90% industrial grade cardanol oil, which is amber-yellow, with a smaller percentage of cardol and methylcardol [22].

12.2.2.1 The Structure of CNSL

The chief constituents of CNSL have been found to be anacardic acid, cardanol and cardol. Minor components include 2-methyl cardol and a small amount of polymeric materials. The first four components of CNSL have been found to comprise mixtures of four constituents differing in side-chain unsaturation, namely fully saturated, monoene, diene and triene. CNSL extracted by a cold-solvent process is called natural CNSL and hot-oil and/or roasting processed CNSL is called technical CNSL (**Table 12.1**).

Quantifying the composition detailed in **Table 12.1** was set by gas-liquid chromatography (GLC), however other chromatographic techniques such as thin layer chromatography (TLC) and high performance liquid chromatography (HPLC) can be used. Although natural CNSL has more anacardic acid than cardanol it is decarboxylated to produce

Table 12.1 Composition of phenol components (%) in natural and technical CNSL [4, 23]		
Components	Natural CNSL (%)	Technical CNSL (%)
Anacardic acid	77.02	-
Cardanol	2.37	82.15
Cardol	16.77	13.71
2 - Methyl cardol	2.83	4.10

anacardol which when hydrogenated yields cardanol (Equation 12.11), the chief constituent of technical CNSL [24].

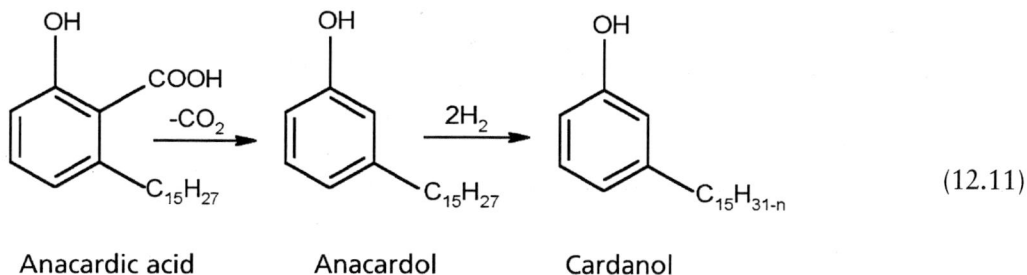

$$(12.11)$$

| Anacardic acid | Anacardol | Cardanol |

12.2.2.1 Polymerisation of CNSL

Heating of phenol to temperatures of between 160-300 °C in the presence of a catalyst is known to enhance three-dimensional crosslinking of the polymer network, resulting in a hard, infusible and insoluble thermosetting polymer. All the constituents of CNSL are typically phenol compounds. Due to the presence of the hydroxyl (-OH) group, the carboxyl (-COOH) group, and variable aliphatic unsaturation in the side chain, CNSL is able to take part in several chemical reactions. The long aliphatic chains present in CNSL impart flexibility due to internal plasticising, resulting in the formation of soft resins at elevated temperatures, unlike phenol-formaldehyde resins, which are hard [21]. The determination of the reaction mechanism of CNSL is complicated by the presence of its constituents, all of which have four different chemical structures including some amount of polymer materials [4, 21]. This implies that an understanding of the chemical reactions and, therefore, structure of one of the CNSL constituents may explain the chemical structure behaviour of cashew nut shell liquid. Since cardanol is the major component of technical CNSL its applications in resin production has been of great interest [25-28]. It seems logical, therefore, to present the chemical structure of cardanol (**Figure 12.6**), its reactions and preparation into resin. The four constituents of cardanol are given in **Figure 12.7**.

In alkaline media, cardanol takes part in ether formation, where an alkyl group such as diethylsulfate replaces the hydrogen of the hydroxyl group, and a pale yellow, high boiling point, liquid of ethyl cardanol (**Figure 12.8**) is produced. Other ethers include cardanoxy acetic acid (**Figure 12.9**) produced by reaction with monochlorinated acetic acid in acetone.

Both these materials are used widely in the manufacture of paints, binders, varnishes and the formation of alkyd resins [29]. The reaction mechanism involved in alkaline catalysis involving cardanol-formaldehyde is presented in **Figure 12.10**.

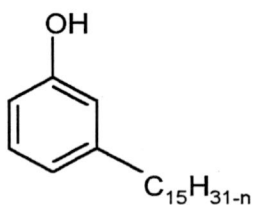

Figure 12.6 The general structure of cardanol (n = 0, 2, 4, and 6) [4, 21, 25]

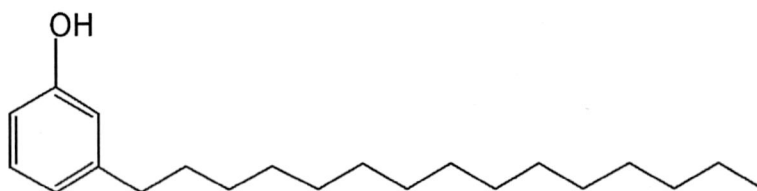

(a) 3-(pentadecyl)-phenol n = 0

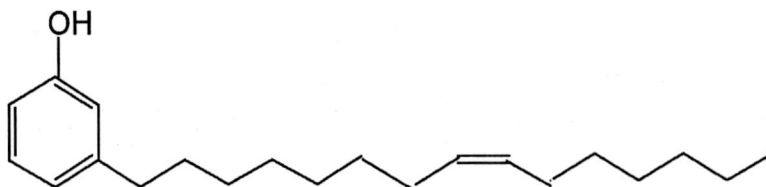

(b) 3-(8-pentadecenyl)-phenol n = 2

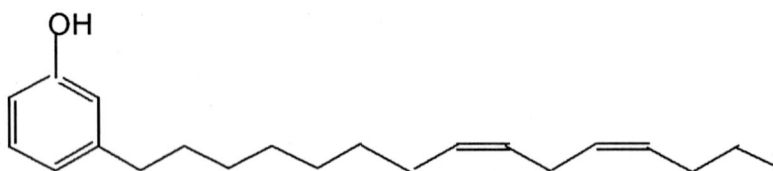

(c) 3-(8, 11-pentadecadienyl)-phenol n = 4

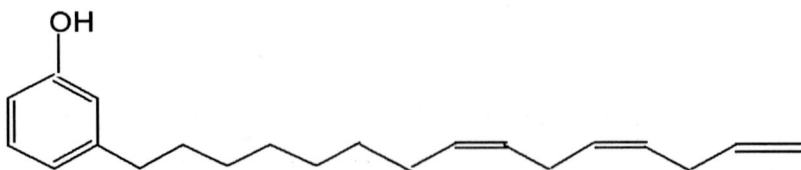

(d) 3-(8, 11, 14-pentadecatrienyl)-phenol n = 6

Figure 12.7 The four chemical constituents of cardanol [4, 21, 25]

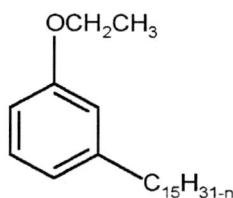

Figure 12.8 Ethyl cardanol

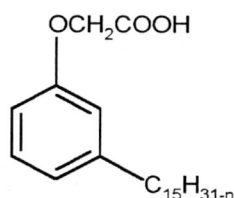

Figure 12.9 Cardanoxy acetic acid

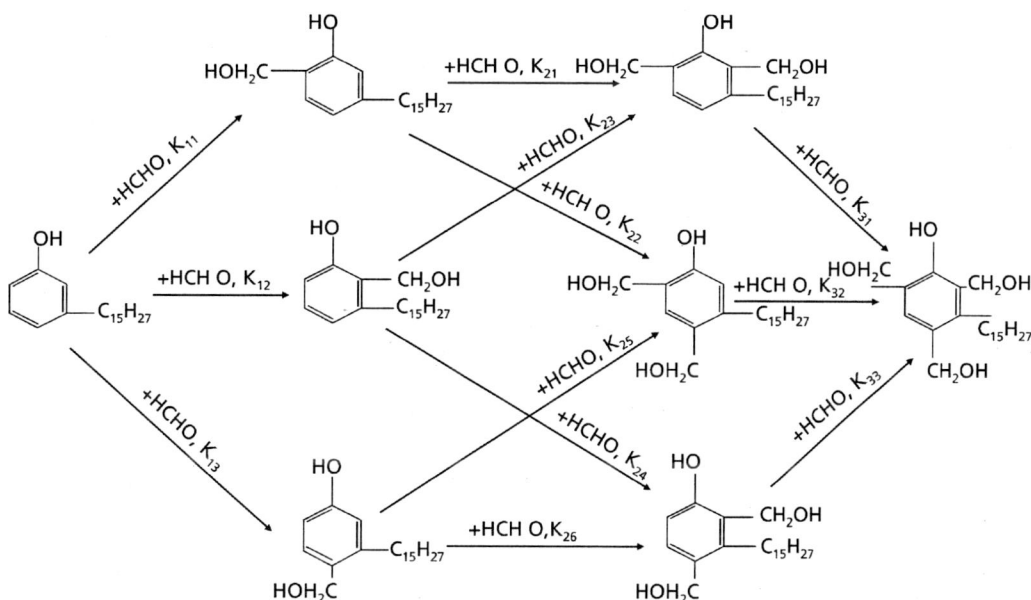

Figure 12.10 Initial reactions of less than or equal to 1 mol formaldehyde and 1 mol cardanol mixture [26]

In **Figure 12.10**, the K_{11} route is characterised by *ortho-ortho* methylol substitution, the K_{12} route is characterised by the *ortho, ortho-para* then *ortho-ortho* and *ortho-para* methylol substituted phenolic ring and route K_{13} is characterised by the *para* and *ortho-para* methylol substituted phenolic ring. Recent studies have shown that cardanol can be polymerised with maleic anhydride (MA) under acid catalysts and cured with HMTA [30]. It is reported that cardanol alkyl chain double bonds are not very reactive and that except for the synthesis of resole or novolak type formaldehyde resins the cardanol molecules do not have enough reactive sites [30]. The cardanol-MA ratios used were 1:1, 1:0.5 and 1:0.33. In the study it was observed that by taking into consideration only the 60% of cardanol species (**Figures 12.7b, 12.7c and 12.7d**) which contain double bonds and are readily available for reactions thus suggesting the following possible structures (**Figure 12.11**) and (**Figure 12.12**).

However, the direct cardanol-cardanol bonding (**Figure 12.12**) was viewed to be less likely to happen because after MA vanished from the reaction mixture the concentration of cardanol remained almost unchanged [30].

Misra and Pandey [25] reported the effect of process parameters such as mole ratio of reactants, catalyst concentration and temperature in the production of resins by the reaction of cardanol with formaldehyde using sodium hydroxide (NaOH) as a catalyst. In their work it was found that the cardanol-formaldehyde reaction did not follow a first-order scheme but was a second-order reaction when tested for its conformity at

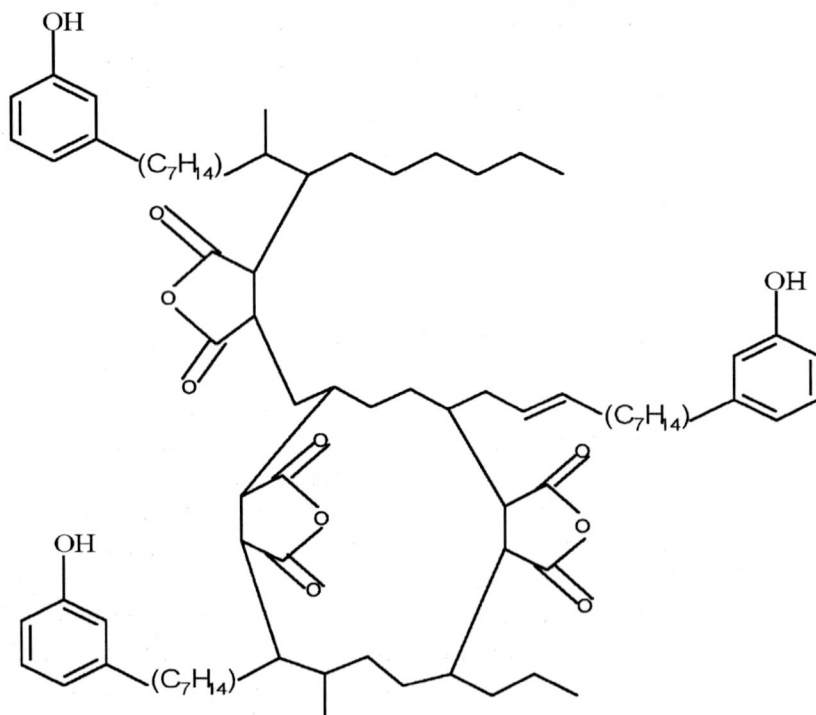

Figure 12.11 Possible cardanol-MA crosslinked structure [30]

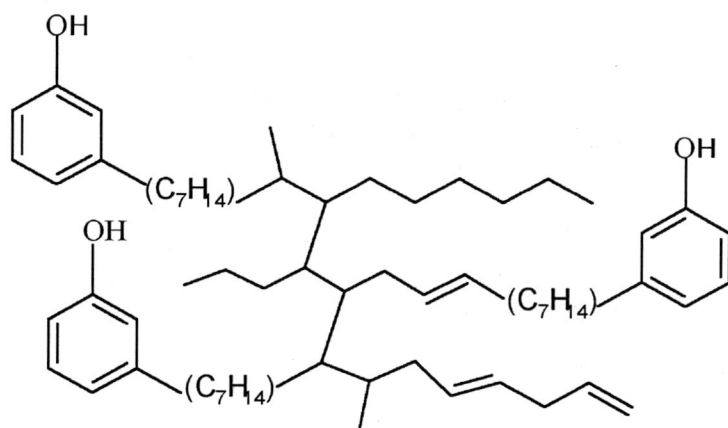

Figure 12.12 Cardanol-MA possible linear structure [30]

various formaldehyde-cardanol molar ratios. However, both first and second order reaction mechanisms have been observed when reacting phenol-formaldehyde by using weak and strong base chemicals such as ammonia and NaOH, respectively, as catalysts. The cure characteristics of thermoset resins are based on the principle that the heat flow in an isothermal reaction is proportional to the reaction rate.

12.2.2.3 Characterisation of Cashew Nut Shell Liquid

It was shown in **Table 12.1** that crude CNSL contains four major components with cardanol being the largest constituent. The chemistry of all the four components is quite similar as they are phenolic in nature and all have an aliphatic chain with 15 carbons and up to 31 hydrogens. The chain is either fully saturated or unsaturated thus rendering the structure susceptible to various reactions that can happen on the side chain and the phenol groups. The reaction of CNSL is complicated in the sense that unless it is distilled to obtain one of the components it is difficult to accurately follow the reaction mechanism. However, economics will favour the direct polymerisation of CNSL rather than using its components such as cardanol, which will require a costly distillation process. It is with such economies in mind that a simple but useful method of following the polymerisation of CNSL should be adopted to study the curing of CNSL-formaldehyde by following changes in the glass transition temperature. In this article characterisation of CNSL by differential scanning calorimetry (DSC) and thermogravimetric analysis techniques are presented.

12.2.2.4 Characterisation of the Neat CNSL by DSC

Figure 12.13 shows the DSC thermogram of neat CNSL subjected to two consecutive dynamic runs. The thermogram marked first run has a wide endotherm with a minimum at around 100 °C, which may be due to moisture encapsulated during the sample preparation time.

After this endotherm a wide and shallow exotherm is observed at around 220 °C beyond which there is a slight drop after which it rises between 250-300 °C to coincide with the thermogram for the second run. The shallow peak observed is caused by residual cure of the polymeric materials acquired either naturally or during the CNSL extraction period. The second run shows a nearly straight inclined thermogram indicating that residual moisture has been evaporated and that curing of the polymeric materials occurring naturally has been completed. Similar observations were reported when curing polyester-based kapok-cotton fabric composites whereby after the removal of water molecules the composite thermogram developed in the same manner as **Figure 12.13** [31]. The crude CNSL was assumed to be 'pure' in the sense that it contained the major components mentioned in previous sections, was polymerised using formaldehyde and catalysed using sodium hydroxide. The progress of cure was characterised and it is discussed in the following sections.

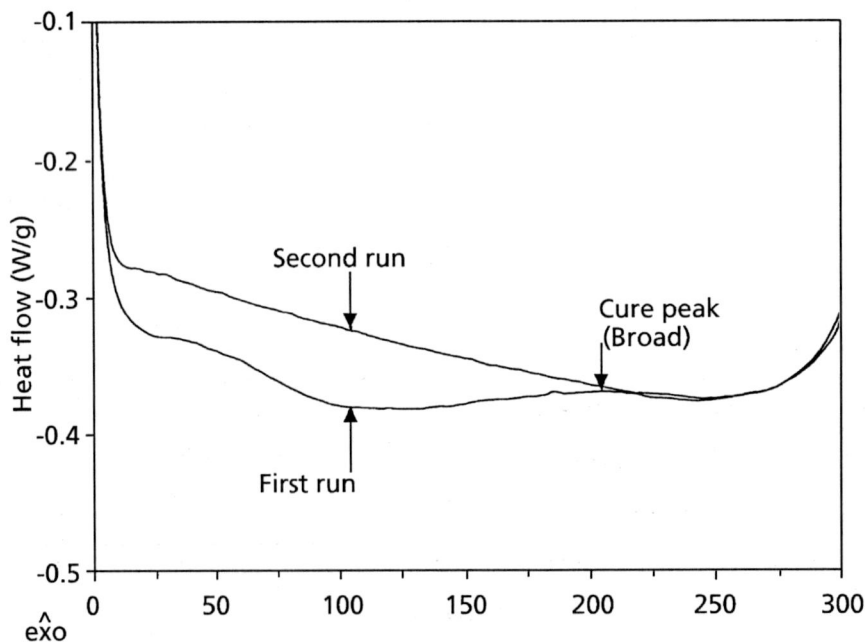

Figure 12.13 DSC thermograms of neat CNSL

12.2.2.5 Cure Characteristics of CNSL-Formaldehyde by Oven and DSC Method

Figure 12.14 shows a thermogram of a sample of oven-cured CNSL-formaldehyde resin analysed using the DSC in a similar way as for the neat CNSL. The resin was subjected to two consecutive DSC dynamic runs. The first run shows an endotherm at around 73 °C and a sharp exotherm at around 138 °C followed by a small and wide exotherm at around 221 °C. The endotherm is due to residual water molecules, entrapped whilst mixing the reactants, the sodium hydroxide catalyst and by-products of the CNSL-formaldehyde poly-condensation reaction. The sharp exotherm indicates that the oven curing was not complete and that residual curing continued during the first DSC run. This is a peak curing temperature, due to the cardanol, which is the most predominant component of crude CNSL.

The second exotherm is due to materials that are more thermally stable than cardanol. According to unpublished work by Mwaikambo and co-workers [32], when crude CNSL is distilled to obtain cardanol, the remaining residue is mainly composed of cardol and exhibits higher thermal stability than cardanol. The second run in **Figure 12.14** shows

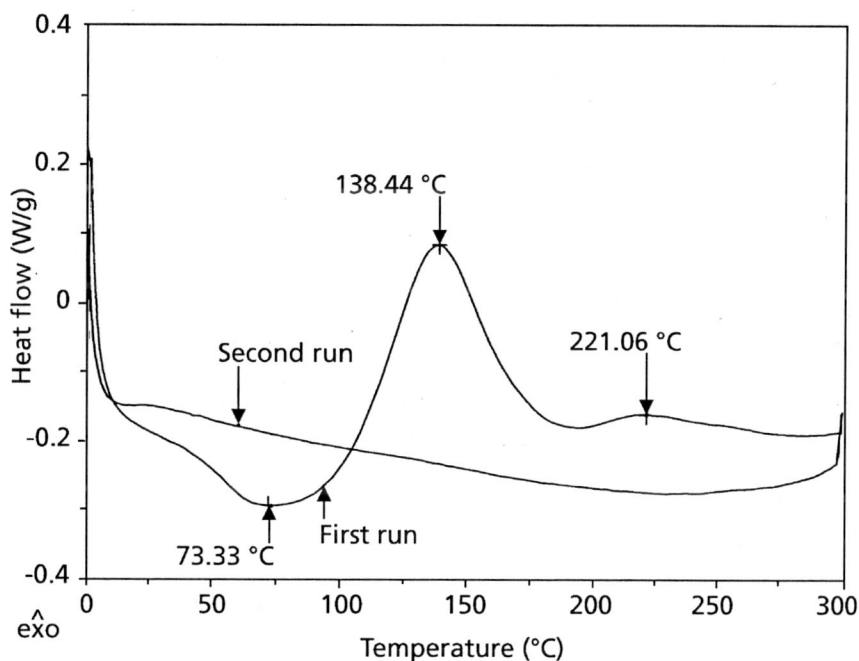

Figure 12.14 DSC thermograms of oven cured CNSL-formaldehyde resin

no endotherm and exotherm peaks indicating that the resin has fully crosslinked. The most common method to determine whether the crosslinking has occurred is by determining the rate at which the reactants diminish, and hence the rate of formation of the products called the rate of reaction, α. In an earlier publication [20] it was demonstrated that since crosslinking results in molecular rigidity, determining the glass transition temperature (T_g), would provide the most reliable information on the progress of cure of the thermosets. It was also observed by using the DSC technique that the sodium hydroxide catalysed CNSL-formaldehyde resin attains a maximum T_g of around 65 °C (**Table 12.2**). Maffezzoli and De Matteis [33] obtained similar T_g values for cardanol, which is a main component in crude CNSL.

The thermal characteristics of CNSL as a function of time represent those of its components shown in **Table 12.1**. The graphs of the T_g *versus* the natural log of time were plotted and are shown in **Figure 12.15**.

Figure 12.15 shows a rapid initial change in the T_g, which appears to rapidly stabilise as the reaction approaches completion. This implies that the speed at which the resin cures and, therefore the rate at which, crosslinking proceeds depends on the amount of residual formaldehyde present in the reaction. It was observed that the T_g reaches a maximum following a curing time of 10 minutes (ln t of around 2.3) at CNSL-formaldehyde concentration ratio of 1:2 and 140 °C when the resin is cured with and without HMTA (**Figure 12.15**).

Table 12.2 Variations in T_g with time of cure for sodium hydroxide catalysed CNSL-formaldehyde at a ratio of 1:2 and 140 °C with and without HMTA

With HMTA			Without HMTA		
Time, t (min)	ln t	T_g (°C)	Time (min)	ln t	T_g (°C)
2.5	0.92	34.67	5	1.61	54.92
5	1.61	49.96	10	2.30	64.56
15	2.71	52.63	15	2.71	64.38
60	4.09	53.06	60	4.09	54.97
120	4.79	53.16	120	4.79	59.00
180	5.19	50.45	180	5.19	55.90
240	5.48	54.34	240	5.48	62.09

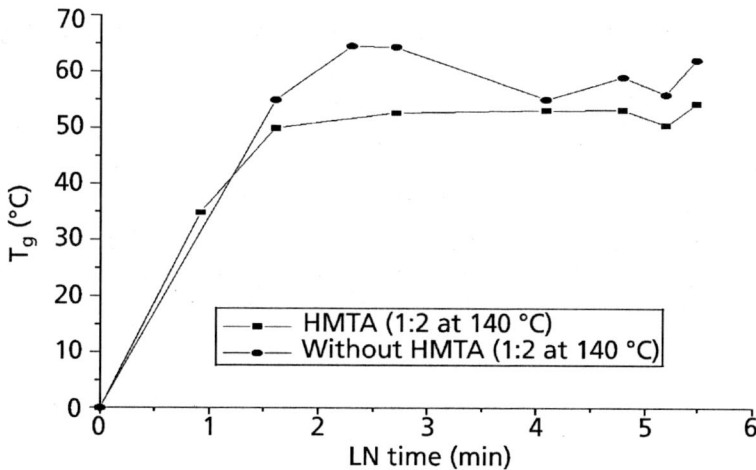

Figure 12.15 Glass transition temperature *versus* ln time for CNSL–formaldehyde (1:2 ratio) with and without HMTA at 140 °C

The link between T_g and time of cure is more relevant for thermoset resins used for composite manufacture since matrix rigidity is critical in end uses where strength and toughness is required.

12.2.2.6 Thermogravimetric Analysis (TGA)

Thermogravimetric analysis determines the loss of mass as a function of temperature. This makes it an invaluable method for evaluating thermoset resins used in high temperature components such as brake pads and gaskets where any loss in mass will result in poor performance. **Figure 12.16** shows the TGA thermogravimetric weight loss curves for distilled CNSL, crude CNSL and CNSL residue.

The samples were heated in nitrogen at a heating rate of 10 °C/min. Crude CNSL shows slight weight loss between 100-200 °C due to decomposition of low molecular weight impurities with mass loss of around 1.5%. Three degradation phases are exhibited in the range 200-700 °C. The degradation that begins at around 296 °C is that of cardanol and the degradation at around 524 °C is that of the cardol and the polymeric materials. **Table 12.3** shows T_g, the degradation temperature and the mass loss of distilled (cardanol) and crude CNSL, and CNSL residue. Cardanol and CNSL residue (cardol) exhibits the same first degradation temperature (296 °C) and it is lower than that of crude CNSL (335 °C).

271

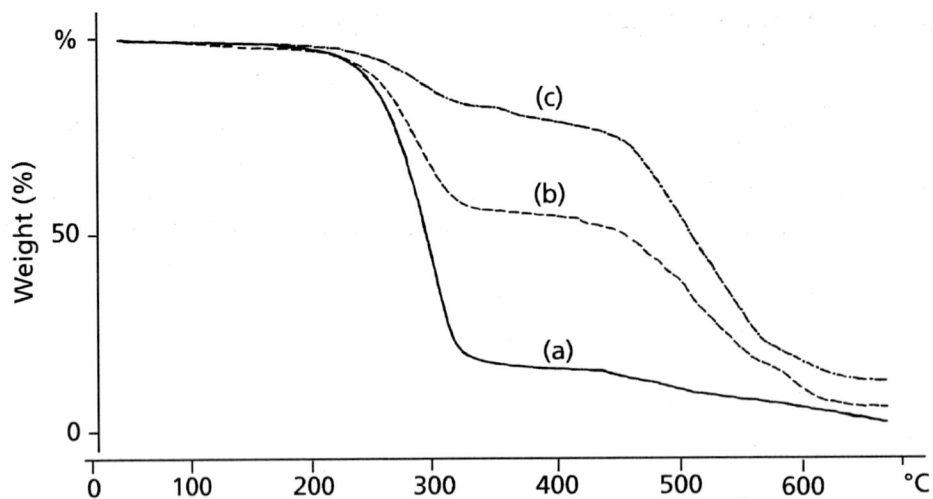

Figure 12.16 TGA thermograms for (a) distilled CNSL (cardanol), (b) crude CNSL and (c) CNSL residue

Table 12.3 The first and second degradation temperatures and mass loss of crude CNSL, distilled CNSL and CNSL residue measured by the TGA method			
Material		Degradation temperature (°C)	Mass loss (%)
Crude CNSL	1	335	44.6
	2	522	89
CNSL residue	1	296	10
	2	524	50
Distilled CNSL	1	296	91.7
	2	-	-

The CNSL residue exhibits the highest first degradation temperature followed by crude CNSL and distilled CNSL, respectively. The highest second degradation temperature was observed in the CNSL residue followed by crude CNSL. Distilled CNSL did not show significant second degradation temperature.

12.2.2.7 Nuclear Magnetic Resonance (NMR) Analysis of Neat CNSL

The NMR spectra of the neat CNSL were recorded in deuterochloroform (CDCl$_3$). The spectrum (**Figure 12.17**) showed the presence of peaks corresponding to aromatic protons, olefinic protons, methylenic protons and methyl protons and is in agreement with the work reported by Varma and co-workers [34] and Nayak and co-workers [27]. Aromatic protons are observed between 6.53-7.20 ppm having an estimate number of 4-hydrogen. Olefinic protons appear as multi peaks between 4.90-5.10 ppm having an estimated number of 4-hydrogen. The methylene protons are seen between 0.80-2.80 ppm with an estimated number of 34-hydrogen [27, 34].

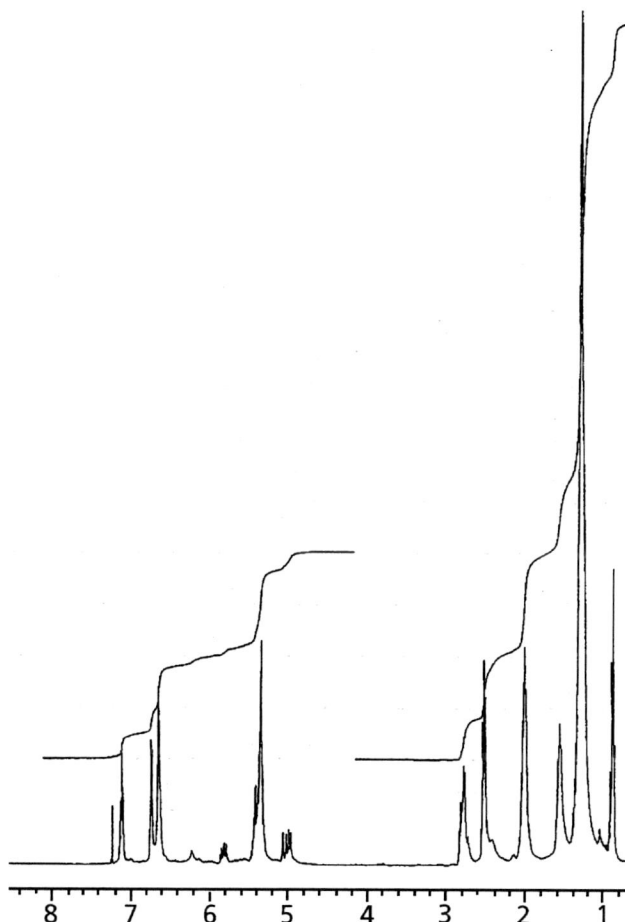

Figure 12.17 H-NMR spectrum of neat CNSL (the numbers in the spectrum are ppm)

The signal at 2.72-2.84 ppm is due to $-CH_2$ groups believed to due to be the triene component of cardanol and the triplet at 2.52-2.56 ppm is due to diene in cardanol. The signal observed at 2.0 ppm is due to $-(CH_2-CH=CH)$ in cardanol. The $-(CH_2)_2$ group appears at 1.26 ppm and the $-CH_3$ terminal group is found between 0.68-0.79 ppm. There are no hydroxyl protons of cardol, which would have appeared as a singlet at 6.50 ppm, and in 6-methyl cardol, normally seen at 4.65 ppm. Hydroxyl protons of cardanol are seen at 5.80 ppm. This implies that the crude CNSL used in this analysis is predominately composed of cardanol.

Figure 12.18 shows the proton NMR spectra of *in situ* polymerisation of CNSL-formaldehyde resin. Comparing the signals identified in **Figure 12.17** as being due to the cardanol component of CNSL with those observed in **Figure 12.18** it is clear that chemical reaction has happened. For instance, the signals due to the hydroxyl protons of cardanol observed at 5.80 ppm in **Figure 12.17** are not seen in **Figure 12.18**. Similarly the signals

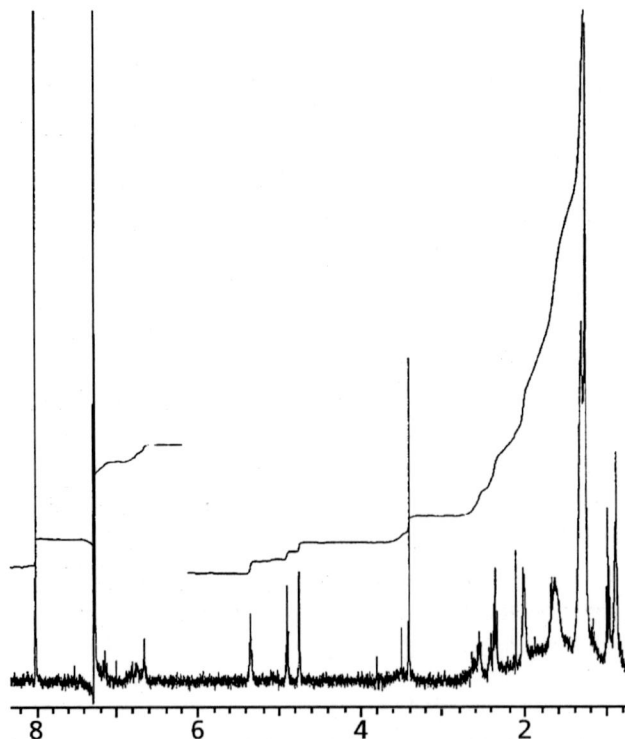

Figure 12.18 H-NMR spectrum of CNSL-formaldehyde resin

of the triene component of cardanol observed between 2.72-2.84 ppm and the signals due to the diene component of cardanol between 2.52-2.56 ppm have diminished in **Figure 12.18**.

The broad base observed between 1.80-2.80 ppm (**Figure 12.18**) is caused by the presence of water molecules resulting from the polycondensation reaction of CNSL-formaldehyde, from sodium hydroxide and formaldehyde solutions. It can be surmised from these results that CNSL-formaldehyde reactions is mainly that of cardanol.

Figure 12.18 also shows signals at 6.63 ppm and at 2.35 ppm and 0.86 ppm caused by the carbon atom at the *para* position and the methylene linkage of the novolak type resin hardened using HMTA.

12.2.2.8 Applications

Cashew nut shell liquid and its by-products have found a wide range of end uses. Typical applications include oil-seals and filling compounds in electrical components, flooring compositions and chemically resistant cements. CNSL-based products are used as additives for antioxidants, stabilisers and demulsifiers for petroleum products. Sulfonated cardanol or partially hydrogenated cardanol, when added to mineral oil or lubricating oil improves the resistance of the latter towards oxidation and corrosion.

CNSL resins have been used as binders for friction materials obtained from completely cured resins. They possess excellent friction-resistant properties because of their plastic nature due to the side chain and resistance to chemical change and degradation in the presence of heat generated by friction. A composition, containing the dust of polymerised and cured CNSL, a thermosetting binder such as phenol-, cresol- or urea-formaldehyde resin and fillers such as the fibres of cotton, sisal, leather and granulated cork, has been found useful in the preparation of wet clutch facings. Fully cured CNSL-based polymers are also used as friction dust in brake-lining compositions, where they reduce the fall in coefficient of friction with increase in temperatures. Particulate CNSL-MA-formaldehyde condensate products have been used as automotive disk-brake pads with improved high-temperature performance. Railroad brake shoes exhibiting good frictional properties have been made from a composition containing CNSL resins. Such brake materials show no cracking, delamination or voids.

CNSL is used in rubber compounds, where they act as reinforcing fillers and improve tensile strength, hardness and abrasion-resistance. The addition of cardanol to phenol-formaldehyde composition, incorporated into rubber, either prior to or after vulcanisation, improves the resistance of the rubber material ozone cracking. The alkyl ethers of CNSL or cardanol have been used as plasticisers for various types of rubbers such as neoprene, styrene-butadiene and polybutadiene. The incorporation of cardanol improves the durability and flexibility of rubber even at low temperatures. Products formulated using copolymerised 1,3-butadiene-acrylonitrile rubber, CNSL, sulfur and hexamine provide good lining materials for tanks.

A research group at the University of Bath has been conducting research on the formulations of alkali catalysed CNSL-formaldehyde resin and has used it as binder for plant fibres such as sisal, hemp and kenaf [20, 35]. The results show that CNSL-based composites can be used in buildings as interior panels and roofing materials.

12.3 Conclusions

Plant oils offer important sources of naturally occurring phenolic-based monomers, which can be formulated to produce thermoset resins. The most abundantly available of all the natural oils is CNSL, which is has a phenolic structure. The presence of four different phenolic components and the side chain unsaturation allows the use of different synthesis and polymerisation pathways using acid and alkali catalysts. The resulting polymers are highly crosslinked thermoset resins, which exhibit good thermal and frictional properties. The use of crude CNSL as the feedstock for the preparation of thermoset resins has economic advantages over the distilled components, as the latter will require extra investment for the distillation processes. The cost disadvantage of the distillates is offset by the requirement for pure products with appealing aesthetic properties.

The availability of CNSL in the tropics and its potential use as a natural replacement for fossil origin products offers a cheap and renewable natural resource which can be used for the manufacture of plant fibre reinforced composites. The formulation of CNSL-formaldehyde resin can be used as adhesives for plant fibres such as hemp, flax, jute, sisal and kenaf for the manufacture of composites.

The use of plant-based renewable resources for the productions of materials for industrial applications has economic as well as environmental benefits appropriate to the 21st century.

Acknowledgement

The authors wish to thank the Sokoine University of Agriculture for availing financial support in a form of scholarship under the Tanzanian-Norwegian technical co-operation programme.

References

1.   A.A.K. Whitehouse, E.G.K. Pritchett and G. Barnett, *Phenolic Resins*, 2nd Edition, revised, The Plastic Institute, Iliffe Books Ltd, London, UK, 1967, 1.

2.   D.J. Walton and J.P. Lorimer in *Polymers*, Oxford University Press, Oxford, UK, 2001, 111.

3.   P.C. Charterjee and B.S. Sitaramam, *Journal of Applied Polymer Science*, 1989, **37**, 33.

4.   P.H. Gedam and P.S. Sampathkumaran, *Progress in Organic Coatings*, 1986, **14**, 115.

5.   J.V. Crivello and R. Narayan, *Chemistry of Materials*, 1992, **4**, 692.

6.   J.A. Sherringham, A.J. Clark, and B.R.T. Keene, *Lipid Technology*, 2000, **12**, 129.

7.   F.D. Blum, B.R. Sinha and D. O'Connor, *Journal of Applied Science*, 1989, 38, 163.

8.   J.C. Caygill in *Secondary Plant Products, Antinutritional and Beneficial Actions in Animal Feeding*, Eds., J.C. Caygill and I. Muller-Harvey, Nottingham University Press, Nottingham, UK, 1999, 1-6.

9.   I. Mueller-Harvey and A.B. McAllen in *Advances in Plant Cell Biochemistry and Biotechnology*, Volume 1, Ed., I.M. Morrison, JAI Press Ltd, London, UK, 1992, 151-217.

10. I. Mueller-Harvey in *Secondary Plant Products, Antinutritional and Beneficial Actions in Animal Feeding*, Eds., J.C. Caygill and I. Muller-Harvey, Nottingham University Press, Nottingham, UK, 1999.

11. *Chemistry and Significance of Condensed Tannins*, Eds., R.W. Hemingway and J.J. Karchesy, Plenum Press, New York, NY, USA, 1989.

12. A. Pizzi and A. Stephanou, *Journal of Applied Polymer Science*, 1993, 50, 2105.

13. J.F. Shackelford, *Introduction to Materials Science for Engineers*, 2nd Edition, Macmillan Publishing Company, London, UK, 1988, 403.

14. S.G. Kuzak, J.A. Hiltz and P.A. Waitkus, *Journal of Applied Polymer Science*, 1998, **67**, 349.

15. B. Golding, *Polymers and Resins: Their Chemistry and Chemical Engineering*, D Van Norstrand Company, Inc., Princeton, NJ, USA, 1969, 303.

16. *Textbook of Polymer Science*, 3rd Edition, Ed., F.W. Billmeyer, Wiley-Interscience, New York, NY, USA, 1969.

17. M.F. Grenier-Loustalot, S. Larroque, D. Grande, P. Grenier and D. Bedel, *Polymer*, 1996, **37**, 6, 955.

18. N.J.L. Megson, *Phenolic Resin Chemistry*, Butterworths Scientific, London, UK, 1958.

19. M.F. Grenier-Loustalot, S. Larroque, D. Grande and P. Grenier, *Polymer*, 1996, **37**, 4, 639.

20. L.Y. Mwaikambo and M.P. Ansell, *Journal of Materials Science*, 2001, **36**, 15, 3693.

21. J.H.P. Tyman, *Chemical Society Review*, 1979, **8**, 499.

22. R. Amorati, G.F. Pedulli, L. Valgimigli, O.A. Attanasi, P. Filippone, C. Fiorucci and R. Saladino, *Journal of the Chemical Society Perkin Transactions 2*, 2001, 2142.

23. A.R.R. Menon, C.K.S. Pillai, J.D. Sudha and A.G. Mathew, *Journal of Scientific and Industrial Research*, 1985, **44**, 324.

24. V.S. Pansare and A.B. Kulkarni, *Journal of Indian Chemical Society*, 1964, **41**, 4, 251.

25. A.K. Misra and G.N. Pandey, *Journal of Applied Polymer Science*, 1984, **29**, 361.

26. A.K. Misra and G.N. Pandey, *Journal of Applied Polymer Science*, 1985, **30**, 969.

27. P.L. Nayak, S.K. Swain, S. Sahao, D.K. Mohapatra, B.K. Mishra and S. Lenka, *Journal of Applied Polymer Science*, 1994, **54**, 1413.

28. C.K.S. Pillai and R. Antony, *Journal of Applied Polymer Science*, 1990, **41**, 1765.

29. A Whelan and J.A. Brydson, *Developments with Thermosetting Plastics*, Applied Science Publishers, London, 1975.

30. M. Joksič in *Proceedings of an International UNIDO workshop on Materials Selection and Design for Low-Cost Housing in Developing Countries*, AREA Science Park, Trieste, Italy, 1999, 45-55.

31. L.Y. Mwaikambo and E.T.N. Bisanda, *Polymer Testing*, 1999, **18**, 3, 181.

32. L.Y. Mwaikambo, M. Avella and E. Martuscelli, *Thermal Characterisation of Cashew Nut Shell Liquid by DSC and TGA Methods*, un-published work, 1998.

33. A. Maffezzoli and F. De Matteis in *Proceedings of an International UNIDO workshop on Materials Selection and Design for Low-Cost Housing in Developing Countries*, AREA Science Park, Trieste, Italy, 1999, 35.

34. I.K. Varma, S.K. Dhara, M. Varma and T.S. Biddapa, *Die Angewandte Makromolekulare Chemie*, 1987, **154**, 2504, 67.

35. E.T.N. Bisanda and M.P. Ansell, *Composite Science and Technology*, 1991, **41**, 2, 163.

13 Commercially Available Low Environmental Impact Polymers

Roger Van Erven

In a field as volatile as biopolymers, any printed information is likely to date quite rapidly, therefore the reader is encouraged to look at www.Biopolymer.net as a good starting point to get more information and news about biopolymers and related subjects and materials. The site was started by BiPP because there was nothing like it, and information was hard to find. BiPP (www.BiPP.nl) is a company dedicated to working with biopolymers, providing all kinds of services, like creating concepts, designing products, producing products and testing materials. This chapter summarises the results of a survey of available biopolymers.

This chapter is based on information obtained by conducting a questionnaire-based survey. Companies who responded have detailed profiles which in some cases have been complimented by additional information from trade directories and web sites. Thus there is some bias toward responsive companies with a significant web presence. We make no claim that this list is definitive, however, it is representative of the scope of commercially available polymers at the time of publication.

Cargill Dow LLC
15305 Minnetonka Boulevard
Minnetonka   Tel: +1 952 742 0400
MN 55345   Fax: +1 952 742 0477
USA   www.cargilldow.com

Material	NatureWorks PLA and Ingeo PLA fibres
Material type	Polylactic acid (PLA). Derived entirely from annually renewable resources like corn or wheat.
Properties	High gloss, clarity, stiffness, good twist retention and deadfold (it stays wrapped around items without needing further sealing).
Grades	Grades available include extrusion thermoforming, injection moulding, biaxially orientated films, injection blow moulding, and foaming.

Typical uses	*Business segment*	Commercially available applications
	Rigid thermoforms	Clear, short shelf life trays and lids. Opaque dairy containers. Consumer displays and electronics packaging. Disposable articles. Cold drink cups.
	Biaxially-oriented films	Shrink wrap for consumer goods packaging. Twist wrap candy and flower wrap. Windows for envelopes, bags and cartons.
	Bottles	Short, shelf-life milk and oil packaging.
	Apparel	Sport, active and underwear. Fashion.
	Non-wovens	Agricultural and geo textiles. Hygiene products (diapers and feminine hygiene). Wipes. Shoe liners. Blends with natural fibers – hemp, sisal and flax.
	Household, industrial and institutional fabrics	Bedding, drapery, table cloths, curtains, mattress ticking. Wall and cubicle fabrics, upholstery.
	Carpet	Surface yarns and fibres.

Fibrefill	Pillows.
	Comforters.
	Mattresses.
	Duvets.
Production volume	The full capacity of the plant in Blair, Nebraska, USA is 140,000 tonnes per year

Biomer
Forst-Kasten-Strasse 15
Krailling   Tel: +49 89 8572665
D-82152   Fax: +49 89 8572792
Germany   www.biomer.de

Material	Biomer® P and Biomer® L
Material type	Biopolyesters; polyhydroxybutyrate (PHB) and poly-L-lactate (PLLA).
Grades	Articles made of Biomer resins are waterproof.

Biomer resins are inert towards water, even after extended exposure. So articles can be produced which can come into contact with water. They also are fairly resistant towards fats and oils. Thus articles made of Biomer resins protect sensitive products against moisture and oily soiling.

Items made of Biomer resins have attractive surfaces and are transparent. Injection moulded products made of certain Biomer formulations have a shiny surface. Thus optically attractive products can be created. The articles made of PLLA formulations are transparent.

Articles made of Biomer materials can be autoclaved.

Articles made of PHB thermoplastics retain their original shape at 121 °C or 132 °C. Because of this perishable goods can be canned into packages produced of Biomer resins and preserved by steam sterilisation.

Articles made of Biomer materials are toxicologically safe.

The monomer as well as the polymer are natural components and metabolites of human cells. They are not toxic even at high concentrations. Thus Biomer formulations can be used for articles which come into contact with skin, feed or food (registration for food contact in progress).

Articles made of Biomer materials are fully biodegradable.

In composts where nitrogen and phosphates are abundant, items made with Biomer resins are taken up by bacteria and fungi as nutrient. No toxic byproducts are accumulated. Under optimal conditions thin walled injection moulded articles are composted within two months. Therefore articles made with Biomer resins can be disposed of with ordinary kitchen wastes. Without composting conditions they remain intact for years.

Properties

Type	*Biomer® P209*	*Biomer® P240*	*Biomer® P226*	*Biomer® L9000*	*Biomer® L1000*
Modulus (MPa)	900-1200	1000-1200	1700-2000	3600	-
Tensile strength (MPa)	15-20	18-20	24-27	70	-

Properties Cont'd	Type	Biomer® P209	Biomer® P240	Biomer® P226	Biomer® L9000	Biomer® L1000
	Elongation (%)	4-18	10-17	6-9	2.4	-
	Flexural strength (N/mm²)	18	17	35	98	-
	Deformation at bending break (%)	4.7	6.5	6.6	2.8	-
	Flexural strength at 3.5% (N/mm²)	16	17	29	-	-
	Impact strength (KJ/m²)	21	80	30	16.5	-
	Notched impact strength (KJ/m²)	2.1	8.6	2.7	3.3	-
	MFI (g/10 min)	17-20	5-7	9-13	3-6	>200
	Vicat-Temp. (°C)	57	53	96	56	-
	Density (g/cm³)	1.20	1.17	1.25	1.25	-
	Moisture absorption (%)	0.75	0.4	0.4	0.3	-
	Hardness (Shore D)	57	56	67	na	-
	Shrinkage	1.3	1.3	1.3	0.12	

Tests done at least 4 week after the manufacture of the test specimens.
na: not available
MFI: melt flow index

Typical uses	High temperature, or wet conditions (PHB) and medical or food use (PLLA).
Production volume	Biomer can supply up to 5 tonnes/month of the P-series and about 1 tonne/month for the L-one. This should increase to 5 tonnes/month later in 2004.

BASF Aktiengesellschaft
Carl-Bosch Strasse 38
Ludwigshafen
D-67056
Rheinlandpfalz
Germany

Tel: +49(0)621 60 49978
Fax: +49(0)621 60 22356
www.basf-ag.de

Material	Ecoflex
Material type	Aliphatic aromatic copolyester.
Properties	The mechanical properties of Ecoflex are comparable with low density polyethylene (LDPE). The films are tear resistant and flexible, as well as being resistant to water and humidity fluctuations. The very high toughness and failure energy represent product characteristics of Ecoflex, which significantly exceed typical properties of LDPE films. The barrier properties differ from LDPE. Ecoflex films are breathable because of their moderate water vapour permeability, that can be adjusted from batch to batch.
Grades	Ecoflex S (for starch blend), Ecoflex F (for film application), Ecoflex P (for paper coating).
Typical uses	Shopping/composting bags, packaging, fibre, non-wovens, coatings, agriculture.
Production volume	2000: 4000 tonnes
Additional information	A new plant (40,000 tonnes/year) is planned.

Hycail BV
Industrieweg 24-1
Noordhorn
9804 TG
Groningen   Tel: +31 594 50 57 69
Netherlands   Fax: +31 594 50 62 53

Material	Hycail HM 1010, poly lactic acid for extrusion thermoforming and injection moulding
	Hycail LM1, hot melt adhesive based on lactic acid
Material type	PLA
Grades	Low and high molecular weight variants, All Hycail ® products will be certified according to NEN 13432 [1], and can be provided in custom colours.
Typical uses	Film blowing, thermo-forming and injection moulding, hot melt adhesives, emulsions and chewing gum base
Production volume	Current capacity is around 1000 tonnes/year, although the capacity is being expanded
Additional information	Hycail BV originally started in 1997, as a spin-off of the University of Groningen, The Netherlands. Its technology was initially based on the work of A.J. Pennings, who in the 1970s and 1980s made numerous fundamental contributions in the field of PLA molecular structure/ property relationships.
	In 1998 Dairy Farmers of America (DFA), the world's largest dairy company, gained shares in Hycail. DFA's goal is to add value to the whey permeate which is released by the preparation of cheese in their numerous cheese factories.

Midwest Grain Products Inc.
1300 Main Street
Atchison
66002-0130   Tel: +01 913 367 1480
Kansas   Fax: +01 913 367 1838
USA   www.midwestgrain.com

Material	Polytriticum 200 and Polytriticum 2000
Material type	Wheat gluten (P200), wheat starch/biodegradable polyester compound (P2000).
Properties	Polytriticum 200 and 2000. These biodegradable base resins are made for injection moulding processes. Advantages over other biodegradable resins, including improved performance and cost effectiveness are claimed.
Grades	**Polytriticum 200** is derived from wheat protein, Because of its natural origin, it is readily compostable and degradable in a soil environment. The moulding of Polytriticum 200 is similar to that of a thermoset resin. However, the scrap reclaim can be cycled back to the virgin resin. Polytriticum 200 is opaque and is best processed at a barrel temperature of 60-80 °C. The mould temperature ranges between 120-155 °C. Under normal storage conditions, the moulded article possesses mechanical properties similar to polypropylene.
	Polytriticum 2000 is a specialty blend of wheat starches, wheat protein and biodegradable polyester. It has applications in types of disposable products such as utensils, tent stakes, golf tees, etc. The base resin is light brown and is fully compostable within six months as required by current DIN standards. The moulding of Polytriticum 2000 is similar to that of thermoplastics, with 100% of the scrap reclaim able to be cycled back to the virgin resin. The processing temperature for Polytriticum 2000 is 100-150 °C. As a result of the high content of wheat starch and wheat protein (60%-70%) in Polytriticum 2000, the moulded article is rigid. The mechanical properties ranges between those of LDPE and polystyrene, except in elongation and moisture absorption. The tensile strength ranges from 17-35 MPa, and the flex strength from 30-60 MPa. In a spiral mould test, the flow length ranges between 13-64 cm. The heat distortion temperature of Polytriticum 2000 is between 75-105 °C, depending on the type of biodegradable polyester used in the blend.
Typical uses	P200 is used in pet treats, land-related products and disposable products. P2000 – example applications are tent pegs and golf tees.

Groen Granulaat
Maccallastraat 32a
Noord-Brabant
5708 KR   Tel: +31 0492 590 612
Helmond   Fax: +31 0492 565 716
The Netherlands   www.groengranulaat.nl/home2.html

Material	Ecoplast
Material type	Ecoplast is composed of wood dust, starch, a binder, an internal mould release agent and water.

Properties

Moisture, %	Tensile strength, MPa	Rigidity, MPa	Elongation at break, %	Notch impact strength, kJ/m²
6.0	33.5	1,400	6.1	1.8
5.7	32.8	1,668	5.7	1.5
0	23.7	3,024	3.4	1.3

(Ecoplast recipe Sj0)

Grades	Injection moulding and hot forming granulates.
Typical uses	Construction, thick section mouldings.
Production volume	Plant capacity is 1200 kg/h
Additional information	Ecoplast is producing biodegradable granules based on two recipes resulting from research at Fontys University, Eindhoven, The Netherlands.

Idroplast Srl
Via Palazzaccio 59
Altopascio
55011   Tel: +39 0583 21 65 35
Lucca   Fax: +39 0583 21 65 38
Italy   www.idroplast.com

Material	Hydrolene
Material type	Polyvinyl alcohol-based (PVOH) plastic available as the following grades: Bubble extrusion of film. Injection moulding. Co-extrusion. Film on reel. Plane, side-folded and tube film. Width ranging between 25 mm and 1500 mm. Bags in various dimensions and thicknesses. Glue having various viscosities and pH. Water soluble and self-adhesive labels.
Properties	Non-toxic, biodegradable, water soluble, good barrier properties.
Grades	Grades with varying solubility in water are available. The company claims that it is possible to modify the water solubility and also other characteristics to obtain a product complying with specific requirements. The standard grades of Hydrolene are:

* Hydrolene LTF type + 15 °C
* Hydrolene LTT type + 15 °C - suitable for thermoforming packaging
* Hydrolene LTS type + 30 °C
* Hydrolene HT type + 45 °C
* Hydrolene VHT type + 65 °C

Typical uses

Medical field:
* Bags for hospital laundries.
* One-dose packaging of enzymes and chemical substances where the packaging is subsequently thrown in septic tanks.

Packaging:
* Packaging of toilet paper, tissue and so on.
* Packaging of chemical substances to be dispersed in water.
* Packaging of bait for fishing.
* Packaging of roots of plants, to avoid ground scattering during transportation and planting operations.

Textile field:
- Packaging of colouring substances used by dye-works.
- Temporary support in industrial and craft embroidery.

Packaging of detergents:
- One-dose packaging of powder and compact detergents through traditional sealing machines or also through thermoforming.
- One-dose packaging of non-water-based liquid detergents.

Agriculture:
- Packaging of antiparasitics, herbicides, harsh and aggressive chemicals in general.
- Packaging of rat poison.

Building industry:
- Packaging of fibreglass and other substances to be added to concrete.
- Packaging of antifreeze and other substances to be thrown directly in the concrete mixer.

Co-extrusion:
- Hydrolene can be coextruded with other traditional plastics to reduce the thicknesses of the latter, as Hydrolene has remarkable barrier properties.

Injection moulding:
- Pellets for injection moulding for the production of rigid items.

Additional information

The properties of non-toxicity, biodegradability and compostability, water solubility and the chemical and physical properties of Hydrolene have been certified by some of the most important European research institutes such as Department of Chemistry and Industrial Chemistry at the University of Pisa (Italy); the University of Essen (Germany); CIS - Italian Experimental Centre of Milan (Italy); and the Polytechnic Institute of Milan (Italy).

Lethal Dose 50 (LD_{50})test and ichthyotoxicity test to evaluate compatibility with Italian Law Merli 319/76 have been successfully carried out by the Local Sanitary Unit of Pisa (Italy).

Kuraray Specialities Europe GmbH
Industriepark Höchst
65926 Frankfurt am Main
Germany

Telephone: +49 69 3 05 853 00
Fax: +49 69 3 05 85399
E-mail: marketing@kuraray-kse.com
www.kuraray-kse.com

Material	Mowiol, Mowital
Material type	PVOH, Polyvinylbutyral (PVB)
Properties	Water-based, solvent-based.
Grades	**Mowital** - Properties and applications depend mainly on degree of acetalisation and viscosity of the Mowital grade. The numerical suffix of the grade designation (20, 30, 45, 60 and 75) indicates the increasing degree of polymerisation (and thus the increasing viscosity) of the various grades. The designation T, H and HH indicate different degrees of acetalisation.
	Mowiol - PVOH contain vinyl alcohol and vinyl acetate units. In partially hydrolysed grades the vinyl alcohol content is such that the entire molecule is freely soluble in water. A high degree of hydrolysis increases the crystallisation tendency and crystallinity of polyvinyl alcohols. Fully hydrolysed grades are therefore less soluble in water
Typical uses	Coatings, adhesives, printing inks, paper.
Production volume	50,000 tonnes per year
Additional information	PVOH are soluble in water, react with many substances, and are not health-damaging. Although renewable origin is not claimed, they are biodegradable.

Biotool Naturnahe Kunststoffe GmbH
Reitweg 5
90587 Veitsbronn   Tel: +49 (0) 911 7530162
Germany   Fax: +49 (0) 911 7530262

Material	Megithan, Megidur, Megicoat, Megivest, Meprolastik
Material type	Thermoset resin systems based on sugar-starch-, sorbitol-, glycerin-, oleo-, ricinus-, soyapolyoles and caprolactames.
Properties	1- or 2-component thermoset resin systems. Durable, but not biodegradable! Non-toxic, absolutely non hazardous system. High impact strength, good heat resistance, low viscosity.
Grades	Hard to flexible (80 Shore D - 40 Shore A). Filled or high transparent casting resins. Tooling resins for thermoforming and sheet metal deep drawing. Resins for laminating (RTM) or bonding.
Typical uses	Transparent cast parts (small to very big). Illuminated ornaments, active light systems, tools. In combination with carbon, glass or natural fibres, e.g., construction parts, coach parts, sports articles, etc.
Production volume	About 200 tonnes per year
Additional information	All resins have a positive life cycle analysis and are CO_2 neutral. Room temperature vulcanising silicones with very good shear strength also belong to the Biotool programme.

PVAXX Limited
Warner Building
85 Reid Street
Hamilton HM12
Bermuda

Telephone: + 1 441 296 5072
Fax : +1 441 296 4306
www.pvaxx.com
E-mail: sales@pvaxx.com
technicalsupport@pvaxx.com
investorrelations@pvaxx.com

Material	PVAXX biocompostable polymers and PVAXX Unifier
Material type	PVA-based polymer - hot and cold water soluble. Unifier allows the conversion of industrial by-products (such as ash, tyres and agricultural biomass) into highly saleable and useable materials. These materials can be processed on standard moulding or extrusion equipment as replacements for traditional wood or plastic products with significant cost savings.
Properties	Blown film, sheet, injection, blow moulding. Water soluble at 5, 55 and 80 °C.
Grades	Injection grades: C63, C60, C66, C73, C70, H64, H63, H60, H76 Film grades: C33, C30, C20, C21, H23, H20 Pharmaceutical Extrusion Grade. Extrusion W42
Typical uses	Compostable bags, all injection items, thermoforming, expanded foam products, cold soluble films for single dosage. Profile and tube extrusion.
Production volume	10,000 tonnes per year
Additional information	

Eastman Chemical BV
Weena 159-161
Rotterdam
3013 CK
The Netherlands

Tel: +31 102402265
Fax: +31 102402100
www.eastman.com

Material	Eastar Bio
Material type	Aliphatic aromatic copolyester, produced from conventional diacids and glycols.
Properties	Eastar Bio copolyester resin can be cast or blown into tough translucent film, extrusion coated or laminated onto paper or other substrates, and can be spun into fibres as well as spunbonded and melt blown non-wovens.
It is easy to process on existing polyethylene (PE) equipment and exhibits a soft flexible hand.
It has tensile properties similar to LDPE, and offers superior adhesion and is compatible with other natural polymers.
Grades	Eastar Bio GP
Eastar Bio Ultra
Eastar Bio additives and colour concentrates
Typical uses	Yard and food waste/compost bags.
Food packaging.
Fast-food serviceware.
Coated paper packaging.
Agricultural films.
Fabrics and non-wovens.
Medical disposables.
Disposable and limited use wipes.
Additional information:	• Eastar Bio meets FDA and EU Foodcontact Standards
• Eastar Bio's compostability is certified by DIN CERTCO
• Eastman is a member of the IBAW and BPI and is active in several scientific committees.
• Production is located in Hartlepool, UK and has a capacity of 15,000 tonnes.

Novamont SpA
via Fauser, 8
Novara
28100
Novara
Italy

Tel: +39.0321.699611
Fax: +39.0321.699600
www.materbi.com

Material	Mater-Bi
Material type	Starch-based biodegradable materials.
Properties	Ranging from flexible film (similar to LDPE) to rigid articles [similar to polystyrene (PS)].
Grades	Film extrusion. Sheet extrusion. Injection moulding. Foams.
Typical uses	Trash and carrier bags, mulch film, hygiene, catering, flexible and rigid packaging, pet toys, loosefillers, expanded sheet.
Additional information	Products certified as biodegradable and compostable according to international standards Complete (from cradle to grave) LCA available for specific film and foam products

Metabolix, Inc.
303 Third Street
Cambridge
MA 02142
USA

Tel: +1 617 492 0505
Fax: +1 617 492 1996
www.metabolix.com

Material	Metabolix PHA
Material type	Polyhydroxyalkanoates
Properties	Metabolix PHA cover an enormous range of physical properties. The following chart shows the range of tensile strength and elongation at break for unformulated PHA when tested according to ASTM D638-03 [2]. PHA can behave both as traditional thermoplastic polymers and as elastomers. While some polymers such as polyethylene, flexible PVC, and thermoplastic elastomers have high elongation at break figures, they yield irreversibly at high levels of extension. Metabolix has developed truly elastomeric grades that have high levels of recovery (typically >80-90%), even under high levels of deformation, (e.g., > 500% ultimate elongation at break). These materials can be used for adhesives, stretch coatings, and fibres, and have properties competing with some vulcanised rubbers

Typical Uses

Market	*Applications*
Polymer Performance Improvers	PLA
	Cellulosics
	Starch
	Synthetic Biodegradable Polyesters
Fibers and Nonwovens	Textiles
	Wipes
	Personal Hygiene Products
	Filtration Products
	Carpets
Adhesives	Hot Melts
	Pressure Sensitives
	Tie Layers
	Heat Seal Resins
Film (including multi-layer)	Barrier Layers
	Packaging
	Agricultural Mulch
	Film
	Trash Bags
	Lawn and Leaf Bags

Typical Uses	*Market*	*Applications*
Cont'd	Functional Specialties	Controlled Release Binders for Metal & Ceramic Powder
	Moulding Resins	Consumer
		Disposables
		Containers
		Appliances
		Toys
		Automotive
	Coatings	Water Resistant
		Paper and Board Coatings
		Architectural Paints
		Wood Products
	Other	Solvents
		Coalescing

Additional information

Metabolix produces a wide variety of bioplastics through the fermentation of natural sugars and oils using microbial biofactories. These materials range in properties from stiff thermoplastics suitable for moulded goods, to highly elastic grades, to grades suitable for adhesives and coatings. In some cases, bioplastics offer combinations of properties not available in synthetic materials. For example, bioplastics, have excellent water resistance with biodegradability, which allows production of flushable personal hygiene products and wet wipes. In the future, bioplastics will be produced directly in plants, making them cost-competitive with even general purpose resins such as polyethylene. PHA will serve as environmentally friendly alternatives to over half of the plastics used today.

Low Environmental Impact Polymers

Tenaro GmbH
Am Goldberg 2
D-99817 Eisenach/Stedtfeld
Germany

Tel: +49 (0) 3691621-320
Fax: +49 (0) 3691621-329
www.Tecnaro.de

Material	Arboform			
Material type	Lignin, natural fibres			
Properties	*Mechanical Properties*	*Standards*	*Unit*	*Range**
	Tensile strength	DIN EN ISO 527-1 [3] DIN EN ISO 527-2 [4]	N/mm^2	10-22
	Ultimate elongation	DIN EN ISO 527-1 [3] DIN EN ISO 527-2 [4]	%	0.3-0.7
	Tensile modulus	DIN EN ISO 527-1 [3] DIN EN ISO 527-2 [4] DIN EN ISO 178 [5] DIN EN ISO 604 [6]	N/mm^2	1000-5000
	Flexural modulus	EN ISO 178 [7]	N/mm^2	1000-5000
	Bending stress	EN ISO 178 [7]	N/mm^2	10-50
	Impact strength	EN ISO 179 [8]	KJ/m^2	2-5
	Hardness	DIN 53505 [9]	Shore D	50-80
	Ball indentation hardness	DIN 53456 [10]	N/mm^2	20-70
	Thermal Properties			
	Expansion coefficient	DIN 53752 [11]	I/°C	$1 \times 10^{-5} - 5 \times 10^{-5}$
	Vicat temperature	DIN EN ISO 306 [12]	°C	80
	Martens temperature		°C	54
	Thermal conductivity	DIN EN 12939 [14] DIN EN 12664 [14] DIN EN 12667 [14]	W/(m*K)	0.384
	Hot-wire test	DIN EN 60669 [17]	—	650 °C, passed
	Electrical Properties			
	Conductivity, surface	DIN IEC 60039 [18] DIN IEC 60167 [19]	G Ohm	5
	Conductivity, continuity	DIN IEC 60039 [18] DIN IEC 60167 [19]	G Ohm	3

Other Properties

Mould shrinkage	-	%	0.1-0.3
Density (in compact moulding)	-	g/cm³	1.3-1.4
Water content	-	%	2-8
Bleeding of various elements	DIN EN 71-3 [20]	-	passed
Saliva and sweat fastness	DIN V 53160-1 [21] DIN V 53160-2 [22]	-	passed
Fire performance	DIN 4102-2 [23]; DIN EN 1363-1 [24]; DIN EN 1364-1 [25]; DIN EN 1364-2 [26]; DIN EN 1365-1 [27]; DIN EN 1365-2 [28]; DIN EN 1365-3 [29]; DIN EN 1365-4 [30]	-	passed

* depending on formulation

Grades	Standard and also to the customers formulation, adjusted (with pigments and/or odour).
Typical uses	Furniture, automotive, toys, musical instruments, loudspeakers.
Production volume	300 tonnes per year.
Additional information	Arboform is thermoplastic material made exclusively from renewable resources based on lignin, a natural polymer which is formed by photosynthesis and makes up about 30 % of the substance of every tree and every woody plant. Lignin is second to cellulose as the most abundant natural polymer and, in the tree trunk for example, forms a three-dimensional supporting structure around the cellulose fibres. Lignin gives natural, as-grown wood with the necessary compressive strength, which cellulose cannot provide. The cellulose fibre can only provide tensile strength, which means a compound of natural fibres (cellulose) and lignin makes a material that, just like natural wood, can withstand tensile and compressive loads in combination. By varying the composition of Arboform® both in terms of quality and quantity it is possible to adjust strength, rigidity, dimensional stability with varying temperatures, and other material properties so as to suit specific product requirements

Borregaard Lignotech
DEA-Scholven - Strasse 9
Karlsruhe
D-76187
Germany

Tel: +49 721 555 910
Fax: +49 721 555 9110
lignopol.com

Material	LignoPol
Material type	LignoPol is a thermoplastic composite based on lignin. The formulations also contain natural fibres and some natural additives. The know-how is protected by patent EP/PCT No. EP 0 720 634 [31].

Properties

- Good mechanical properties, comparable to filled polypropylene and acrylonitrile-butadiene-styrene.
- Based on renewable raw materials: CO_2 neutral, compostable and recyclable.
- Very good flow properties in injection moulding.
- Very low mould shrinkage.
- Good dimensional stability, passes climate tests, e.g., by the automotive industry.
- Natural flame retardancy without additives.
- Good affinity to adhesives and paints.

Grades   Customised formulations for different applications

Typical uses

- Durable injection moulded articles for automotive, building, electrical and general applications.
- Ideal as backing material for veneer: same coefficient of thermal expansion as wood and compatible with simple adhesives.
- Matrix for composites with natural fibre mats for compression moulding, e.g., of interior panels for vehicles.

Additional information

LignoPol is compostable, it biodegrades at approximately the same rate as wood. Hence it is suitable for durable articles (not rapidly biodegradable).

Lignin is the only renewable polymer, suitable as a base for durable thermoplastic materials, which is available in large quantities at prices that are competitive with synthetics.

Environmental Polymers Group plc
Polymer House
9 Woodrow Way
Fairhills Industrial Estate
Irlan
Manchester   Tel: +44 (0)161 777 4830
M44 6ZQ   Fax: +44 (0)161 777 4846
UK   www.epgplc.com

Material	Depart
Material type	PVOH-based biodegradable polymer
Properties	Temperature controllable water solubility, biodegradability, high tensile and puncture strength, flexibility, high gloss and transparency. Static dissipative. Excellent gas barrier properties.
Grades	Hot, warm and cold water soluble films. Looking for partners to develop thermoformed sheet and injection moulded products.
Typical uses	Packaging, compost waste bags, laundry bags.
Additional information	Scale up to production of 10,000 tonnes per year of pellets by 2004 underway

Low Environmental Impact Polymers

Rodenburg Biopolymers
Denariusstraat 19
Oosterhout NB
4903 RC   Tel: +31 162 49 70 40
Noord Brabant   Fax: +31 162 49 70 41
The Netherlands   www.biopolymers.nl

Material	Solanyl
Material type	Starch-based, optionally cellulose fibre reinforced or blended with commercially available biodegradable polyesters
Properties	Tailormade to customer specification. General properties, fully biodegradable under composting and in soil. Mechanical properties are equivalent to those of PE or PS. Speed of degradation can be modified by blending with other bioplastic.
Grades	Solanyl IM (injection moulding) Solanyl EX (pipe extrusion) Solanyl CR (controlled release) Solanyl BM (blend) Solanyl CM (composite materials)
Typical uses	Injection moulding applications in packaging, agriculture, disposables, technical products.
Additional information	Food contact under investigation. Sheet extrusion and vacuum moulding under development.

Biologische Naturverpackungen GmbH

Werner-Heisenberg-Strasse 32	Tel: +49 (0)2822-92510
D 46446 Emmerich	Fax: +49 (0)2822-51840
Germany	www.biotec.de

Material	BIOPLAST
Material type	Completely biodegradable blends based on renewable resources as for example, thermoplastic starch (TPS).
Properties	• Extrudable on standard machinery, e.g., for converting of LDPE • Completely biodegradable, compostable depending on material thickness • Some suitable for food contact • High water vapour transmission ratre
Grades	BIOPLAST GF 103/11 - film blowing BIOPLAST GF 103/51 - film blowing BIOPLAST GF 105/30 - film blowing BIOPLAST GF 105/50 - sheet extrusion, thermoforming, injection moulding
Typical uses	Garbage bags, nets, trays, agricultural films – for further applications see website.

Additional information

The rest of the material suppliers did not reply to our questionnaire. For these suppliers we have included the name of the company and material, and contact details.

Material	Company	Address	Telephone/Fax	Website Address
Biopar (starch)	Biop	BIOP Biopolymer Technologies AG Technologie – Zentrum Gostritzer Strasse 61-63 01217 Dresden	Tel: + 49 210 322 695 Fax: + 49 210 322 695	www.biopag.de
Paragon (starch)	Avebe	AVEBE Postbus 15 9640 AA VEENDAM The Netherlands	Tel: +31 598 669111 Fax: +31 598 664368	www.avebe.com
Placorn (starch)	Nihon Shokuhin Kako Co. Ltd.	33-8, Sendagaya 5-Chome Shibuya-Ku Tokyo 151-0051 Japan	Tel: +81 3 53604411 Fax: +81 3 53604423	
ReNEW resin (starch)	StarchTech Inc.	720 Florida Avenue South Golden Valley MN 55426 USA	Tel: +1 612 545-5400 Fax: +1 612 545-9450	www.starchtech.com
Supol (starch)	Supol GmbH	Mittagstrasse 24-25 D-39124 Magdeburg Germany	Tel: + 49 (0) 391 25 26 917 Fax: + 49 (0) 391 25 26 916	www.supo.de
Lacty (PLA)	Shimadzu Corporation			www.shimadzu.com
Aquadro (Polyvinyl) & Enviroplastic-Z (wood) CA 92131	Planet Polymer Technologies, Inc.	Planet Polymer Technologies, Inc. 9985 Businesspark Avenue San Diego USA	Tel: + 1 858 549 5130 Fax: +1 858 549 5133	www.planetpolymer.com

Material	Company	Address	Telephone/Fax	Website Address
Gohsenol (polyvinyl)	The Nippon Synthetic Chemical Industry Co., Ltd. (NIPPON GOHSEI)	Umeda Sky Building Tower East 1-88, Oyodonaka 1-chome, Kita-ku Osaka 531-0076 Japan	Tel: +81 6 6440 5300 Fax: +81 6 6440 5330	www.nichigo.co.jp
Poval (polyvinyl)	Kuraray Co., Ltd.	Shin-Hankyu Building, 1-12-39 Umeda Kita-ku Osaka 530-8611 Japan		www.kuraray.co.jp
Solplax (polyvinyl)	Solplax	6265 Stevenson Las Vegas NV 89120 USA	Tel: +1 702 4542121 Fax: +1 702 4347594	www.solplax.com
Fasal (wood)	Austel: subsidiary of IFA Tulln	Franz Josef Straße 33 A-5020 Salzburg Austria	Tel: +43 662 640 3652	www.austel.au
Treeplast (wood)	PE Design and Engineering BV	PO Box 3051 NL 26021 DB Delft The Netherlands	Tel: +31 15 214 8903 Fax: +31 15 214 3323	www.treeplast.com
Absorbable (medical grade)	Absorbable Polymer Technologies, Inc.	2683 Pelham Parkway Pelham AL 35124 USA	Tel: +1 205-620-0025 Fax: +1 205-620-9888	www.absorbables.com
Lactel (medical grade)	Birmingham Polymers, Inc.	756 Tom Martin Drive Birmingham AL 35211-4467 USA	Tel: +1 205 917-2231, Fax: +1 205 917-2245	www.birminghampolymers.com

Material	Company	Address	Telephone/Fax	Website Address
Purasorb (medical grade)	Purac	PURAC biochem Arkelsedijk 46 PO Box 21 4200 AA Gorinchem The Netherlands	Tel: +31 183 695 695 Fax: +31 183 695 600	www.purac.com
Resomer (medical grade)	Boehringer Ingelheim GmbH	Dr. Boehringer Gasse 5-11 1121 Vienna Austria	Tel: +431 80 10 50 Fax: +431 804 08 23	www.boehringer-ingelheim.com
Bionolle	Showa Highpolymer	3-1-35 Hakozakihuto Higashi-ku Fukuoka Japan	Tel:+092-651-2932 Fax: +092-651-2934	www.showa.co.jp
Biomax	DuPont	DuPont Building 1007 Market Street Wilmington DE 19898 USA	Tel: + 1-302-774-1000	www.dupont.com
Greenpol	SK Chemicals:	SK Corporation Chemical Business Department 99 Seorin-dong Jongro-gu Seoul 110-11 Korea	Tel: +82 2 2121 6487 Fax:+82 2 2121 6937	www.skchem.com
Lunare SE	Nippon Shokubai	Kogin Building 4-1-1 Koraibashi Chuo-ku Osaka 451-0043 Japan	Tel: + 81 6 6223 9111 Fax: + 81 6 6201 3716	www.shokubai.co.jp

Material	Company	Address	Telephone/Fax	Website Address
Mazin	Mazin			www.angelfire.com/ne/mazin
Napac	Napac	33, Boulevard du Général Martial Valin 75015 Paris France	Tel: + 33 (0)1 44 25 20 40 Fax: +33 (0)1 44 25 20 50	www.napac.fr
Pullulan	Hayashibara Biochemical Labs.		Tel: + 81 86-224-4315 Fax: + 81 86-224-4577	www.hayashibara.co.jp
Capa	Solvay Capralactones	Baronet Road Warrington Cheshire WA4 6HB UK	Tel: +44 (0) 1925 651277 Fax: +44 (0) 1925 232207	www.solvaycaprolactones.com
Celgreen	Daicel Chemical Industries, Ltd	2-5, Kasumigaseki 3-chome Chiyoda-ku Tokyo Japan		www.daicel.co.jp
Unitika Poval (polyvinyl)	Unitika Ltd.	JP Building 3-4-4 Nihonbashi-Muromachi Chuo-ku Tokyo Japan 103-8321	Tel: +81-3-3246-7536	www.unitika.co.jp
Vegemat (starch)	Vegemat	Rue de la menoue 32400 RISCLE France	Tel: 05 62 69 71 72 Fax: 05 62 69 92 72	www.vegemat.com

References

1. NEN EN 13432, *Packaging - Requirements for Packaging Recoverable Through Composting and Biodegradation - Test Scheme and Evaluation Criteria for the Final Acceptance of Packaging*, 2000.

2. ASTM D638-03, *Standard Test Method for Tensile Properties of Plastics*, 2003.

3. DIN EN ISO 527-1, *Plastics - Determination of Tensile Properties - Part 1: General principles*, 1996.

4. DIN EN ISO 527-2, *Plastics - Determination of Tensile Properties - Part 2: Test Conditions for Moulding and Extrusion Plastics*, 1996.

5. DIN EN ISO 178, *Plastics - Determination of Flexural Properties*, 2003.

6. DIN EN ISO 604, *Plastics - Determination of Compressive Properties*, 2003.

7. EN ISO 178, *Plastics - Determination of Flexural Properties*, 2003.

8. EN ISO 179, *Plastics - Determination of Charpy Impact Strength*, 1996.

9. DIN 53505, *Shore A and Shore D Hardness Testing of Rubber*, 2000.

11. DIN 53456, *Testing of Rubber; Impact Test for the determination of the Low-Temperature Brittleness Point*, 1980.

10. DIN 53752, *Testing of Plastics; Determination of the Coefficient of Linear Thermal Expansion*, 1980.

12. DIN EN ISO 306, *Plastics - Thermoplastic Materials - Determination of Vicat Softening Temperature (Vst)*, 1997

13. DIN 52612-1, *Testing of Thermal Insulating Materials; Determination of Thermal Conductivity by the Guarded Hot Plate Apparatus; Test Procedure and Evaluation of Results*, 1979.

14. DIN EN 12939, *Thermal Performance of Building Materials and Products - Determination of Thermal Resistance by Means of Guarded Hot Plate and Heat Flow Meter Methods - Thick Products of High and Medium Thermal Resistance*, 2001.

15. DIN EN 12664, *Thermal Performance of Building Materials and Products - Determination of Thermal Resistance by Means of Guarded Hot Plate and Heat Flow Meter Methods - Dry and Moist Products with Medium and Low Thermal Resistance Thermal Resistance*, 2001.

16. DIN EN 12667, *Thermal Performance of Building Materials and Products - Determination of Thermal Resistance by Means of Guarded Hot Plate and Heat Flow Meter Methods - Products of High and Medium Thermal Resistance*, 2001.

17. DIN EN 60669-1, *Switches for Household and Similar Fixed Electrical Installations - Part 1: General Requirements*, 2003.

18. DIN IEC 60093, *Methods of Test for Insulating Materials for Electrical Purposes; Volume Resistivity and Surface Resistivity of Solid Electrical Insulating Materials,*

19. DIN IEC 6017, *Methods of Test for Insulating Materials for Electrical Purposes; Insulation Resistance of Solid Materials*, 1993

20. DIN EN 71-3, *Safety of Toys - Part 3: Migration of Certain Elements*, 2002.

21. DIN V 53160-1, *Determination of the Colourfastness of Articles for Common Use - Part 1: Resistance to Artificial Saliva*, 2002.

22. DIN V 53160-2, *Determination of the Colourfastness of Articles in Common Use - Part 2: Resistance to Artificial Sweat*, 2002.

23. DIN 4102-2, *Fire Behaviour of Building Materials and Building Components; Building Components; Definitions, Requirements and Tests*, 1977

24. DIN EN1363-1, *Fire Resistance Tests - Part 1: General Requirements*, 1999.

25. DIN EN 1364-1, *Fire Resistance Tests on Non-Loadbearing Elements - Part 1: Walls*, 1999.

26. DIN EN 1364-2, *Fire Resistance Tests for Non-Loadbearing Elements - Part 2: Ceilings*, 1999.

27. DIN EN 1365-1, *Fire Resistance Tests on Loadbearing Elements - Part 1: Walls*, 1999.

28. DIN EN 1365-2, *Fire Resistance Tests for Loadbearing Elements - Part 2: Floors and Roofs*, 2000.

29. DIN EN 1365-3, *Fire Resistance Tests for Loadbearing Elements - Part 3: Beams*, 2000.

30. DIN EN 1365-4, *Fire Resistance Tests for Load-Bearing Elements - Part 4: Columns*, 1999.

31. P. Wunning and M. Wunning, inventors; Lignopol Polymere Stoffe GmbH, assignee; EP 0720634B1, 1998.

Abbreviations

3HB	3-Hydroxybutyrate
3HHp	3-Hydroxyheptanoate
3HHx	3-Hydroxyhexanoate
3HO	3-Hydroxyoctanoate
3HV	3-Hydroxyvalerate
AERT	Advanced Environmental Recycling Technologies
AIDS	Acquired immunodeficiency syndrome
aPECH	Atactic PECH
ASTM	American Society for Testing and Materials
BAT	British American Tobacco
BMC	Bulk moulding compound
BOD	Biological oxygen demand
BPI	British Plastics Institute
BSI	British Standards Institute
BTA	Co-polyester of terephthalic and adipic acids with 1,4-butandiol
CAB	Cellulose acetate butyrate
CC	Cellulose carbamate
CEN	European Committee for Standardisation
CHP	Combined heat and power
CL	Caprolactone
CNSL	Cashew nut shell liquid
CRPA	Centro Ricerche Produzioni Animali
DCPO	Dicumylperoxide
DEEDS	Design for Environment Decision Support

ASR	Automobile shredder waste
ASTM	American Society for Testing Materials
ATPISO2	Acetylene-terminated polyimidesulfone
ATPS	Aqueous two-phase systems
ATR	Attenuated total reflectance
AU	Allylurea
B	Butadiene
BACA	*N,N´*-bis(acryloyl)cystamine
B-AN	Butadiene-acrylonitrile
BB-PPO	Butylbenzoyl poly (phenylene oxide)
BC	Block copolymer
BCF	Bulk continuous filament
BD	1,4-Butandiol
BET	Braunauer-Emmett-Teller equation
BF	Bicomponent fibre
BMA	Butyl methacrylate
BMA-*co*-MMA	Poly(butyl methacrylate-*co*-ethyl methacrylate
BMEP	Bis(2-methacryloxyethyl) phosphate
BMI	Bismaleimide
BOD	Biochemical oxygen demand
BP	Benzyl peroxide
BPO	Benzoyl peroxide
B-PPO	Benzoyl poly (phenylene oxide)
BR	Butyl rubber
BSA	Bovine serum albumin
BTX	Benzene-toluene-xylene aromatics
BVPy	Poly(*n*-butylmethacrylate-*co*-4-vinylpiridine)
C	Collagen
C/C	Cover-core structure
C/S	Sheath-core structure
CB	Carbon black

DFA	Dairy Farmers of America
DFE	Design for Environment
DIN	Deutsches Institüt für Normung
DIN CERTCO	DIN certification company
DMA	Dynamic mechanical analysis
DML	Deflection of maximum load
DP	Degree of polymerisation
DS	Degree of substitution
DSC	Differential scanning calorimetry
DTI	Department of Trade and Industry
E´	Storage modulus
ECD	Ecodesign
EDTA	Ethylene diamine tetra acetic acid
ELV	End-of-life vehicle
EML	Energy-to-maximum load
EPSRC	Engineering and Physical Sciences Research Council
ESCA	Electron spectroscopy for chemical analysis
EU	European Union
FDA	Food and Drug Administration
GLC	Gas liquid chromatography
GM	Genetically modified
GMO	Genetically modified organism
GMT	Glass mat thermoplastics
GRP	Glass reinforced plastic
HB	Hydroxybutyrate
HDPE	High density PE
HPLC	High pressure liquid chromatography
HTMA	Hexamethylenetetramine
HV	Hydroxyvalerate(s)

IBAW	International Biodegradable Polymer Association and Working Groups
ICI	Imperial Chemical Industries
IR	Infra-red
ISO	International Organisation for Standardisation
LCA	Life cycle analysis
LD_{50}	Lethal dose which causes death of 50% of the test population
LDPE	Low-density polyethylene
LEIP	Low environmental impact polymers
LiDS	Life Cycle Design Strategies
LVL	Laminated veneer lumber
MA	Maleic anhydride
MAPP	Maleated PP
mcl-PHA	Medium-chain-length PHA
MDI	Di-isocyanate (4, 4′-methylene bis(diphenyl di-isocyanate)
MFI	Melt flow index
MSW	Municipal solid waste
M_w	Molecular weight(s)
NADPH	Nicotinamide adenine dinucleotide phosphate
NIRA	Near Infra-Red Analysis
NMMO	*N*-methylmorpholine-*N*-oxide
OECD	Organisation for Economic Co-operation and Development
PA-6	Nylon-6
PBA	Polybutylacrylate
PCL	Polycaprolactone
PDP	Product development process
PE	Polyethylene
PECH	Poly(epichlorohydrin)
PEO	Polyethylene oxide

PEST	Political Economic Social Technical
PET	Polyethylene terephthalate
PHA	Polyhydroxyalkanoate(s)
PHB	Polyhydroxybutyrate
PHBH	Copolymer of 3HB and 3HHx
PHBV	Polyhydroxybutyrate-*co*-3-hydroxyvalerate
PHBV4	PHBV containg 4% valerate
PHV	Polyhydroxyvalerate(s)
PHVB	Polyhydroxybutyrate-hydroxyvalerate
PLA	Polylactic acid
PLA	Polylactide
PLLA	Poly-L-actide
PMC	Polymer matrix composite(s)
PP	Polypropylene
PS	Polystyrene
PU	Polyurethane(s)
PVA	Polyvinyl acetate
PVB	Polyvinylbutyral
PVC	Polyvinyl chloride
PVOH	Polyvinyl alcohol(s)
R&D	Research & Development
RH	Relative humidity
RIM	Reaction injection moulding
RMIT	Royal Melbourne Institute of Technology
rpm	Revolutions per minute
RRIM	Reinforced reaction injection moulding
RTM	Resin transfer moulding
SD	Standard deviation
SMC	Sheet moulding compound

SRIM	Structural reaction injection moulding
STI	Sustainable Technologies Initiative
TDI	Toluene 2, 4 di-isocyanate
T_g	Glass transition temperature
TGA	Thermogravimetric analysis
TLC	Thin layer chromatography
T_m	Melting point
TPS	Thermoplastic starch
T_s	Softening temperature
TV	Television(s)
USDA	United States Department of Agriculture
UV	Ultraviolet

Author Index

Company Index

Index

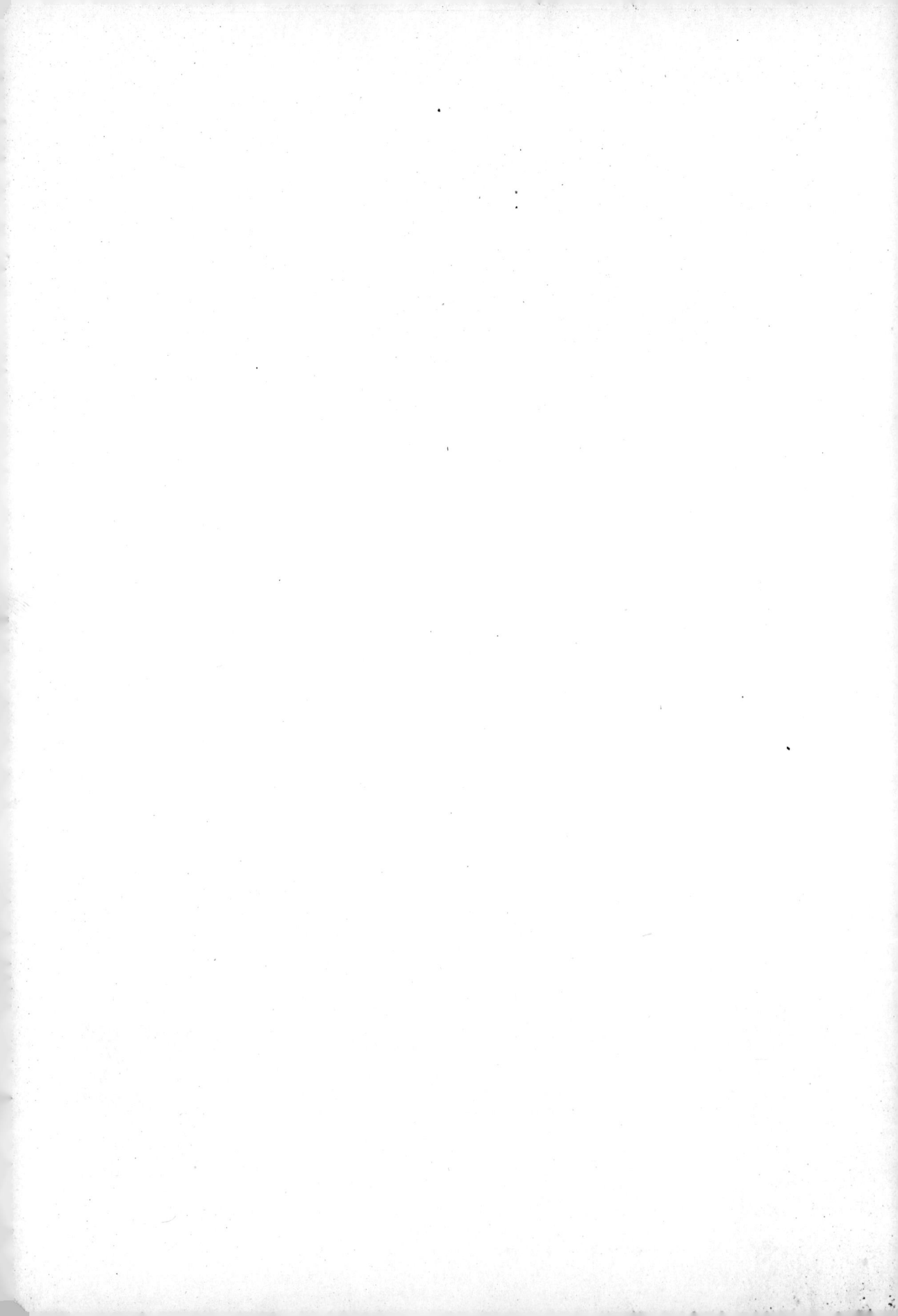